Suping Peng

Jincai Zhang

Engineering Geology for Underground Rocks

Suping Peng
Jincai Zhang

Engineering Geology for Underground Rocks

with 191 Figures

Springer

Professor Suping Peng
China University
of Mining and Technology
11 Ding Xueyuan Road
Haidian District
100083 Beijing
China

Dr. Jincai Zhang
Knowledge Systems, Inc.
One Sugar Creek Center Blvd.
Suite 1100
Sugar Land TX 77478
USA

Library of Congress Control Number: 2007929428

ISBN 978-3-540-73294-5 Springer Berlin Heidelberg New York

Springer is a part of Springer Science+Business Media
springer.com

Cover design: deblik, Berlin
Production: Almas Schimmel
Typesetting: Camera-ready by Authors

Printed on acid-free paper 30/3180/as 5 4 3 2 1 0

Foreword

As underground engineering projects are going deeper and are being constructed in difficult formations, there is an ever greater need for a better understanding of engineering behaviors and geological characteristics of rocks. Conventional engineering geology can provide fundamental knowledge and solutions to rock engineering. However, today underground structures are being built in geologically more complicated formations, for instance,

- at great depth with high temperature, high in-situ stress, and high pore pressure;
- in geologically young formations with unconsolidated rocks;
- in tectonically active faulted zones with abnormally high stresses;
- in highly fractured formations with high permeability and low rock strength; and
- in salt formations with relatively low viscosity and the attendant creep behavior, etc.

For underground engineering projects to be built in these formations a multi-disciplinary approach is required for the engineering design as well as for the modeling. Engineering Geology for Underground Rocks written by Prof. Peng and Dr. Zhang aims to solve the multi-disciplinary coupling problems in underground engineering.

Engineering Geology for Underground Rocks deals with physical, geomechanical, and geophysical properties of underground rocks. It considers the coupled processes and diverse behaviors of underground rocks under various stresses and different fluid pressure environments. This book mainly focuses on underground engineering geological problems and solutions as they are encountered in the energy industry with emphasis on both theoretical study and practical application. As underground engineering projects advance to ever greater depths, increases of the in-situ stress and the pore pressure as well as the temperature present significant challenges for geological engineering. Examples of these challenges during excavation include wellbore and tunnel instabilities, casing failures, water inrushes into active mining regions, rock and coal bursts, mining induced seismicity and gas blowouts. I feel that the authors are making a significant contribution to the resolution of this class of problems. This book is the result of intensive research and shows a dedicated effort of the authors.

Hartmut A. Spetzler, Ph.D.

Professor and Past Chair of Department of Geological Sciences,
Fellow and Past Associate Director of Cooperative Institute for Research in Environmental Sciences,
University of Colorado at Boulder, Colorado, USA

Dedication

"If I have seen further it is by standing on ye shoulders of Giants."
Sir Isaac Newton

To all who have helped us.

Preface

Engineering Geology is devoted to the investigation, study and solution of the engineering and environmental problems which may arise as the result of the interaction between geology and the works and activities of man as well as to the prediction and the development of measures for prevention or remediation of geological hazards (The International Association of Engineering Geology (IAEG) Statutes, 1992). Engineering Geology for Underground Rocks is mainly focused on underground works and activities related to engineering geology such as underground mining, oil and gas drilling exploration and production, and underground excavations etc.

People have known about fossil energy for thousand years. In the Bronze Age, people in Wales began to use coal. Coal is mentioned in the Bible and in the writings of the ancient Greeks. Three thousand years ago, the Chinese began mining and burning coal instead of wood. Coal burned much hotter, and this allowed people to melt soft metals such as copper they mined to make bronze drill bits for oil and gas drilling. In ancient inland China, where salt was scarce and consequently expensive, briny wells were deliberately developed by muscle power and bamboo rig. As long ago as 900 B.C., ancient Chinese found oil and gas in the saltwater wells they drilled. The oil and gas were sent through bamboo pipes for heating and cooking. More than 1,500 years ago, the Chinese drilled for oil to depths of over 900 m using bronze and bamboo tubes. In 1859 the world's first modern oil well was drilled in Titusville, Pennsylvania. Petroleum industry was founded then, and engineering geology started to serve to the oil and gas industry.

Nowadays, due to rocketing energy and mineral requirements, mining and oil and gas development must continue in ever more difficult and complex geological settings, challenging traditional technology and often bringing more geological problems, even disasters. Mining and oil and gas development in these difficult settings require much more innovative theoretical research, specific techniques, and practical technology in both engineering and geology. For example, underground mining has excavated as deep as 3,000 m; borehole drilling has reached to 12,000 m below the Earth; the deepest offshore drilling exceeds 4,000 m below the water surface in the Gulf of Mexico. One of the authors had participated pore pressure, wellbore and casing stability analyses in the world deepest oil well (34,070 feet or 10,384 m in true vertical depth from the sea level) in the deep water of the Gulf of Mexico in 2005. As the depth of underground engineering increases, in-situ stress, pore pressure, and temperature increase, causing various engineering and geological problems, such as excavation and wellbore instability, water inrush and influx, rock and coal burst, mining-induced seismicity, gas blowout and explosion, etc. This book attempts to solve some problems mentioned above.

The aims of this book are to:

- introduce basic principles of engineering geology in underground mining and oil and gas development,
- discuss methods to determine in-situ stress and pore pressure using geophysical means.
- focus on applications of rock mechanics and poromechanics in excavation and wellbore instability, and
- emphasize coupled rock stress/deformation, pore pressure, and fluid flow in porous rocks and fractured porous formations.

The authors would like to thank the following organizations for their supports:

The State Key Laboratory of Coal Resource and Safe Mining, China University of Mining and Technology

Knowledge Systems, Inc., U.S.A.

CIRES, the University of Colorado at Boulder, U.S.A.

This book was supported by the National Science Foundation of China (grant No. 50221402, 50490271, and 50025413), the China National Program on Key Basic Research Project (grant No. 2002CB211707 and 2005CB221500), the key project of the Ministry of Education (No. 306002) and the Program for Changjiang Scholars and Innovative Research Team in University of MOE，PRC (IRT0408).

China Coal Research Institute and some coal mining bureaus in China, such as Datong, Daliuta, Datun, Huainan, Huaibei, Handan, Fengfeng, Feicheng, Jiaozhuo, Kailuan, Lianshao, Shuangyashan, Yanzhou, Xingtai, Zaozhuang, Zibo, etc. gave the authors access to their database, which is gratefully acknowledged.

Authors appreciate the help from Hal H. Zhang and Justin W. Lee.

The authors would also like to thank the supports and encouragement to this book from the following individuals:

Academicians of Chinese Academy of Engineering: Profs. Dexin Han, Weitang Fan, Minggao Qian, Heping Xie, Shining Zhou, Jishan He, Desheng Gu, Sijing Wang.

Academicians of Chinese Academy of Sciences: Profs. Lianjun Ye and Yinfo Chang.

Prof. J. -C. Roegiers in Mewbourne School of Petroleum and Geological Engineering at the University of Oklahoma.

Prof. H. A. Spetzler at the University of Colorado at Boulder.

The authors also thank the following professors in the State Key Laboratory of Coal Resource and Safe Mining at China University of Mining and Technology for providing related materials: Pengfei Zhang, Daiyong Cao, Zhaoping Meng, Guowei Zhu, Lianying Sun, Yangbing Li. Some graduate students of the first author also provided help, which is grateful.

The authors would like to memorize Prof. Tianquan Liu, Academicians of Chinese Academy of Engineering.

Contents

List of contributors

Peng, Suping, Ph.D.
Yangtze Chair Professor,
China University of Mining and Technology
Director,
China State Key Laboratory of Coal Resource and Safe Mining
Beijing 100083, China

Zhang, Jincai, Ph.D.
Senior Geomechanics Engineer and Geomechanics Manager,
Knowledge Systems, Inc.
One Sugar Creek Center Blvd., Suite 1100, Sugar Land, TX 77478
USA

1 Rock properties and mechanical behaviors

Rock physical and mechanical properties are very important parameters for geological engineering design and construction. For instance, in coal mining industry many geological disasters induced by mining were associated with misunderstanding of rock mechanical properties (Peng 1998a, Han and Peng 2002). Rock physical properties include density, porosity, and permeability, etc. Rock mechanical properties mainly include elastic modulus, Poisson's ratio, and rock strength. These parameters can be obtained by lab experiments of core samples or by in-situ tests. The other characteristics of rocks include time-dependent rheological and creep behaviors (Wang 1981). When the rock samples are not available, such as oil and gas drilling and mining at deep depth, the well log data and geophysical data can be used to analyze and interpret rock physical and mechanical parameters (Peng 1997, Peng 1999).

1.1 Density

Rock density (ρ) is a measure of mass of the rock contained in a given unit volume (density = mass/volume). It is usually expressed in g/cm^3. Rock density is also called bulk density. Since most of rocks are porous media, bulk density of porous rocks depends on not only the density of the solid matrix material, but also the density of pore fluids as well as saturation, as shown in the following equation:

$$\rho_b = (1-\phi)\rho_{ma} + \phi\rho_f \tag{1.1}$$

where ρ_b is the rock or bulk density; ρ_{ma} is the matrix or solid density, ρ_f is the fluid density, ϕ is the porosity.

Tables 1.1 and 1.2 list matrix density for common minerals (Schön 1996) and bulk density for various rock types.

For most rocks and soils, the particle density is about 2.65 g/cm^3, roughly the density of quartz and clay minerals. The density of water is 1 g/cm^3, thus, the bulk density of the rock can be calculated when the formation porosity is known. Figure 1.1 shows density changes with depth below the sea floor in the Gulf of Mexico. The density in the figure is calculated from the core porosity data (Ostermeier et al. 2001) by using Eq.1.1. It shows that the density of formations increases as the burial depth increases.

Table 1.1. Matrix density for common minerals

Minerals	Density ρ_{ma} (g/cm^3)
Quartz	2.65
Calcite	2.71
Dolomite	2.87
Montmorillonite	2.06
Illite	2.64
Kaolinite	2.59
Chlorite	2.88
K-Feldspar	2.56
Biotite	2.90
Halite	2.165
Anhydrite	2.96
Sylvite	1.99
Plagioclase (Na)	2.62
Plagioclase (Ca)	2.76
Barite	4.48

Table 1.2. Bulk density and porosity for various rock types

Rock type	Density ρ_b (g/cm^3)	Porosity ϕ (%)
Granite	2.5-2.8	0.5-1.5
Dolerite	3.0-3.1	0.1-0.5
Limestone	2.5-2.8	5-20
Dolomite	2.5-2.6	1-5
Quartzite	2.65	0.1-0.5
Sandstone	2.0-2.6	5-25
Shale	2.0-2.7	10-30
Coal - Anthracite	1.3-1.6	
Coal - Bituminous	1.1-1.4	
Sediments	1.7-2.3	
Metamorphic rocks	2.6-3.0	

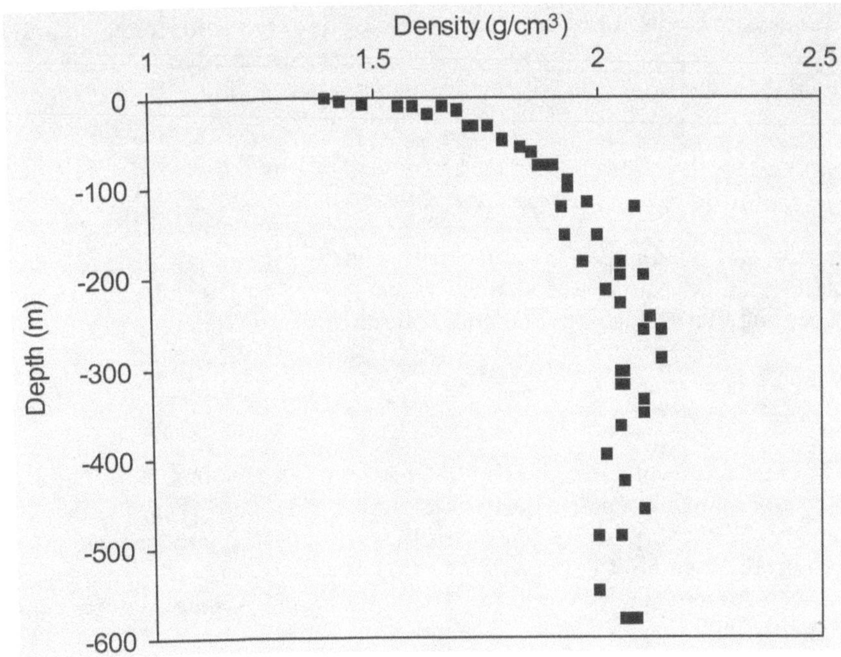

Fig. 1.1. Density variations with depth below the sea floor in the Gulf of Mexico.

Rock density can be measured in the lab using core samples. For underground rocks, it is more convenient to determine rock bulk density by geophysical methods. Gardner et al. (1974) presented the following relationship between bulk density and seismic wave velocity in sedimentary rocks:

$$\rho_b = 1.741 v_p^{0.25} \tag{1.2}$$

where ρ_b is the bulk density, in g/cm^3; v_p is the velocity of the compressional wave, in km/s.

This equation represents a fair average for a large number of laboratory and field observations of different brine-saturated rock types.

Rock density is controlled by densities and volumetric fractions of components which the rock is composed. Therefore, density differences between minerals, fluids, and gases cause a strong correlation between rock density and porosity. For underground rocks, as depth increases the rock compaction increases, causing porosity reduction. This induces an increase in rock density as depth increases, as shown in Fig. 1.1.

It needs to mention that specific gravity (γ) is very useful for calculating rock overburden stress. Specific density is dimensionless, equal to the density of the material divided by the density of water.

1.2 Porosity

Porosity (ϕ) is an important property to analyze oil and gas reservoir and aquifer storage. Porosity is defined to be the ratio of a volume of void spaces within a rock to the total bulk volume of the rock, i.e.:

$$\phi = \frac{V_{pores}}{V_{rock}} \tag{1.3}$$

In engineering practice, porosity has several descriptions, but, the two most common are total porosity (defined as the above) and effective porosity. Effective porosity represents the ratio of the interconnected pore space to the total bulk volume of the rock. Therefore, effective porosity is also called interconnected porosity. The fluids in interconnected pores contribute to fluid flow.

Porosity is primarily controlled by the shape, size and arrangement of the rock grains (Peng and Meng 2002). It also depends on rock mechanical processes (such as compaction, deformation, fracture evaluation) and geochemical processes (e.g. dissolution, precipitation, mineralogical changes). Table 1.3 shows different packing effects on porosity for identical spherical rock grains.

Table 1.3. Porosity of different packing of spherical rock grains

Packing	ϕ
Cubic	0.48
Orthorhombic	0.40
Tetragonal	0.30
Rhombohedral	0.26

Porosity can be determined by lab experiments in core samples and by well logs, such as density, resistivity, neutron, and NMR (Nuclear Magnetic Resonance) logs. The density-porosity equation (Eq. 1.1) can be rewritten for calculating porosity purpose:

$$\phi = \frac{\rho_{ma} - \rho_b}{\rho_{ma} - \rho_f} \tag{1.4}$$

For clean sandstones at high pressures and with moderate porosities, porosity is often estimated by the empirical time average formula (Wyllie et al. 1956).

$$\phi = \frac{1/v_p - 1/v_{ma}}{1/v_f - 1/v_{ma}} \tag{1.5}$$

where v_p is the formation or rock compressional velocity; v_{ma} is the grain or matrix compressional velocity; and v_f is the fluid velocity. The equation can be rewritten by using transit time terminology in the following form:

$$\phi = \frac{\Delta t - \Delta t_{ma}}{\Delta t_f - \Delta t_{ma}} \tag{1.6}$$

where Δt is the acoustic transit time; Δt_{ma} is the acoustic transit time of the rock matrix; and Δt_f is the acoustic transit time of interstitial fluids.

The Wyllie equation represents consolidated and compacted formations, generally for porosity of less than 0.25 in sandstones. Unconsolidated sandstones, such as those in U.S. Gulf Coast, Nigeria, and Venezuela, often have much higher porosity (0.28 to 0.50). If the Wyllie equation is used in unconsolidated sandstones, correction for compaction is necessary. Generally, such sandstones can be identified by observing Δt in adjoining shale beds. Formations not subjected to sufficient overburden stress do not have enough compaction and rigidity for proper transmission of an acoustic wave, and as a result, with decreasing overburden pressure, Δt through unconsolidated formations increases disproportionately (Baker Atlas 2002). An empirical equation can be used to calculate porosity in unconsolidated sandstones:

$$\phi = \frac{\Delta t - \Delta t_{ma}}{\Delta t_f - \Delta t_{ma}} \frac{1}{C_p} \tag{1.7}$$

where the compaction correction factor can be expressed as (Baker Atlas 2002):

$$C_p = \frac{\Delta t_{sh} C}{100}$$

where Δt_{sh} is the specific acoustic transit time in adjacent shales (μsec/ft); C is the shale compaction coefficient, ranging from 0.8 to 1.3.

The Raymer-Hunt-Gardner correlation is another empirical fit of independently determined porosity to observe interval times (Raymer et al. 1980). It is expressed as:

$$\phi = c\frac{\Delta t - \Delta t_{ma}}{\Delta t} \tag{1.8}$$

where c is a constant and varies typically from 0.625 to 0.7.

Field tests and lab experiments have shown that rock porosity decreases as the burial depth increases as shown in Fig. 1.2. The following equation is the most commonly used to describe porosity and depth relationship:

$$\phi = ae^{-bZ} \tag{1.9}$$

where a and b are constants; Z is the depth. In the Northern North Sea, porosity and depth have the following relationship (Schön 1996):

For sandstone

$$\phi = 0.49e^{-2.7\times10^{-4}Z} \tag{1.10}$$

For shale

$$\phi = 0.803e^{-5.1\times10^{-4}Z} \tag{1.11}$$

where Z is the depth in meters.

Similarly, the relationship between porosity and effective stress can be expressed as the follows:

$$\phi = ae^{-b\sigma'} \tag{1.12}$$

where a, b are the constant; σ' is the effective compressive stress.

It can be seen that porosity decreases as compressive effective stress (σ') increases, causing formation more compacted and consolidated.

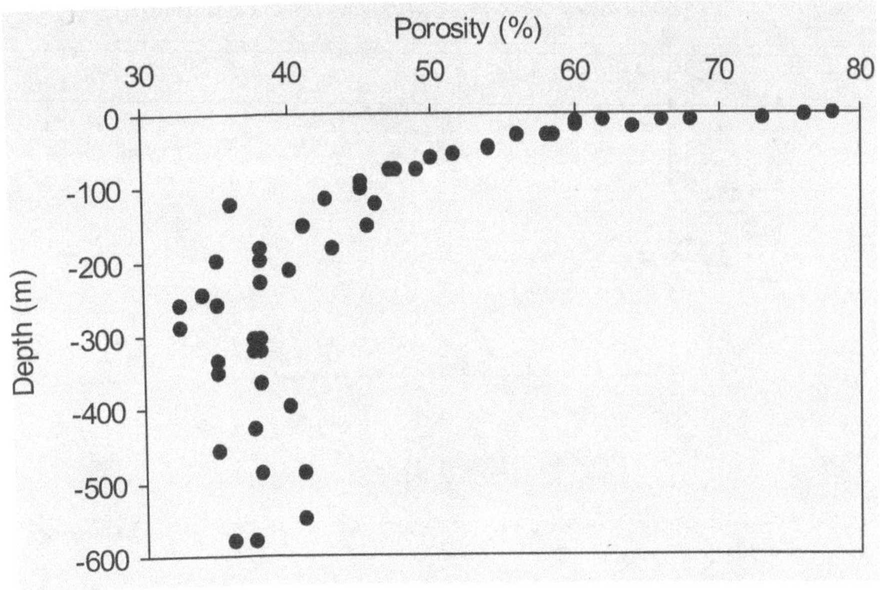

Fig. 1.2. Porosity decrease with depth (starting from the sea floor) in the central deepwater Gulf of Mexico (plotted from the core test data presented by Ostermeier et al. 2001).

Lab experimental tests show that rock porosity decreases as the applied stress increases. Figure 1.3 presents porosity and axial stress relationships under uniaxial strain conditions for two sandstone specimens cored 1000 m below the sea floor. The experiments show that the stress and porosity have negative exponential relationships, i.e.:

$$\phi = 33.6e^{-0.0023\sigma} \tag{1.13}$$

$$\phi = 36.7e^{-0.0027\sigma} \tag{1.14}$$

where ϕ is the porosity (%); σ is the axial stress (MPa).

The stress and porosity can be expressed as the following general form:

$$\phi = \phi_0 e^{-c\sigma} \tag{1.15}$$

where ϕ_0 is the initial porosity; c is the constant; and σ is the axial stress.

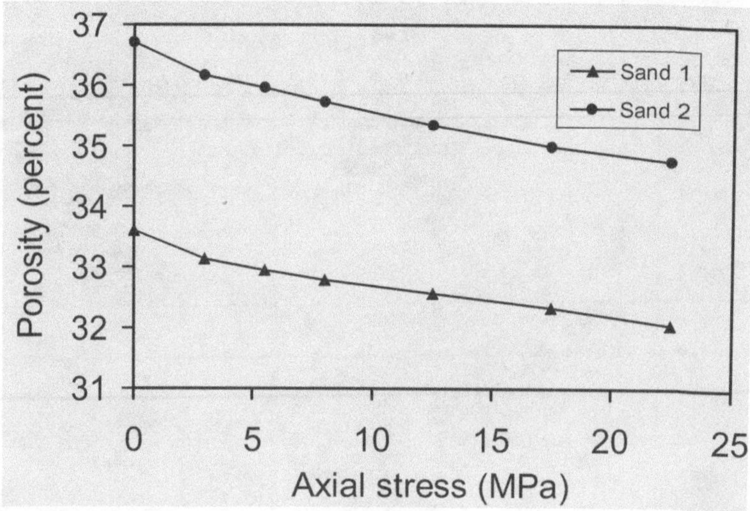

Fig. 1.3. Porosity decreases with the increase of the axial stress in uniaxial strain tests for sand specimens.

1.3 Permeability

1.3.1 Darcy's law and hydraulic conductivity

In 1856, Henry Darcy investigated the flow of water in vertical homogeneous sand filters in connection with the fountains of the city of Dijon. From his experiments, Darcy concluded that the rate of flow q is proportional to the constant cross-sectional area A, proportional to the difference of hydraulic head $(h_1 - h_2)$, and inversely proportional to the distance of the two hydraulic heads L. When combined, these conclusions give the famous Darcy formula:

$$q = KA\frac{h_1 - h_2}{L} \tag{1.16}$$

where K is the hydraulic conductivity. Hydraulic conductivity is commonly used in hydrogeology (Bear 1972) and is a measure in how easily a particular fluid passes through a particular earth material. The following equation gives permeability and hydraulic conductivity relationship (Zhang et al. 1997):

$$K = \frac{\rho_f g}{\mu} k \tag{1.17}$$

where ρ_f is the fluid density; μ is the dynamic viscosity of fluid; and k is the permeability.

1.3.2 Permeability tensor

Permeability is the most important physical property of a porous medium. Permeability measures quantitatively the ability of a porous medium to conduct fluid flow. Permeability has dimensions of an area, and it is measured in units of Darcy (1 Darcy = 0.98×10^{-12} m^2), or more commonly the milli-darcy or md.

Permeability of the rock depends largely on the connectedness of the void spaces, the grain size of the rock, and cementation between rock grains. A rock could be extremely porous, but if each pore is isolated from the others, the rock would be impermeable. If rock grain size is small, then void spaces of the rock are small. Therefore, the surface film of the fluid can actually choke the movement of additional fluids through the small spaces. This is why clays are so impermeable, even though their porosities can be as high as 50%.

Permeability is a tensor in a manner similar to stress tensor, and in Cartesian coordinates it can be expressed as:

$$k = \begin{bmatrix} k_{11} & k_{12} & k_{13} \\ k_{21} & k_{22} & k_{23} \\ k_{31} & k_{32} & k_{33} \end{bmatrix} \tag{1.18}$$

More often the permeability is isotropic in the direction of the bedding plane but anisotropic in the direction of perpendicular to the bedding plane. In this case, if two coordinates are in the plane of the bedding, the permeability tensor is diagonal with the coefficients in the plane of the bedding having equal values and denoted as horizontal permeability, k_h, and the coefficient in the direction perpendicular to the bedding denoted as vertical permeability, k_v. In this case the permeability tensor can be expressed as follows:

$$k = \begin{bmatrix} k_h & 0 & 0 \\ 0 & k_h & 0 \\ 0 & 0 & k_v \end{bmatrix} \tag{1.19}$$

In the sedimentary rocks, usually horizontal permeability has a large value than vertical permeability, depending on porosity and grain size and grain packing. For example, a rock consisted of large flat grains has horizontal permeability of 2000 md and vertical permeability of 800 md. However, for a rock with small flat grains, it has horizontal permeability of 800 md and vertical permeability of 50 md (Baker Atlas 2002).

1.3.3 Permeability and porosity correlation

For clastic rocks general correlation between permeability and porosity follows a power law, i.e.:

$$k = a\phi^b \tag{1.20}$$

where a and b are empirical constants.

Permeability often increases with porosity. However, rocks with very low porosity have exhibited high permeability, and some high porosity rocks have very low matrix permeability (Baker Atlas 2002). Permeability can be determined by several means (Peng and Wang 2001); e.g., well test, wireline formation tests, or core tests.

Permeability also shows strong correlations with burial depth, stress, and rock properties (Peng et al. 2000). The in-situ stress and stress changes caused by underground engineering have important effects on permeability (Wang et al. 2001, Zhang et al. 2007). More detail discussions in stress-dependent permeability will be given in Chap. 8.

1.4 Elastic modulus

Elastic modulus (also called Young's modulus) is an important parameter to describe stress and strain relationship. For most rocks, the uniaxial stress-strain curve before failure takes approximately the linear form (Fig. 1.4). This can be presented by (Jaeger and Cook 1979):

$$\sigma = E\varepsilon \tag{1.21}$$

where σ is the stress; ε is the strain; the constant, E, is called elastic modulus. A material is linearly elastic if the above relation holds accurately.

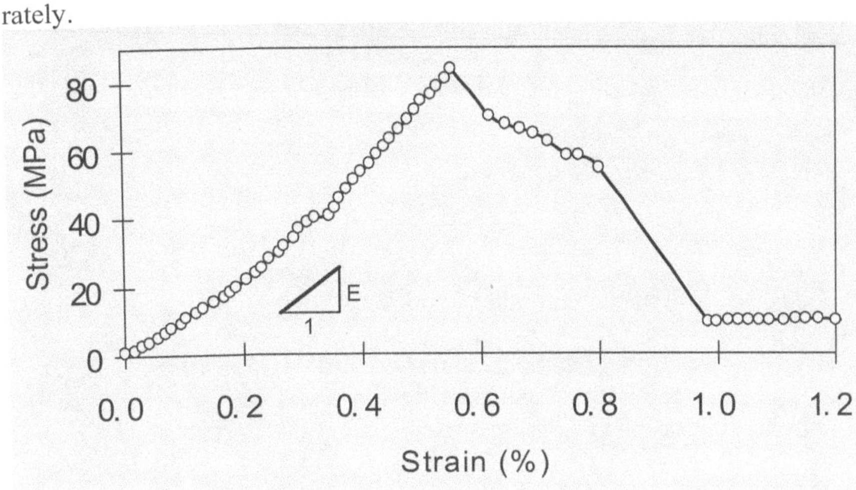

Fig. 1.4. Elastic modulus and stress-strain relationship for a medium-grained sandstone under uniaxial compressive test.

Elastic modulus describes the capacity of rock deformation, or the stiffness of a rock. For a high elastic modulus rock, it is less deformable (i.e. stiff). The initial part of the complete stress-strain curve will be steep. For a low elastic modulus (soft) rock, it is more deformable, and the initial part of the complete stress-strain curve will be gentle (Hudson and Harrison 1997).

Rock static elastic modulus can be obtained from lab core tests of either uniaxial or triaxial compressive experiment (Meng et al. 2002). Rock dynamic elastic modulus (E_d) can be solved from the following equations by knowing the rock elastic compressional and shear wave velocities:

$$v_p = \sqrt{\frac{E_d}{\rho_b} \frac{(1-\nu)}{(1+\nu)(1-2\nu)}} \tag{1.22}$$

$$v_s = \sqrt{\frac{E_d}{\rho_b} \frac{1}{2(1+\nu)}} \tag{1.23}$$

where v_p and v_s are the compressional and shear wave velocities of the rock, respectively; ρ_b is the rock bulk density; ν is the Poisson's ratio of the rock. Rock elastic velocities can be obtained by seismic survey or sonic well log. The dynamic elastic modulus can be converted into static modulus by empirical correlations (Du et al. 2001). The dynamic modulus also can be expressed as the following forms if one knows the interval transit time:

$$E_d = \frac{\rho_b}{t_p^2} \frac{(1+\nu)(1-2\nu)}{(1-\nu)} \tag{1.24}$$

$$E_d = \frac{2\rho_b(1+\nu)}{t_s^2} \tag{1.25}$$

where t_p and t_s are the compressional and shear interval transit time, respectively.

From lab tests in over 400 core samples from the Gulf of Mexico, Lacy (1997) derived the following correlation between rock dynamic elastic modulus and ultrasonic compressional velocity:

$$E_d = 0.265 v_p^{2.04} \tag{1.26}$$

where the dynamic modulus (E_d) is measured in Mpsi; the compressional velocity v_p is in km/s and ranges from 1 to 6 km/s.

In general, the dynamic values of elastic modulus have been found to be significantly greater than the static values. It was also noted that the discrepancy was far greater for soft rocks, such as sandstone, than for hard rocks, such as granite (Howarth 1984). The discrepancies between the dynamic and static moduli have been widely attributed to microcracks and pores in the rocks.

From ultrasonic test data of 600 core samples in the Gulf of Mexico, Lacy (1997) obtained the following correlation for sandstones:

$$E_s = 0.0293 E_d^2 + 0.4533 E_d \tag{1.27}$$

where E_s is the static elastic modulus.

The similar correction exists for shales (Lacy 1997):

$$E_s = 0.0428 E_d^2 + 0.2334 E_d \tag{1.28}$$

From sandstones, shales, limestones, and dolomites, the generalized correlation can be expressed as follows (Lacy 1997):

$$E_s = 0.018 E_d^2 + 0.422 E_d \tag{1.29}$$

where E_s and E_d in Eqs. 1.27 to 1.29 are the static and dynamic elastic moduli with units of one million psi (Mpsi).

Ohen (2003) gave the following relationship between the dynamic and static moduli for the rocks in the Gulf of Mexico, i.e. the dynamic Young's modulus is about 18 times of the static Young's modulus.

$$E_s = 0.0158 E_d^{2.74} \tag{1.30}$$

where the elastic moduli E_s and E_d are in Mpsi.

1.5 Poisson's ratio

Poisson's ratio is the ratio of transverse strain to corresponding axial strain on a material stressed along one axis. For a rock core subjected to axial load, Poisson's ratio (ν) can be expressed as:

$$\nu = -\frac{\varepsilon_l}{\varepsilon_a} \tag{1.31}$$

where ε_l is the lateral strain; ε_a is the axial strain. Therefore, Poisson's ratio can be determined by measuring the lateral and axial deformations of the uniaxial compressive test in rock samples.

Poisson's ratio can also be calculated from the velocities of the elastic wave:

$$\nu = \frac{\frac{1}{2}(v_p / v_s)^2 - 1}{(v_p / v_s)^2 - 1} \tag{1.32}$$

where v_p and v_s are the compressional and shear velocities, respectively.

Figure 1.5 shows the calculated Poisson's ratio by Eq. 1.32 from sonic compressional and shear velocities in an offshore well at a water depth of 264 m. It shows that Poisson's ratio depends on both lithology and depth and is larger in shales and smaller in sandstones.

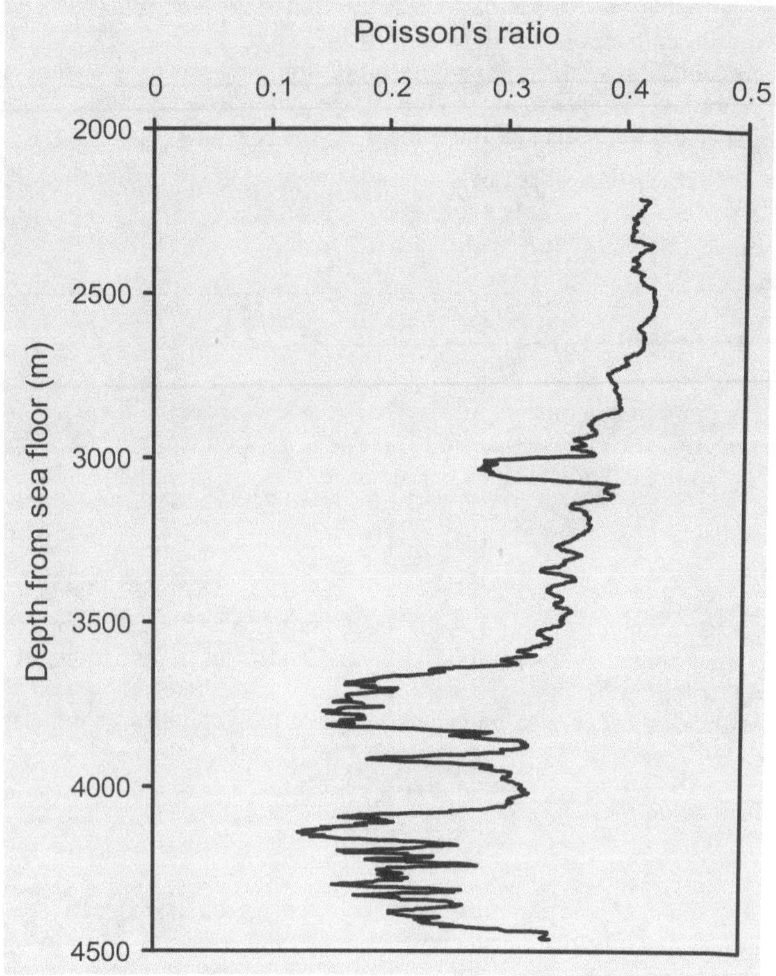

Fig.1.5. Calculated Poisson's ratio from sonic compressional and shear velocities obtained from wireline log in an offshore borehole.

Experimental results show that Poisson's ratio in a rock depends on lithology, confining stress, pore pressure, and porosity of the rock. Normally Poisson's ratio is about 0.2 for sandstones, about 0.3 for carbonate rocks, and greater than 0.3 for shales.

It is noticed that Poisson's ratio varies significantly with the burial depth. Figure 1.6 shows that Poisson's ratio increases as the increase of the depth below the sea floor in the Gulf of Mexico, where most formations are shales. The following correlation between Poisson's ratio and the depth can be obtained from Fig. 1.6:

$$\nu = 0.0582 \ln Z - 0.0174 \tag{1.33}$$

where Z is the burial depth measured in meters.

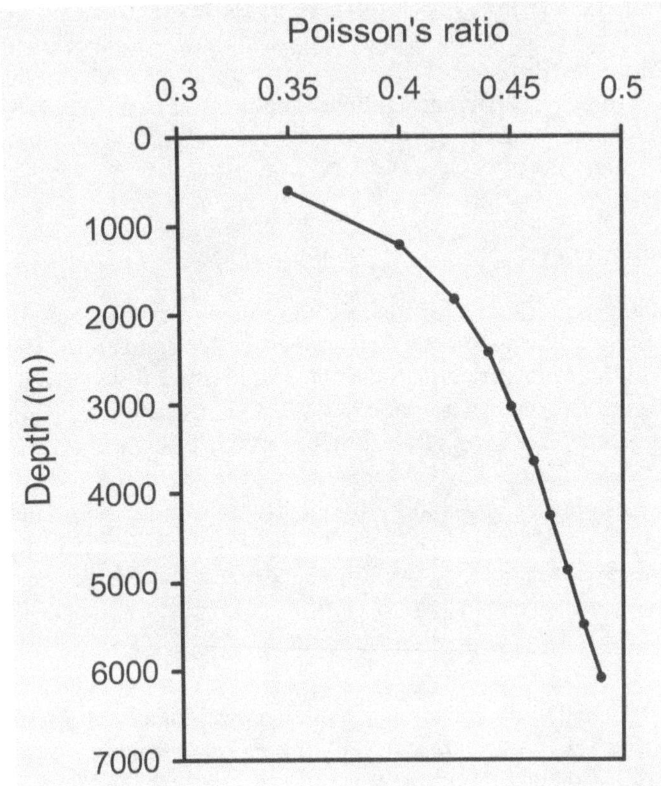

Fig.1.6. Poisson's ratio as a function of burial depth in the Gulf of Mexico.

1.6 Rock strength estimation from physical properties

Compressive strength and tensile strength parameters are the widely used rock strength parameters for geomechanical analyses, particularly the uniaxial compressive strength (UCS), rock cohesion (c), and angle of internal friction (φ). Rock strength parameters can be obtained from core sample tests in lab experiments (Peng 1998b, Peng et al. 2002a, Peng et al. 2002c). However, in most case the core samples are unavailable for laboratory testing, particularly in oil and gas industry, where the formations are very deep. In this case the empirical corrections between rock strength and geophysical well log data can be used for estimation (Peng et al. 2001). These correlations usually are developed for some specific rock formations based on relationships of laboratory core tests and geophysical data. Because there are multiple choices of strength models and correlations for various rock types in different geological settings, it is necessary to understand the characteristics of the model and their range of applicability prior to utilizing them (Chang et al. 2006).

1.6.1 Sandstones

The following empirical relationships were given between uniaxial compressive strength and other physical properties for sandstones (Change et al. 2006). In the following equations sonic or seismic P-wave velocity (v_p) or equivalently interval transit time ($\Delta t = 1/v_p$) is used for determination of rock strength (UCS) (Peng et al. 2002b). In the following equations, the unit of UCS is MPa, the unit of v_p m/s. However, the unit of Δt used in the context is μs/ft, because it is more frequently used in the petroleum industry.

Freyburg (1972) developed the following relationship based on the data in Thuringa, Germany (Chang et al. 2006):

$$UCS = 0.035 v_p - 31.5 \tag{1.34}$$

For fine grained, both consolidated and unconsolidated sandstones in the Bowen Basin of Australia, McNally (1987) presented the following strength estimation using sonic logs:

$$UCS = 1200 \exp(-0.036 \Delta t) \tag{1.35}$$

For weak and unconsolidated sandstones in the U.S. Gulf Coast the following empirical equation exists (Chang et al. 2006):

$$UCS = 1.4138 \times 10^7 \Delta t^{-3} \tag{1.36}$$

Fjær et al. (1992) proposed the following relationship that is applicable to sandstone with UCS > 30 MPa:

$$UCS = 3.3\times10^{-14}\rho^2 v_p^2[(1+\nu)/(1-\nu)]^2(1-2\nu)(1+0.78V_{clay}) \quad (1.37)$$

where ρ is the density of the rock, in g/cm^3; and V_{clay} is the friction of clay in the sandstone.

In the Gulf of Mexico, the following relationship was developed (Chang et al. 2006):

$$UCS = 3.87\exp(1.14\times10^{-10}\rho v_p^2) \quad (1.38)$$

where ρ is the density of the rock, in g/cm^3.

Except for Eq. 1.34, all of the equations appear to badly under-predict the strength data for high travel time ($\Delta t > 100$ μs/ft), or very low velocity ($v_p < 3000$ m/s) as reported by Chang et al. (2006). Therefore, calibration is extremely important before using any of these strength correlations. For very weak North Sea sandstone provided by Bradford et al. (1998), Equations 1.36 and 1.37 fit the data fairly well (Chang et al. 2006).

Weingarten and Perkins (1992) presented the following relation to predict sandstone internal friction angle, φ (degrees) using porosity ϕ (fraction):

$$\varphi = 57.8 - 105\phi \quad (1.39)$$

1.6.2 Shales

Lal (1999) developed the following shale strength correlations tied only to compressional sonic velocity in shales. The relations were developed using an extensive shale database. The following relations show the internal friction angle, φ (degrees), cohesion, c (MPa), and UCS (MPa) as a function of compressional sonic velocity v_p (km/s):

$$\sin\varphi = (v_p - 1)/(v_p + 1) \quad (1.40)$$

$$c = 5(v_p - 1)/\sqrt{v_p} \quad (1.41)$$

$$UCS = 10(304.8/\Delta t - 1) \quad (1.42)$$

Through laboratory triaxial compressive tests of shale samples obtained from deep boreholes from the North Sea, Horsrud (2001) proposed the following correlations:

$$UCS = 0.77 v_p^{2.93} \tag{1.43}$$

$$UCS = 243.6 \phi^{-0.96} \tag{1.44}$$

$$E = 0.076 v_p^{3.23} \tag{1.45}$$

where ϕ is porosity in percent; v_p in km/s; and E is the Young's modulus in MPa. Using the P-wave interval transit time from the sonic log, Δt, in μs/ft, the correlations can be expressed as (Horsrud 2001):

$$UCS = 0.77(304.8 / \Delta t)^{2.93} \tag{1.46}$$

$$E = 0.076(304.8 / \Delta t)^{3.23} \tag{1.47}$$

The Lal's and Horsrud's correlations predict strength fairly well for the North Sea and the Gulf of Mexico where high porosity, unconsolidated Tertiary or younger shales are dominant. Therefore, these correlations can be used to estimate shale strength in weak formations. In these strength correlations Horsrud's correlation predicts a higher shale strength than Lal's for the shale with high sonic velocity (or low transit time). Figure 1.7 presents the rock cohesion calculated from sonic well log data by using Lal's and Horsrid's correlations in an oil exploration well in Mexico.

1.6.3 Limestone and Dolomite

Milizer and Stoll (1973) presented the following strength and interval transit time relationship for limestone and dolomite (Chang et al. 2006):

$$UCS = (7682 / \Delta t)^{1.82} / 145 \tag{1.48}$$

Golubev and Rabinovich (1976) proposed the following empirical equation for limestone and dolomite (Chang et al. 2006):

$$UCS = 10^{(2.44 + 109.14 / \Delta t)} / 145 \tag{1.49}$$

Equation 1.48 predicts a much lower strength than Eq. 1.49.

For the limestone and dolomite with low to moderate porosity ($0.05 < \phi < 0.2$) and high UCS ($30 < UCS < 150$ MPa) in Middle East, the following empirical equation was obtained (Chang et al. 2006):

$$UCS = 143.8 \exp(-6.95 \phi) \tag{1.50}$$

where ϕ is porosity in fraction.

When laboratory tests are not possible, the field estimates of rock strength can be conducted by examining rock samples and drilling cuttings. Table 1.4 presents the field methods to estimate rock uniaxial compressive strength proposed by Hoek and Brown (1997).

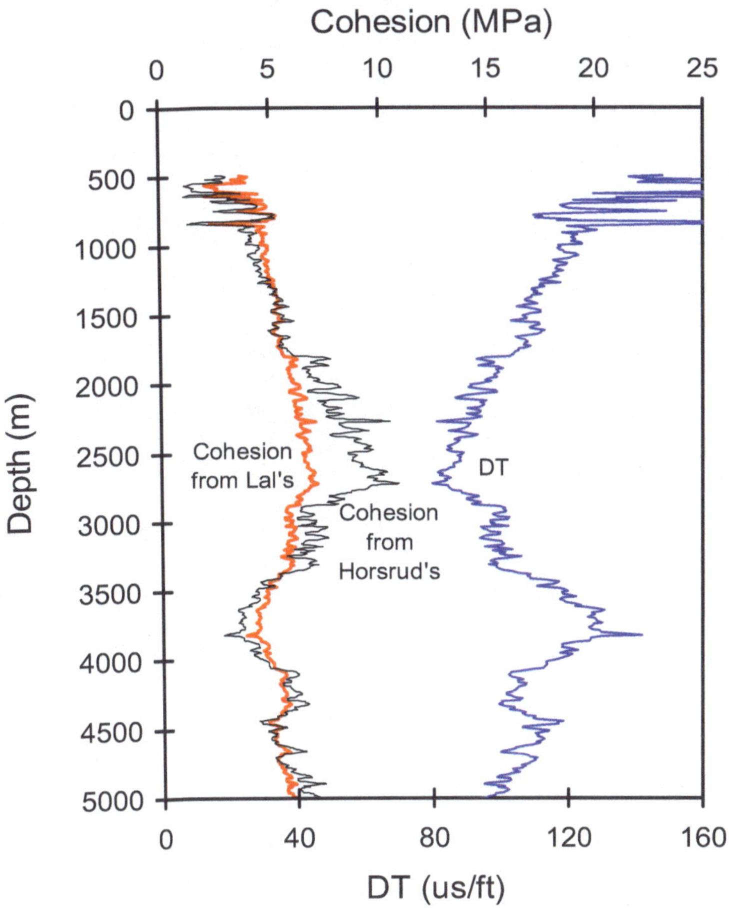

Fig.1.7. Rock cohesion calculated from sonic transit time (Δt or DT) by using Horsrud's and Lal's correlations.

Table 1.4. Field estimates of rock uniaxial compressive strength (UCS)

	UCS (MPa)	Field characteristics	Examples
Extremely strong	> 250	Specimen can only be chipped with a geologic hammer	Fresh basalt, chert, diabase, gneiss, granite, quartzite
Very strong	100-250	Specimen requires many blows of a geological hammer to fracture it	Amphibolite, sandstone, basalt, gabbro, gneiss, granodiorite, limestone, marble, rhyolite, tuff
Strong	50-100	Specimen requires more than one blow to fracture it	Limestone, marble, phyllite, sandstone, schist, shale
Medium strong	25-50	Cannot be scraped or peeled with a pocket knife. Specimen can be fractured with a single blow from a geological hammer.	Claystone, coal, concrete, schist, shale, siltstone
Weak	2-25	Can be peeled with a pocket knife with difficulty. Shallow indentation made by firm below with point of a geological hammer.	Chalk, rocksalt, potash
Very weak	1-5	Crumbles under firm blows with point of a geological hammer, can be peeled by a pocket knife	Highly weathered or altered rock
Extremely weak	0.25-1	Indented by thumbnail	Stiff fault gouge

Partially reproduced from Hoek and Brown 1997 with permission of Elsevier

1.7 Elastic stress-strain relationship

For linear elastic isotropic materials the three-dimensional stress and strain follow the Hooke's law:

$$\varepsilon_x = \frac{1}{E}[\sigma_x - \nu(\sigma_y + \sigma_z)] \tag{1.51}$$

$$\varepsilon_y = \frac{1}{E}[\sigma_y - \nu(\sigma_x + \sigma_z)]$$

$$\varepsilon_z = \frac{1}{E}[\sigma_z - \nu(\sigma_x + \sigma_y)]$$

$$\Gamma_{xy} = \frac{1}{2G}\tau_{xy}$$

$$\Gamma_{xz} = \frac{1}{2G}\tau_{xz}$$

$$\Gamma_{yz} = \frac{1}{2G}\tau_{yz}$$

where σ, τ are the normal and shear stresses, respectively; ε, Γ are the normal and shear strains, respectively; G is the shear modulus.

In plane stress (biaxial stress) state, $\sigma_z = \tau_{yz} = \tau_{zx} = 0$, the strain-stress has the following forms:

$$\varepsilon_x = \frac{1}{E}(\sigma_x - \nu\sigma_y) \tag{1.52}$$

$$\varepsilon_y = \frac{1}{E}(\sigma_y - \nu\sigma_x)$$

$$\Gamma_{xy} = \frac{1}{2G}\tau_{xy}$$

and stress-strain relation is:

$$\sigma_x = \frac{E}{1-\nu^2}(\varepsilon_x - \nu\varepsilon_y) \tag{1.53}$$

$$\sigma_y = \frac{E}{1-\nu^2}(\varepsilon_y - \nu\varepsilon_x)$$

$$\tau_{xy} = 2G\Gamma_{xy}$$

This case occurs when a thin plate is stressed in its own plane. It also occurs in the analysis at any free surface, if the x- and y-axes are taken in the surface (Jaeger and Cook 1979).

In plane strain (biaxial strain) state, $\varepsilon_z = \Gamma_{yz} = \Gamma_{zx} = 0$, the strain-stress can be expressed as:

$$\varepsilon_x = \frac{1+\nu}{E}[(1-\nu)\sigma_x - \nu\sigma_y] \tag{1.54}$$
$$\varepsilon_y = \frac{1+\nu}{E}[(1-\nu)\sigma_y - \nu\sigma_x]$$
$$\Gamma_{xy} = \frac{1}{2G}\tau_{xy}$$

and stress-strain relation is:

$$\sigma_x = \frac{E}{(1+\nu)(1-2\nu)}[(1-\nu)\varepsilon_x + \nu\varepsilon_y) \tag{1.55}$$
$$\sigma_y = \frac{E}{(1+\nu)(1-2\nu)}[(1-\nu)\varepsilon_y + \nu\varepsilon_x)$$
$$\sigma_z = \frac{\nu}{2}(\sigma_x + \sigma_y)$$
$$\tau_{xy} = 2G\Gamma_{xy}$$

This case is applicable to boreholes, and two-dimensional openings.

For linear elastic anisotropic materials with orthorhombic symmetry, the three-dimensional stress and strain can be expressed as the following relations (Fjaer et al. 1992):

$$\sigma_x = C_{11}\varepsilon_x + C_{12}\varepsilon_y + C_{13}\varepsilon_z \tag{1.56}$$
$$\sigma_y = C_{12}\varepsilon_x + C_{22}\varepsilon_y + C_{23}\varepsilon_z$$
$$\sigma_z = C_{13}\varepsilon_x + C_{23}\varepsilon_y + C_{33}\varepsilon_z$$
$$\tau_{yz} = 2C_{44}\Gamma_{yz}$$
$$\tau_{xz} = 2C_{55}\Gamma_{xz}$$
$$\tau_{xy} = 2C_{66}\Gamma_{xy}$$

where C_{ij} are the elastic constants.

1.8 Effective stress and poroelastic stress-strain relationship

Most underground rocks belong to porous media. For such a medium the solid and fluid parts are deformed independently, and effective stress needs to be considered in a fluid-saturated medium. The relationship between changes in total stress (σ_{ij}) and effective stress (σ'_{ij}) was given by Terzaghi (1943) and can be expressed as:

$$\sigma_{ij} = \sigma'_{ij} + \alpha p \delta_{ij} \tag{1.57}$$

where p is the fluid pressure; δ_{ij} is the Kronecker's delta, $\delta_{ij} = 1$, when $i = j$; and $\delta_{ij} = 0$, when $i \neq j$; α is the Biot's constant (Biot, 1941), and

$$\alpha = 1 - K / K_s \tag{1.58}$$

where K is the bulk modulus of the skeleton; K_s is the bulk modulus of the solid grains. α is restricted to the range of $\phi < \alpha \leq 1$. In unconsolidated or weak rocks, it is close to 1 (Fjaer et al. 1992).

The strain and stress relation can be expressed as:

$$\begin{aligned}
\varepsilon_x &= \frac{1}{E}[\sigma_x - \nu(\sigma_y + \sigma_z) - \alpha(1-2\nu)p] \\
\varepsilon_y &= \frac{1}{E}[\sigma_y - \nu(\sigma_x + \sigma_z) - \alpha(1-2\nu)p] \\
\varepsilon_z &= \frac{1}{E}[\sigma_z - \nu(\sigma_x + \sigma_y) - \alpha(1-2\nu)p] \\
\Gamma_{xy} &= \frac{1}{2G}\tau_{xy} \\
\Gamma_{xz} &= \frac{1}{2G}\tau_{xz} \\
\Gamma_{yz} &= \frac{1}{2G}\tau_{yz}
\end{aligned} \tag{1.59}$$

References

Biot MA (1941) General theory of three-dimensional consolidation. J Appl Phys 12(1):155-164

Baker Atlas (2002) Introduction to wireline log analysis. Baker Hughes

Bear J (1972) Dynamics of Fluids in Porous Media. Dover

Bradford LDR, Fuller J, Thompson PJ, Walsgrove TR (1998) Benefits of assessing the solids production risk in a North Sea reservoir using elastoplastic modeling. SPE/ISRM Eurock'98 held in Trondheim Norway pp261-269

Chang C, Zoback MD, Khaksar A (2006) Empirical relations between rock strength and physical properties in sedimentary rocks. J Petrol Sci Eng 51:223-237

Du L, Du L, Peng S, Wang Y (2001) Back calculations of formation elastic properties in VTI media. World Geology 20(4):396-416

Fjær E, Holt RM, Horsrud P, Raaen AM (1992) Petroleum related rock mechanics. Elsevier

Gardner GHF, Gardner LW, Gregory AR (1974) Formation velocity and density - the diagnostic basics for stratigraphic traps. Geophysics 39(6):770-780

Han D, Peng S (2002) Investigation and future research strategy on engineering disasters induced by coal mining at deep depth in China coal industry. China Coal 28(2):5-9 (in Chinese)

Hoek E, Brown ET (1997) Practical estimates of rock mass. Int J Rock Mech Min Sci 34 (8):1165-1186

Horsrud P (2001) Estimating mechanical properties of shale from empirical correlations. SPE Drilling Completion 16:68-73

Howarth DF (1984) Apparatus to determine static and dynamic elastic moduli. Rock Mech Rock Engng 17:255-264

Hudson JA, Harrison JP (1997) Engineering rock mechanics, an introduction to the principles. Pergamon

Jaeger JC, Cook NGW (1979) Fundamentals of rock mechanics, 3rd edition, Chapman & Hall

Lacy LL (1997) Dynamic rock mechanics testing for optimized fracture designs. Paper SPE38716

Lal M (1999) Shale stability: drilling fluid interaction and shale strength. SPE Latin American and Caribbean Petrol Eng Conf held in Caracas, Venezuela

McNally GH (1987) Estimation of coal measures rock strength using sonic and neutron logs. Geoexploration 24:381-395

Meng Z, Peng S, Fu J (2002) Key factors to control rock mechanical properties in coal measure formations. Chinese J Rock Mech Eng 21(1):102-106 (in Chinese)

Ohen HA (2003) Calibrated wireline mechanical rock properties method for predicting and preventing wellbore collapse and sanding. Paper SPE 82236

Ostermeier RM, Pelletier JH, Winker CD, Nicholson JW (2001) Trends in shallow sediment pore pressure – deepwater Gulf of Mexico. SPE/IADC 67772 presented at SPE/IADC drilling conference held in Amsterdam, The Netherlands

Peng S (1997) Investigation of engineering disasters induced by coal mining at deep depth in China. 4th Ann meeting, Geology Branch, China Coal Industry Tech Committee, Qingdao (in Chinese)

Peng S (1998a) Establishment and implementation of geological support system of high production and high efficiency in China coal mines. Symp 31st Young scientists, China Sci Tech Soc, Coal Industry Press Beijing (in Chinese)

Peng S (1998b) Approach on mechanical properties of clastic rocks and rock micro-structure. Proc. 5th Ann Conf of Young Scientists, China Coal Society. Coal Industry Press (in Chinese)

Peng S (1999) The state-of-art and the future work of engineering disasters induced by coal mining at deep depth in China. Symp Engineering and Disaster, Dept Material Science and Engineering, China NSF, Mt Jiuhua, Jianxi (in Chinese)

Peng S, Qu H, Luo L, Wang L, Duan Y (2000) Complete stress-strain and permeability experiments. J China Coal Society 25(2):113-116 (in Chinese)

Peng S, Wang X, Xiao J, Wang L, Du M (2001) Seismic detection of rockmass damage and failure zone in tunnel. J China Univ Min Tech 30(1):23-26 (in Chinese)

Peng S, Wang J (2001) Safe mining above confined aquifers. Coal Industry Press Beijing (in Chinese)

Peng S, Meng Z (2002) Theory and practice of mining engineering geology. Geological Press (in Chinese)

Peng S, Sun L, Ma C (2002a) Comprehensive analyses of rockmass deformation due to underground mining and their application. J China Min Tech Univ (Natural Sci Ed) 31(1):80-83 (in Chinese)

Peng S, Ling B, Liu D (2002b) Application seismic CT detection technique into roof coal caving comprehensive mechanical longwall mining. Chinese Rock Mech Eng 21(12):1786-1790 (in Chinese)

Peng S, Wang S, Guo J, et al. (2002c) VSP log and its application in Dong #2 borehole of Huainan Coalfield. J China Coal Society 27(6):576-580 (in Chinese)

Raymer LL, Hunt ER, Gardner JS (1980) An improved sonic transit time-to-porosity transform. Paper P, Twenty-first Annual Logging Symp Lafayette Louisiana

Schön JH (1996) Physical properties of rocks: fundamentals and principles of petrophysics. Handbook of geophysical exploration – seismic exploration (Helbig K, Treitel S, ed) Vol18, Pergamon

Terzaghi K (1943) Theory soil mechanics. John Wiley & Sons

Wang J, Peng S, Meng Z (2001) Permeability changes in complete stress-strain test of rock samples. J Beijing Sci Tech Univ 23(6):489-491 (in Chinese)

Wang S (1981) On the mechanism and process of slope deformation in an open pit mine. Rock Mechanics Rock Engng 13(3):145-156

Weingarten JS, Perkins, TK (1992) Prediction of sand production in gas wells: methods and Gulf of Mexico case studies. Paper SPE 24797 Washington DC, 4-7 October

Wyllie MRJ, Gregory AR, Gardner LW (1956) Elastic wave velocities in heterogeneous and porous media. Geophysics 21:41-70

Zhang J, Zhang Y, Liu T (1997) Rock mass permeability and coal mine water inrush. Geological Publ House Beijing (in Chinese)

Zhang J, Standifird WB, Roegiers JC, Zhang Y (2007) Stress-dependent fluid flow and permeability in fractured media: from lab experiments to engineering applications. Rock Mech Rock Engng 40(1):3-21

2 Sedimentary environments and geologic structures

2.1 Sedimentary environments and rock geomechanical behaviors

Sedimentary environments are important for investigating rock physical, geological, and geomechanical behaviors (Peng and Zhang 1995). The brief overview of sedimentary geology is introduced in this section.

Different from other mineral resources, coal-bearing formations mainly formed in different ancient coal accumulation environments (Peng 1991). Its characters are controlled by the ancient geologic environment and its transition when the peat was piled up. Because of different sedimentary environments and sedimentary features, the thicknesses of the coal seam roof and floor formed under different environments changes greatly in both vertical and lateral directions (Peng 1993a). These lead to heterogeneity and discontinuity of coal-bearing formations. During mining process of the coal seams, roof stability becomes worse in these weak formations. Roof caving, bottom heaving, and rock burst accidents often occur in the transition zone between the sandstone and mudstone of the roof. Investigation shows that one of the major reasons for the low production of the fully mechanized mining face in China was the difficulty caused by roof stability (Peng 1994). The non-productive time caused by the roof caving in the mining face occurs about 40% to 60% of the total non-productive time (Peng 1999a). Statistical data from the world's main coal producing countries show that about one third of underground casualty accidents were caused by improper control on roof stability or the earth pressure.

On the other hand, during mine development and production, in order to meet the requirements of transportation and ventilation, a safe, stable, and complete shaft and tunnel network system need to be established. However, because the lithologies of the shaft and tunnel change significantly in the lateral direction, soft or weak formations can be encountered. In this case, shaft and tunnel construction and maintenance become difficult. In the past, mine designers often considered that the rocks distribute constantly in both thickness and lithology in the lateral direction. This design was not always true and led to the failure of the original design. Therefore, it is necessary to study rock sedimentary environments, to understand rock lithology distribution in the studied area, and to make the appropriate design according to geological conditions (Wang 2003, Sun and Wang 2000).

Formations making up the Earth's crust are described by the term, facies. Sedimentary facies is the most interested facies with regard to fluid flow. Broadly divided into sandy facies, shaly facies, and carbonate facies, sedimentary facies is related to the environment in which their sediments are deposited. In general, sedimentary environments include alluvial fan sediments, fluvial deposits, delta deposits, lake deposits, barrier island deposits, and lagoon deposits.

2.1.1 Alluvial fan sediments

Alluvial fan sediments are deposits of sediments in regions of high relief, generally where streams issue from mountains onto a level plain. The fan starts at the apex, the source of sediments from regions of higher relief, as shown in Fig. 2.1. Sediment transport from the apex tends to follow the steepest slope downward, and the sediments, therefore, spread out in a fan. The largest boulders or pebbles are deposited near the apex. Downslope, the fan channel splits up into a number of smaller channels. This reduces the velocity of the water flow, and capacity for carrying sediment is lowered. Therefore, sediments become finer-grained down-slope, even if there is no reduction in steepness of the slope. Further downslope it becomes braided streams and lake deposit. In front of the apex the large faulting sedimentary basin usually are formed. This phenomenon is quite common in the Mesozoic coal-bearing strata in the eastern China.

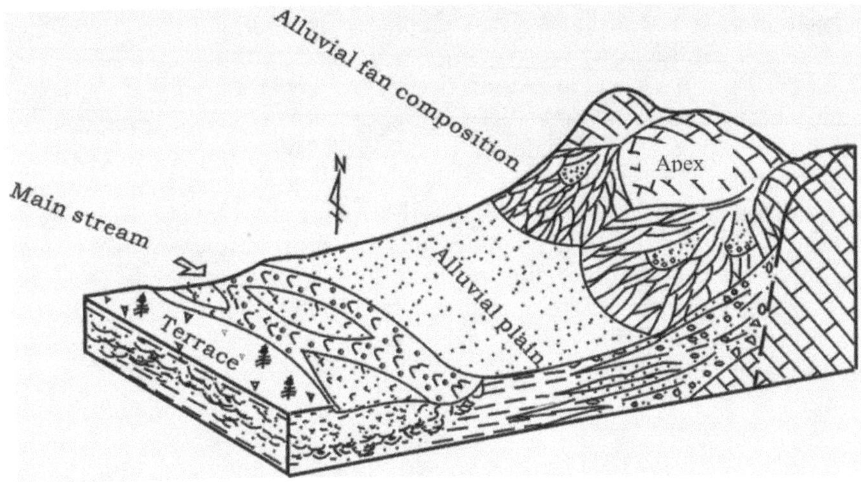

Fig. 2.1. Schematic sediments of an alluvial fan.

Near the apex the sediments are primarily consisted of conglomerates, and their main components are coarse gravels and boulders. The gravels are filled with clay, silt, and sand. The conglomerate layer distributes in strips and is parallel to the direction of the water flow.

Downslope to the middle fan, the sedimentary rocks are mainly comprised of sandstone with some gravels. In the fan tail the sediments are much finer. The sedimentary rocks are consisted of sandstones, siltstones, claystones, and coal seams (Meng et al. 1999). The main geomechanical features of the alluvial fan sediments are as follows:

- The rocks are blocky with great thickness and high strength.
- The sediment lithology is complex with coarse granularity, and the rocks are easily weathered.
- On the end of an alluvial fan, the coarse rocks have high porosity and it may be a good aquifer.

2.1.2 River sediments

Fluvial deposits are sediments deposited as the result of rivers. Fluvial sediments include deposits of braided streams, meandering rivers, and anastomosing streams (Peng 1990). Figure 2.2 shows the fluvial facies of meandering rivers. Braided streams have branched channels because the river channel is not very stable. This usually occurs with steeper stream gradients and an abundant supply of sediments. Braided streams favor the deposition of coarse sediments containing coarse sand and gravel with little clay and silt. Meandering rivers move in loops, with the greatest velocity at the outer bank where erosion occurs and lower velocity at the inner bank where deposition occurs. The fining-upward sequence from sand to silt to clay is typical of meandering river facies. This sequence is the result of the water velocity decreasing as the river, over a given spot, migrates from the outer bank to the inner bank. If the spot is no longer in the river but in the flood plain, then only clay and silt from nearly stagnant water are deposited. An anastomosing stream is defined as a branching, interlacing stream having a netlike appearance. Along the U.S. Gulf Coast is a typical example of bayous and slough in regions of very low stream gradient and with subsidence.

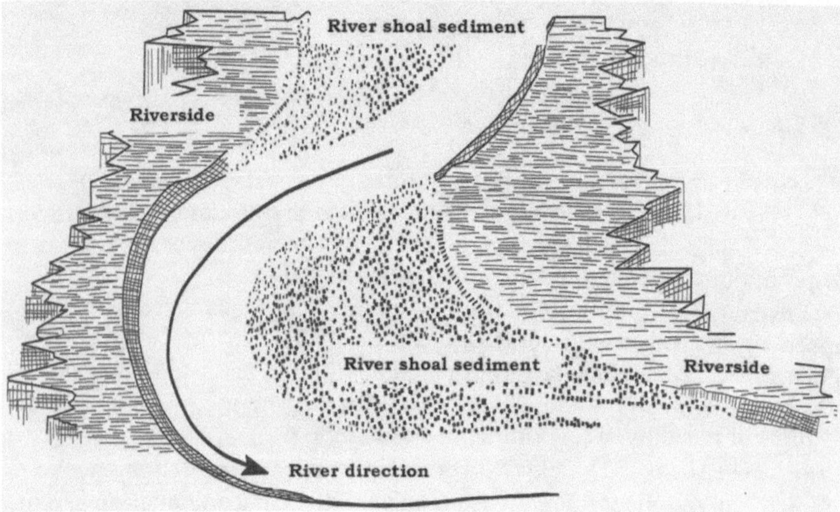

Fig. 2.2. Schematic representation of meandering rivers of fluvial facies.

Meandering river sediments vary from coarse to fine grains upwards, as shown in Fig. 2.3. During the basin deposit process, thick sediments may form due to the continuous development of the meandering rivers. In the meandering river sedimentary system, channel sands act as the skeleton in the rockmass. Generally, it forms laterally many strips of sandstones surrounded by the flood basin deposits. In the lateral direction some layers in the strata were thickened, thinned, or even disappeared. Vertically, the sandstones array or overlap each other in lens shape, and lithology varied cyclically. Therefore, in the meandering river deposit system, fine-grained sediments always surround sand sediments. These made rock mechanical properties anisotropic and heterogeneous (Peng 1996b).

The rocks in fluvial deposits are sandstone, siltstones, shales, mudstones, and claystones and have the following geomechanical behaviors:

- The rocks are stratified layers and interbedded and alternated with soft and hard layers.
- The sandstones are weak in weathering resistance, and the strength changes gradually from the bottom to the top.
- The strength of the sandstone in which the sandstone becomes thinner and near the dead-end is the lowest.

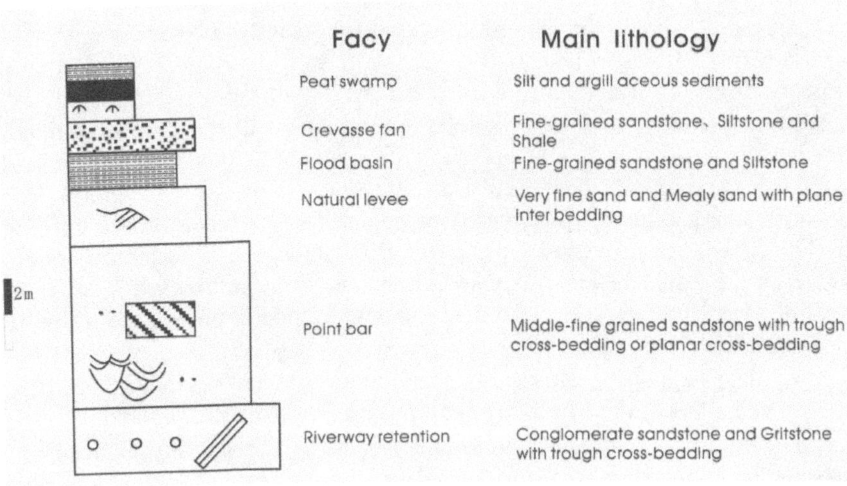

Fig. 2.3. Schematic representation of different types of fluvial facies.

2.1.3 Delta sediments

Deltas form where rivers carrying a large supply of sediments empty into a sea coast where the sediments cannot be transported away as fast as it is deposited (Peng 1996a). Thus, deltas lie in the transit region between the fluvial and marine environments. The Mississippi River Delta is a good example.

The rocks in delta deposits have the following geomechanical behaviors (Peng 1994):

- The rocks are stratified with significant variations of lithology and thickness in the lateral direction.
- The rocks are weak in weathering resistance, and the rock strength

increases from the bottom to the top as the granularity increases.

2.1.4 Lake sediments

Lake sediments are deposited in a lake accumulated on the lake shore and on lake floor. They are deposited in a terrestrial environment and contain organic and inorganic particles, microfossils such as pollen and algae, and macrofossils such as leaves and seeds. Deposit speed in lake environment is faster than that in marine environment, because of a smaller wave in the lake.

Lakeshore deposits are generally well-sorted sands. The sediments load of a stream entering a lake will be dropped as the stream's velocity and transporting ability suddenly decrease. The resulting deposit, which extends outward into the lake, is a delta (Murck and Skinner 1999). Inclined, generally well-sorted layers on the front of a delta pass downward and outward into thinner, finer, evenly laminated layers on the lake floor.

Most lake sediments are layered, in which the layers/strata are defined by color variations. In the deeper parts of the lake, the sedimentary layers are very thin, and deeper-water sediments are fine-grained while those in shallow water are coarse. The strata in lake deposits have the following geomechanical behaviors:

- The rocks have alternately soft and hard layers deposited. Periodic changes of the lake level generate cyclical soft and hard strata.
- The rock layers are continuous with little change in thickness and have low strength.
- Most strata are impermeable layers.

2.1.5 Barrier island sediments

Barrier islands or spits are long, narrow, offshore deposits of sand or sediments that parallel the coast line. The islands are separated from the main land by a shallow sound, bay or lagoon. Barrier islands are often found in chains along the coast line and are separated from each other by narrow tidal inlets. The rising waters carried sediments from those beach ridges and deposited them along shallow areas just off the new coast lines. Waves and currents continued to bring in sediments that built up, forming the barrier islands. In addition, rivers washed sediments from the mainland that settled behind the islands and helped build them up. The sediments of barrier islands are well-sorted, generally medium to fine grained sandstones with silicate cementation. These sandstones have very high strength.

Some strata of barrier island sediments can extend for 160 km or more. They are very hard strata. If a coal seam roof is this kind of strata for longwall caving mining, it is very easy to form a large area of un-caving strata; therefore it is likely to have rock burst.

2.1.6 Lagoon and tidal lagoon sediments

Lagoon, tidal lagoon, and barrier islands or spits are sedimentary elements that parallel to the coast line, as shown in Fig. 2.4. Figure 2.5 shows the littoral sediment of the lagoon and barrier spit of Bolinas Bay of California (Danmeier and Williams 2005).

Lagoon environment belongs to a shallow basin that are separated from the ocean by barrier islands or barrier spits and jointed with the ocean through tidal inlets. In the places where tide develops, a lagoon is a shallow depression full of water even in the period of low tide. If there is sufficient sediment supply, a coastal lagoon can gradually develop into a tidal lagoon or swamp. Therefore, lagoon deposit is closely associated with tidal lagoon and swamp deposits. They transit vertically and are contiguous horizontally.

Lagoon sediments generally are laminated fine-grained sediments, such as clay and silt. In humid and semi-humid regions where coal measures form, these fine sediments often are rich in organic substances. Tidal lagoon is a wide and flat region around the lagoon and depends on the difference of the low and high tides and the ground slope. Near low tide line in the intertidal zone, due to the strong hydraulic activity, flat sand deposits can be formed and developed to be large slaty or sphenoid crossbeddings. Near high tide line, the sediments are mostly mud and silt with horizontal lamination and current lamination.

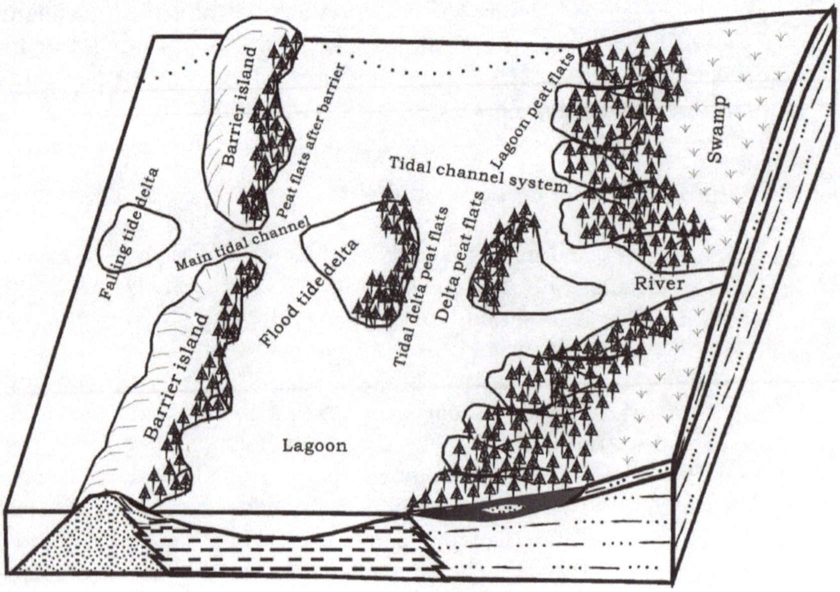

Fig. 2.4. Schematic representation of Lagoon sediments.

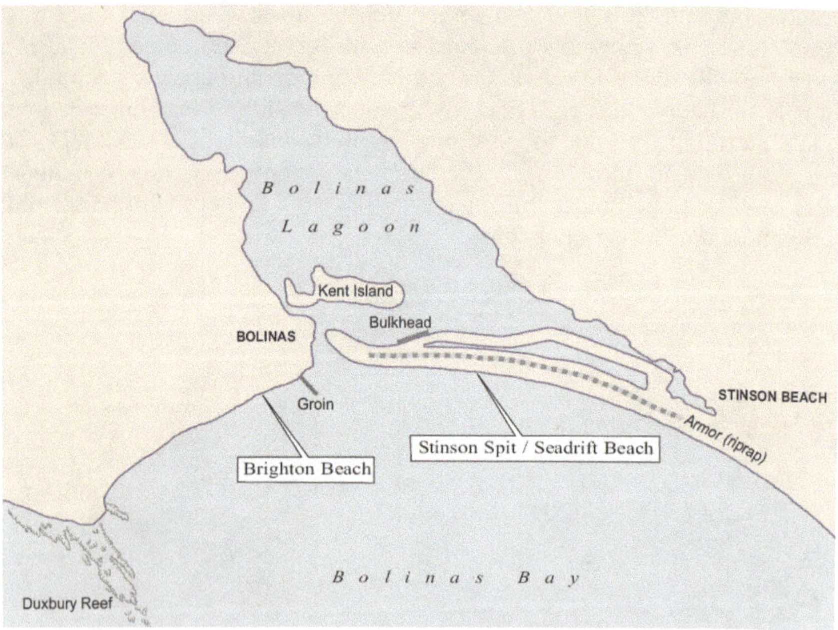

Fig. 2.5. Conceptual representation of Bolinas lagoon and barrier island/spit.

Sediments from the lagoon are uniform in mineralogical and mechanical composition. The geomechanical features of the sediments in this sedimentary mode are as follows:

- The rocks have alternately soft and hard layers deposited. Weak interfaces exist between hard and soft layers.
- The rock layers are continuous with little change in thickness and have low strength.
- The formations usually contain clay minerals, which are most likely to swell and weaken, particularly when they are exposed to water.

Analyses of sedimentary environments are important and applicable to rock engineering. Through investigation of sedimentary environments of the rocks one can understand rock structure, lithologic characteristics, and strata sedimentary sequences. This investigation is beneficial to rock engineering design and construction.

2.2 Sedimentary characteristics and strata lithology

Sedimentary rocks differ according to their sedimentary environments. The mechanical properties of the sedimentary strata depend on the sediment facies characteristics. For instance, in mining engineering, the distribution of facies of sedimentary rocks controls the integrated quality of seam roof stability (Peng 1993b). Therefore, a full understanding of the lithology and thickness of the sedimentary rocks prior to coal mining is important for strata control and management as well as for roof stability of the mining panel. The following case shows the effects of lithologic variations on rock mechanical properties (Peng 1995, Peng and Li 1996).

The Permian coal seams in Huainan Coalfield of Anhui Province, China were deposited in delta-plain environments (Peng et al. 2002). The major coal seams, Seam 13-1 and Seam 11-2, were formed in a braided river system in the compound delta plain of late Permian. The roof strata were formed in the abandoned phase when the shoreline pushed landwards, and the strata are mainly composed of mudstone, silty mudstone, siltstone, and medium- and fine-grained sandstones (Meng et al. 2000). For example, in Xinji coal mine of Huainan Coalfield the immediate roof of Seam13-1 consists of mudstone or sandy mudstone with horizontal bedding, and it sometimes transits into siltstone. The thickness of the immediate roof varies from 0 to 12 m and is instable, varying greatly from east to west (Peng and Meng 2002). Along the east-west direction, the immediate roof alternates between having thick and thin rock layers. These layers extend along the north-south direction of the roof (Figs. 2.6 and 2.7). The sedimentary environment of the formations belonged to the flooding plain facies, which made the immediate roof come in contact with main roof of the sandstone by flooding erosion. The immediate roof was partially scoured, thus the coal seam and the main roof touch directly in some areas, as shown in Fig. 2.7.

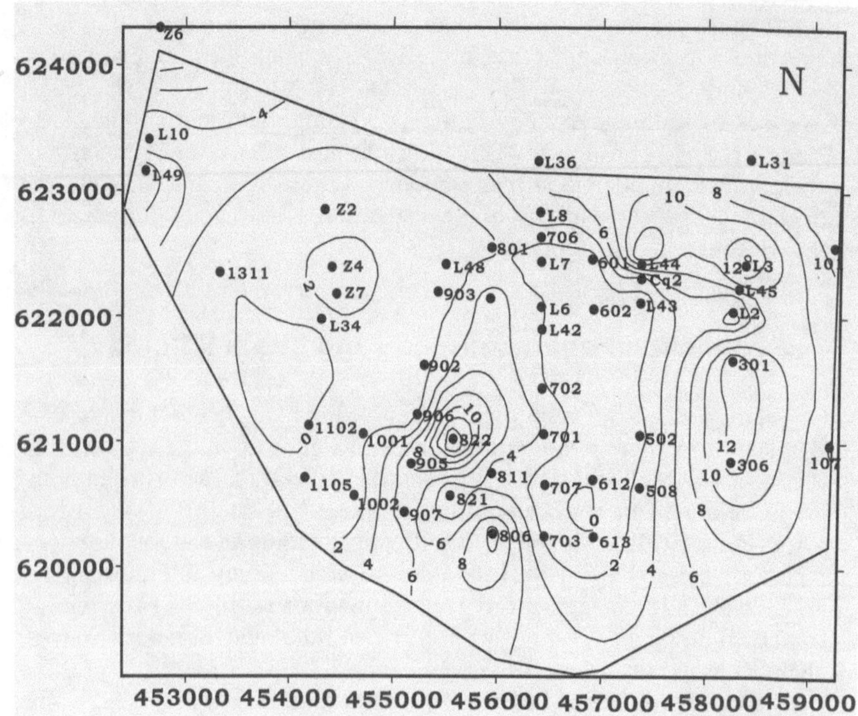

Fig. 2.6. The plan view of the contours of the thicknesses of the immediate roof in the coal seam 13-1 in Xinji coal mine, Huainan Coalfield, China. The solid dots are boreholes with hole names marked, and the unit in the figure is in meters.

The main roof of Seam 13-1 consists of fine- or medium-grained sandstone cemented by siliceous or calcic minerals. Thicker in the west and thinner in the southeast, the thickness of the roof varies from 0 to 24 m, with greater variation occurring along the east-west direction. The sandstone is distributed anastomosingly in the plan view, while in cross-sectional view it is in lentiform, i.e. the sandstone thickness varies greatly (Fig. 2.7). Upwards, at shallower depths, it usually transits into interbedding of sandstone and mudstone, and further upwards a very thin layer of coal seam is frequently found. These characteristic show that the main roof of sandstone of Seam 13-1 was created by an anastomosing river deposit (Peng and Meng 2002).

The lithology and mechanical properties of sedimentary rocks are different and controlled by their sedimentary environments and facies. In the vertical direction, lithology varies cyclically, and in the lateral direction some layers in the strata are thickened, thinned, or even disappear. These variations generate obvious spatial differences in mechanical properties of the rocks. The roof stability in tunnels and mining panels are directly dependent on the sedimentary environments of the rocks, because roof stability depends on mechanical properties of the rocks. For instance, a stronger washing action of the main roof sandstone on its underlying rock creates a more apparent and smoother interface. This is important since when mining under this condition the rocks below the main roof are more likely to collapse. When the thicknesses of sandstone and mudstone in the roof vary, or when lithology changes suddenly, formations usually are in a weak rock zone (Peng and Meng 1999a). This corresponds to a risky zone in the roof.

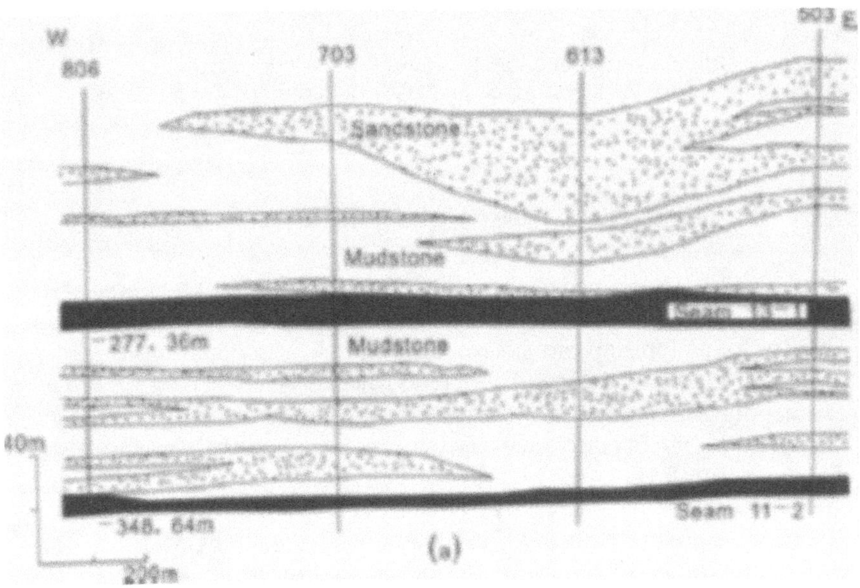

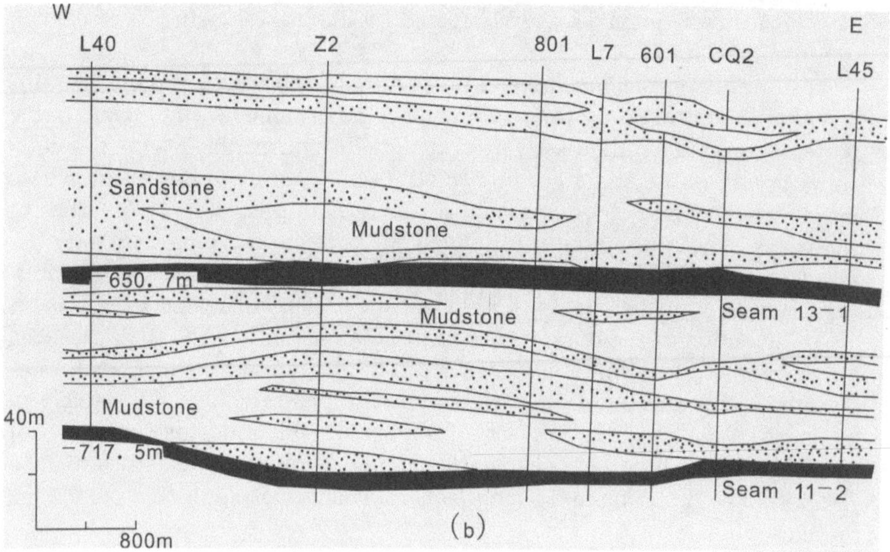

Fig. 2.7. The geologic cross-sections showing the lithology of the strata along formation strike direction in Xinji coal mine, Huainan Coalfield. **a.** Cross-section in the south side of the studied area; **b.** Cross-section in the north side of the studied area.

2.3 Geological structures of sedimentary rock mass

A rock usually is defined as the isotropic and homogeneous aggregation of minerals without macroscopically discontinuity. In geological engineering, most rocks are anisotropic, heterogeneous, and include different scales of discontinuities and geological structures, particularly when the size of rock engineering is large enough to contain these structures' planes. Figure 2.8 shows a heavily fractured and discontinuous rock mass that contains different scales of fractures. In fact, all rock masses are naturally fractured, and it is only in very rare case that the spacings between discontinuities are appreciably greater than the dimensions of the rock engineering project. Thus, formations where rock engineering is constructed can, for the most part, be treated as fractured media with different scales of discontinuities.

The geologic structures of sedimentary rock masses depend on the rock diagenesis, sedimentary environment, and alteration processes, such as tectonics (Peng and Meng 1999b, Meng et al. 1997). According to different sedimentary environments, the rock structure planes can be classified as sedimentary, geologic, and engineering structure planes.

Fig. 2.8. The sedimentary rock masses with different scales of discontinuities (Golden, Colorado).

Sedimentary structure plane

A sedimentary plane is formed during the sedimentary deposit period. An example is the bedding plane which forms at the interface of two rock layers, as shown in Fig. 2.9. The rock mechanical properties are different in the directions that are parallel to and perpendicular to the bedding.

Fig. 2.9. The bedding planes in stratified sedimentary rocks (Golden, Colorado).

Tectonic structure plane

Rock masses have historically experienced many tectonic activities (Ling et al. 2001). Therefore, most rocks are fractured or broken materials with different scales of discontinuities, such as joints, cleavages, folds, and faults. These discontinuities have significant influences on stability of rock engineering. Figure 2.10 shows a rock mass highly deformed and fractured due to tectonic activities. A fault change continuity of the formation, and the strata in both sides of the fault have different thickness and dip angles, as shown in Fig. 2.11. Therefore, faults have an important impact on rock continuity, rock strength, and stability of rock engineering (Peng et al. 2001, Meng et al. 2001).

Fig. 2.10. The geologic structures (fault, fold, and fractures) in stratified sedimentary rock masses (South Dakota).

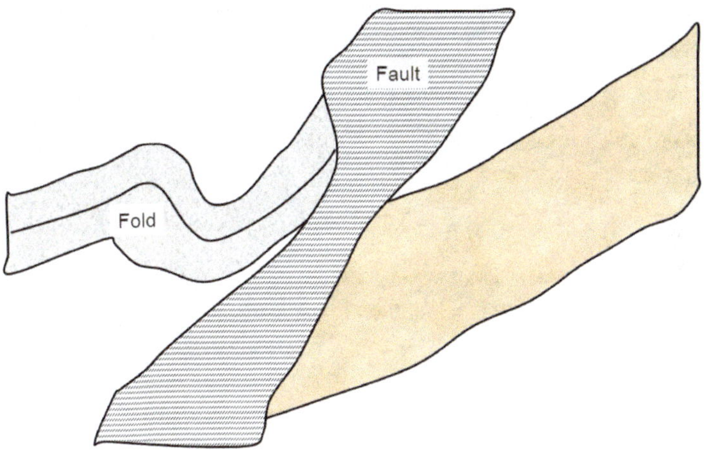

Fig. 2.11. Schematic representation of the fault and fold in the picture as shown in Fig. 2.10.

Figure 2.12 shows that an asymmetric anticline and a syncline cause formation highly deformed and twisted. In this formation the abnormal stress or tectonic stress are expected because of tectonic deformations. Therefore, formation in-situ stress and rock properties differ for different geological structures and tectonic activities. Prior to rock engineering, detection and understanding of formation geologic structures are critical important for engineering design (Peng 1999b). High resolution seismic tools are promising to detect faults in coal fields (Peng et al. 1999c).

Fig. 2.12. The geologic structures in Mountain Kidd showing an asymmetric anticline and a syncline causing formation highly deformed (Kananaskis, Canada).

References

Danmeier D, Williams P (2005) Bolinas lagoon ecosystem restoration feasibility project: final public reports. Philip Williams & Associates

Ling B, Peng S, Meng Z (2001) Overthrust structure and its influence on geological conditions of mining in Xinji coalfield. China Coal Geology 13(4):7-9 (in Chinese)

Meng Z, Peng S, He, R, Yang X, Li Y, Jiang Z (1997) Effects of geologic structures on coal reserves. China Coal Geology 9(3):36-39 (in Chinese)

Meng Z, Peng S, Qu H, Zhao H, Luo L (1999) Influence of stratigraphic facies variations on the roof stability by physical modeling study. Mining Sci and Tech '99, A A Balkema pp203-206

Meng Z, Peng S, Yin S, Wang L (2000) Physical modeling of influence of rock mass structure on roof stability. J China Univ of Min Tech 10(2):172-176 (in Chinese)

Meng Z, Peng S, Li H (2001) Variations of coal physical and mechanical properties near normal faults and their influences on stress distributions of mining. J China Coal Society 26(6): 561-566 (in Chinese)

Murck BW, Skinner BJ (1999) Geology today: understanding our planet. John Wiley & Sons

Peng S, Meng Z (2002) Theory and practice of mining engineering geology. Geological Press (in Chinese)

Peng S (1990) Fundamental characteristics of the anastomosing fluvial system on complex delta plain. Chinese Science Bulletin 35(10):835-839 (in Chinese)

Peng S, Zhang J (1995) Sedimentary environments of coal measure and their effects on mining in Wuda Coalfield. Coal Industry Press (in Chinese)

Peng S (1994) Sedimentary mode of complex delta and its characteristics. J China Coal Society 19(1):89-98 (in Chinese)

Peng S (1995) Geological modeling techniques for longwall mining roof stability: a case study. Proc Rock Mechanics and Strata Control in Mining and Geotech Eng, Beijing.

Peng S, Li Y (1996) Wedge-shaped sandstone roof strata and roof stability, examples from Seam 13-1 in Huainan Coalfield, China. Proc for Rock Mechanics and Strata Control in mining and Geotech Eng Xi'an

Peng S (1991) Paleo-hydrology in Permian No. 3 and 4 coal-bearing formations in Huainan coalfield. J Sedimentary 9(3):1-9 (in Chinese)

Peng S (1993a) State-in-art research in recent decade and future development in China's coal geology. Symp Young Scientists, China Coal Industry Press (in Chinese)

Peng S (1993b) Strata control of roof stability in mining workings. Proc Int Congr on Mining Sci Applied Geology Tech, St. Petersburg

Peng S (1994) Sedimentology of coal measure in some regions of Eastern China. China Sedimentology (Feng Z et al. ed), China Petroleum Industry Press (in Chinese)

Peng S (1996a) Modern peral river delta and Permian Huainan Coalfield, China: a comparative sedimentary facies study. Organic Geochem 24(2):159-179

Peng S (1996b) Mining-induced stress redistribution and roof hazardous characteristics of the channel-filled sandstone roof: a physical modeling study. Rock Mechanics Tools and Technique, A A Balkema pp423-429

Peng S (1999a) Geological problems and approach on high production and high efficiency of deep coal mining in China eastern coal mines. Symp national Mining Conf, Metallurgy Industry Press Beijing (in Chinese)

Peng S (1999b) Detecting instruments and techniques of complex geologic structures. Invited presentation at 4th China's Ann Tech Meeting of Mining Geology, Xi'an (in Chinese)

Peng S, Meng Z (1999a) Numerical modeling on coal roof and floor stability in longwall mining. J China Univ Min Tech 28(1):41-45 (in Chinese)

Peng S, Meng Z (1999b) Classification and its application of roof stability using lithology and geological structure. Symp 46th Young scientists, China Sci Tech Soc, China Sci Tech Press Beijing pp35-42 (in Chinese)

Peng S, Liao Y, Zhang B, Sun M (1999c) A high resolution 3D-seismic technique for distinguishing geological structures in the complex coalfield. Mining Sci Tech '99, A A Balkema

Peng S, Meng Z, Li Y (2001) Physical modeling of fault effects on roof stability. Coal Geology and Exploration 29(3):1-4 (in Chinese)

Peng S, Wang S, Guo J, et al. (2002) VSP log and its application in Dong #2 borehole of Huainan Coalfield. J China Coal Society 27(6):576-580 (in Chinese)

Sun J, Wang S (2000) Rock mechanics and rock engineering in China: development and current state-of-the-art. Int J Rock Mech Min Sci 37:447-465

Wang S (2003) Centurial achievement and tasks in Chinese rock mechanics and engineering. Chinese J Rock Mech Eng 22(6):867-871 (in Chinese)

3 In-situ stress and pore pressure

3.1 In-situ stress regimes

In-situ stress magnitudes and orientations play a very important role in geological engineering, and they are the most basic parameter inputs in design of underground structures. One of the main functions of rock mechanics has been to determine in-situ stress. In-situ stress is characterized by the magnitudes and directions of three principal stresses. Generally, in-situ principal stresses are consisted of three mutually orthogonal stresses, i.e. vertical stress (σ_V), minimum horizontal stress (σ_h), and maximum horizontal stress (σ_H). In different geographic, geologic, and tectonic regions, in-situ stress magnitudes and orientation are very different. The three in-situ stresses correspond to three principal stresses, namely greatest stress (σ_1), intermediate stress (σ_2), and least stress (σ_3). According to the relationships between principal stresses and vertical, minimum and maximum horizontal stresses, three in-situ stress regimes (Fig. 3.1) can be used to describe the stress states, i.e.:

1. Normal faulting stress regime. In this case, gravity or vertical stress drives normal faulting and fault slip occurs when the minimum stress reaches a sufficiently low value (Zoback et al. 2003). In this stress state the vertical stress is the greatest principal stress ($\sigma_V = \sigma_1$), i.e. $\sigma_V \geq \sigma_H \geq \sigma_h$.
2. Strike-slip faulting stress regime. In this case, the vertical stress is the intermediate principal stress ($\sigma_V = \sigma_2$). When the difference between maximum and minimum stresses is sufficiently large, slip faulting occurs. Therefore, this case also represents slip faulting region. In this stress state, one has $\sigma_H \geq \sigma_V \geq \sigma_h$.
3. Reverse (or thrust) faulting stress regime. In this case, the vertical stress is the least principal stress ($\sigma_V = \sigma_3$). When the stress field is very compressive, both horizontal stresses exceed the vertical stress, and folding and reverse faulting could occur when the maximum horizontal principal stress is sufficiently large relative to the vertical

stress. Some thrust fault regions belong to this category. In this stress state, one has $\sigma_H \geq \sigma_h \geq \sigma_V$.

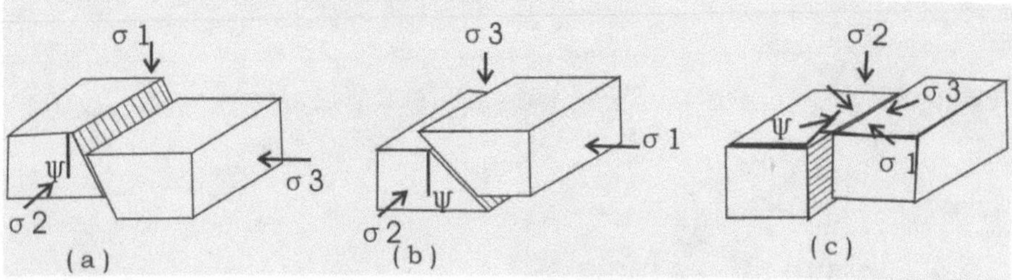

Fig. 3.1. Illustration of different faulting stress regimes: **a.** Normal faulting; **b.** Reverse faulting; and **c.** Strike-slip faulting. In the figure, $\psi = \pi/2 - \varphi_f$.

It can be seen that knowing the stress regime can better constrain in-situ stress magnitudes. Assuming there are critically oriented faults constraining stress magnitudes, the Mohr-Coulomb criterion in faults can be expressed as follows:

$$\sigma_1' \leq C_f + q_f \sigma_3' \tag{3.1}$$

where σ_1' is the maximum effective stress, and $\sigma_1' = \sigma_1 - \alpha p$; σ_3' is the minimum effective stress, $\sigma_3' = \sigma_3 - \alpha p$; C_f is the cohesion of the fault; $q_f = (1+\sin\varphi_f)/(1-\sin\varphi_f)$; φ_f is the angle of internal friction of the fault, ranging from 30 to 45°; p is the pore pressure.

In significant depth, the effect of cohesion of a fault is relatively unimportant in terms of the stress magnitudes, thus the cohesion can be neglected. Consequently the stress relationship in faults can be rewritten as the following form, assuming α=1:

$$\frac{\sigma_1 - p}{\sigma_3 - p} \leq q_f \tag{3.2}$$

In different stress regimes, according to stress magnitudes and Eq. 3.2 the following relationships exist:

1) Normal faulting stress regime

$$\frac{\sigma_V - p}{\sigma_h - p} \leq q_f \tag{3.3}$$

2) Strike-slip faulting stress regime

$$\frac{\sigma_H - p}{\sigma_h - p} \leq q_f \tag{3.4}$$

3) Reverse/thrust faulting stress regime

$$\frac{\sigma_H - p}{\sigma_V - p} \leq q_f \tag{3.5}$$

Equations 3.2 to 3.5 give a bound between in-situ stresses. These equations allow to define the limiting stress magnitudes for different tectonic environments: in normal faulting areas $\sigma_h \cong 0.6\sigma_V$, in reverse faulting areas $\sigma_H \cong 2.3\sigma_V$, and in strike-slip faulting areas if $\sigma_V \cong (\sigma_H + \sigma_h)/2$, $\sigma_H \cong 2.2\sigma_h$ (Zoback and Zoback 1989).

It is noted that the world stress map presents globally measured data of in-situ stresses, where both stress regimes and maximum stress directions can be obtained in some areas (Reinecker et al. 2006).

3.2 Overburden stress

Overburden stress, also called vertical stress, is induced by the weight of the overlying formations. If the overlying formations have an average density of ρ_a, then the overburden stress can be calculated by:

$$\sigma_V = \rho_a g z \tag{3.6}$$

where g is the acceleration due to gravity.

If density varies with depth, the vertical stress can be calculated by integration of rock densities from the surface to the depth of interest, z, i.e.:

$$\sigma_V = \int_0^z \rho_b(z) g dz \tag{3.7}$$

where $\rho_b(z)$ is the rock density as a function of depth.

In offshore areas, water depth (z_w) needs to be considered to calculate the vertical stress, that is:

$$\sigma_V = \rho_w g z_w + \int_{z_w}^z \rho_b(z) g dz \tag{3.8}$$

where ρ_w is the density of water.

Density log can be used to calculate overburden stress. However, density log is usually not recorded at shallow depth. In this case some empirical methods can be used to estimate rock density at shallow depth. Most clastic rocks have an average density of 2.3 g/cm^3, as shown in Chap. 1. Overburden stress gradient is normally in range of 18 to 25 kPa/m (about 1.0 psi/ft). Figure 3.2 gives an example to calculate overburden stress and stress gradient using density log.

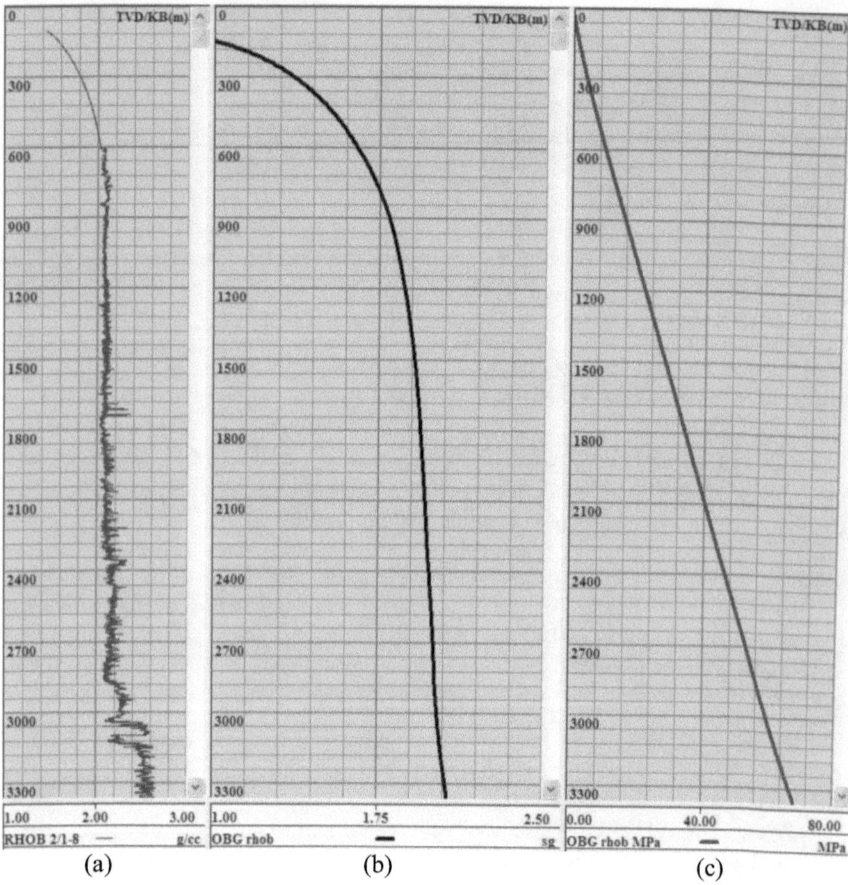

Fig. 3.2. Density log data versus the true vertical depth (TVD in meters) and Drillworks[1] calculation of overburden gradient and overburden stress in offshore

[11] Drillworks is a suite of software developed by Knowledge Systems, Inc. for predicting in-situ stress, pore pressure and wellbore stability.

drilling: **a.** density log data in g/cm^3; **b.** overburden gradient in specific gravity (sg); and **c.** overburden stress in MPa.

Worldwide in-situ stress measurements show that the vertical stress and depth have linear relationships, as shown in the following:

$$\sigma_V = (1.9 \pm 1.26) + (0.0266 \pm 0.0028)z \tag{3.9}$$

$$\sigma_V = 0.027z \tag{3.10}$$

where σ_V is the vertical stress in MPa and z is the depth in m.

3.3 Minimum horizontal stress

In normal faulting stress regime, the minimum horizontal stress is the minimum principal stress and can be calculated by the following equation:

$$\sigma_h = \alpha p + K_0(\sigma_V - \alpha p) \tag{3.11}$$

where p is the pore pressure; α is the Biot's constant; K_0 is the effective stress ratio; and

$$K_0 = \frac{\nu}{1-\nu} = \frac{\sigma_h - \alpha p}{\sigma_V - \alpha p} \tag{3.12}$$

where ν is the Poisson's ratio. The effective stress ratio, K_0, ranges from 0.33 to 1.

According to the joint industry projects conducted at Knowledge Systems (2001), the K_0 of 0.8 is fairly accurate to predict the minimum horizontal stress and fracture gradient in shaly formations in depth of below 3000 m in the Gulf of Mexico.

By considering horizontal strain and deformation effect, Hooke's law can be applied to derive the horizontal minimum stress and strain relationships. The following equation can be obtained, and it can be used to calculate the minimum horizontal stress with tectonic strain effects:

$$\sigma_h = \alpha p + \frac{\nu}{1-\nu}(\sigma_V - \alpha p) + \frac{E\nu}{1-\nu^2}\varepsilon_H + \frac{E}{1-\nu^2}\varepsilon_h \tag{3.13}$$

where E is the elastic modulus of the rock; ε_H and ε_h are the tectonic strains in maximum and minimum horizontal stress directions, respectively.

Using fracture data from the US Gulf Coast, Breckels and van Eekelen (1982) proposed the relationships between minimum horizontal stress and depth. In these relationships the abnormal pore pressure effects were also considered. They can be written as follows:

$$\sigma_h = 0.0053D^{1.145} + 0.46(p - 0.0105D) \quad (D < 3{,}500m) \tag{3.14}$$
$$\sigma_h = 0.0264D - 31.7 + 0.46(p - 0.0105D) \quad (D > 3{,}500m)$$

where D is the depth in meters; p is the pore pressure in MPa. σ_h is the minimum horizontal stress in MPa.

Breckel and van Eekelen pointed out that their minimum stress prediction is fairly reliable in the depth range down to 3500 m, and it can be used with a fair degree of confidence in other tectonically relaxed areas such as the North Sea (Fjær et al. 1992).

Hydraulic fracturing is a common technique for stress measurements in deep boreholes. Hydraulic fracturing stress determination aims to determine both minimum and maximum horizontal stresses. Hydraulic pressurization is conducted with specialized equipment for high pressure injection. During hydraulic fracturing, the borehole is pressured by fluid injection, and the pump injection flow and the well pressure are measured (Rutqvist et al. 2000). A similar test, leak-off test (LOT), has been conducted in drilling industry to evaluate cement job and estimate formation fracture gradient. Hubbert and Willis (1957) concluded that hydraulic fractures in the earth always propagate perpendicular to the orientation of the least principal stress, because it is the least energy configuration. This provides a fundamental principle for conventional hydraulic fracturing measurements and LOT.

Figure 3.3 illustrates the pressure, volume/time plot for a typical hydraulic fracturing or LOT measurement with the first and second pressurization cycles.

In a hydraulic fracture test, an interval of borehole is isolated and sealed with inflatable packers. In a LOT, only the hole below the casing and any new formation drilled prior to the test are exposed (Edwards et al. 2002). The fluid is then pumped into the hole with a constant flowrate. The pressure increase in the hole is typically linear as long as there are no leaks in the system, and the exposed formation is not highly permeable. As the pressure increases at some point, the rate of pressure increase changes such that the pressure-time curve departs from linearity. This departure from linearity is called (Fig. 3.3) fracture initiation pressure (p_i).

After departure from linearity, pressure is typically seen to increase at a lower rate until a maximum pressure is reached, and this pressure is rock breakdown pressure (p_b). After this point the pressure falls rapidly, or remains steady. Horizontal stress magnitudes can be estimated from the fracture breakdown pressure (Haimson and Fairhurst 1970):

$$p_b = 3\sigma_h - \sigma_H - p + T_0 \tag{3.15}$$

where p is the pore pressure; p_b is the fracture breakdown pressure; and T_0 is the tensile strength of the rock.

This relationship assumes that the rock behaves elastic and isotropic. It indicates that fracture breakdown pressure is greater in magnitude than the minimum horizontal stress.

After the rock is broken down (hydraulic fracture created), at some point the pressure in the hole levels off and remains fairly constant (p_{prop}) at the same flowrate. During this stage, the fracture is propagating. When the pumps are turned off, the pressure immediately drops to the instantaneous shut-in pressure (p_{isip}). After the well is shut-in, the pressure begins to drop as fluid leaks into the formation. As the pressure declines, the fracture starts to close. If the fracture is vertical and largely in the far field, then the stress acting to close the fracture is equal to the minimum horizontal stress (σ_h).

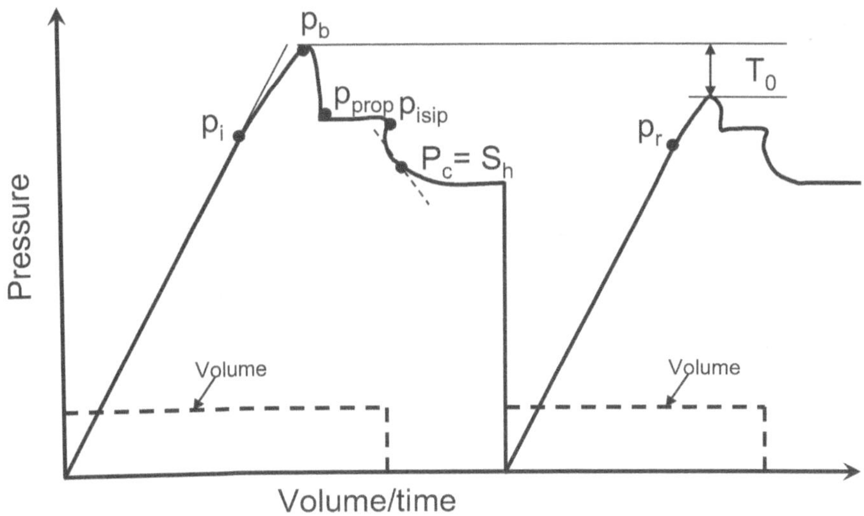

Fig. 3.3. Schematic pressure versus volume/time plot of a hydraulic fracture process. The figure also shows the determination of minimum stress/closure pressure (S_h or σ_h) using the tangent method.

In the LOT test the minimum horizontal stress (σ_h or S_h) is determined from the shut-in pressure. The minimum horizontal stress is equal to the closure pressure (p_c), as shown in Fig. 3.3.

$$\sigma_h = p_c \tag{3.16}$$

3.4 Maximum horizontal stress

3.4.1 Determinations from field tests

The maximum horizontal stress can be estimated from the LOT with fracture reopening test, as shown in Fig. 3.3. In an ideal case, when no fluid penetrates the fracture until fracture reopening, the maximum horizontal stress (σ_H) may be calculated using the following equation for a vertical borehole (Bredehoeft et al. 1976):

$$\sigma_H = 3\sigma_h - p_r - p_0 \tag{3.17}$$

where p_r is the reopening pressure and p_0 is the pore pressure in the fracture.

This equation is derived from the Kirsch solution for a circular hole subjected to an internal pressure in an isotropic, homogeneous, and linear elastic medium. The assumption in the derivation is that the reopening occurs when the fluid pressure applied on the borehole wall is high enough to cancel out the minimum tangential stress at the borehole wall.

If the fracture is slightly conductive, the fluid pressure at the wellbore wall may only partially penetrate, and the pressure inside the fracture at the wellbore wall is equal to the well-pressure. In this case, the maximum horizontal stress can be obtained by (Ito et al. 1997):

$$\sigma_H = 3\sigma_h - 2p_r \tag{3.18}$$

In the extreme case with a more conductive fracture, the fluid pressure may completely penetrate to the crack tip. This implies that the reopening is dominated by total force formed by the fluid pressure on the fracture surfaces, and reopening pressure would be equal to the minimum horizontal stress (Rutqvist et al. 2000):

$$\sigma_h = p_r \tag{3.19}$$

3.4.2 Determinations from various equations

If there is no direct measurement available for determination of the value of the maximum horizontal stress, the following relationship can be used to estimate the maximum horizontal stress:

$$\sigma_H = \sigma_h + m(\sigma_V - \sigma_h) \tag{3.20}$$

where m is a constant, normally $m = 0$ - 2. In the normal faulting stress regime, m can generally be taken as 0.5.

While drilling, when the mud weight in a borehole is high enough, the mud pressure will start to fracture the rock, and tensile failure will occur. At this moment, the minimum tangential stress around wellbore is equal to rock tensile strength. From the Kirsch equation (refer to Chap. 7) the following relationship can be obtained:

$$\sigma_H = 3\sigma_h - p - p_m + T_0 \tag{3.21}$$

where p is the pore pressure; p_m is the mud pressure in the borehole.

Using elasticity theory and the Mohr-Coulomb failure criterion for slippage on the faults, the relationship between horizontal stress and pore pressure during mobilization of normal and thrust faulting are as follows (Addis et al. 1996):

Normal faulting:

$$\sigma_H = \frac{2\nu}{1 - \sin\varphi_f}(\sigma_h - p) + p \tag{3.22}$$

Thrust faulting:

$$\sigma_H = \frac{1}{\nu(q_f + 1)}\{q_f\sigma_h - p[q_f(1-\nu) - \nu]\} \tag{3.23}$$

where is φ_f is the internal friction angle of the fault and $q_f = \dfrac{1 + \sin\varphi_f}{1 - \sin\varphi_f}$.

When the stress concentration around a wellbore exceeds the rock compressive strength, breakouts are induced around the wellbore rocks. Because breakout width is expected to remain stable as breakout growth occurs, Barton et al. (1988) presented a method to determine the maximum horizontal stress utilizing observations of breakout width when the rock uniaxial compressive strength is known, that is:

$$\sigma_H = \frac{UCS + p + p_m + \sigma^{\Delta T} - \sigma_h(1 + 2\cos 2\theta_b)}{1 - 2\cos 2\theta_b} \tag{3.24}$$

where UCS is the uniaxial compressive strength of the rock; θ_b is the breakout angle; $\sigma^{\Delta T}$ is the thermal stress arising from the difference (ΔT) between the mud temperature (T_m) and the formation temperature (T_f) , and

$$\sigma^{\Delta T} = \frac{\alpha_t E(T_m - T_f)}{1 - \nu} \tag{3.25}$$

where α_t is the thermal expansion coefficient of the rock; E is the Young's modulus of the rock.

Worldwide in-situ stress measurements show that the mean horizontal stress components and vertical stress at a depth of ≤ 3000 m have the following relationship (Hoek and Brown 1980):

$$\frac{100}{z} + 0.3 < k < \frac{1500}{z} + 0.5 \tag{3.26}$$

where z is the burial depth in meters; k is defined as the ratio of the average horizontal stress, $(\sigma_H + \sigma_h)/2$, to the vertical stress, σ_V. That is:

$$k = \frac{\sigma_H + \sigma_h}{2\sigma_v}$$

In-situ stress measurements show that the maximum and minimum horizontal stress components have the following relationship:

$$\sigma_H / \sigma_h = 1.4 \text{ to } 3.3$$

3.5 Pore pressure prediction

Pore pressure is one of the most important parameters for geomechanical and geological analyses. One cannot obtain a correct prediction for any geomechanical model without a right pore pressure input for the porous formations (Zhang et al. 2006, Madge et al. 2006). Pore pressure varies from hydrostatic (pressure gradient of about 1.0 MPa/km, or 0.44 psi/ft), which is normal pressure, to severely overpressured (48% to 95% of the overburden stress). The abnormal pore pressure is caused by abnormal formation compaction. When sediments compact normally, their porosity is reduced at the same time as pore fluid is expelled. This causes hydrostatic pore pressure. Non-hydrostatic abnormal pressure has many origins. The overburden stress is a major reason to induce an abnormal pressure. During burial process, increasing overburden and no fluid expulsion from the pores of the formation causes pore pressure increase, producing overpressured pore pressure. The incidents due to inappropriate pore pressure prediction have caused significant non-productive time in the drilling industry. Therefore, various models have been proposed to predict formation pore pressure (Standifird et al. 2004).

Assuming the normally pressured and overpressured formations with identical elastic velocities/resistivities have identical effective stresses, the overpressured pore pressure can be calculated by comparing it to the hydrostatic pressure.

Using well log data, Eaton (1975) established the following empirical relationship between the pore pressure gradient and the formation resistivity:

$$PP = OBG - (OBG - p_n)\left(\frac{R}{R_n}\right)^{1.2} \tag{3.27}$$

where PP is the formation pressure gradient; OBG is the overburden stress gradient; p_n is the hydrostatic pore pressure gradient (0.465 psi/ft or 1.03 MPa/km); R is the shale resistivity; R_n is the shale resistivity at normal hydrostatic pressure ($R_n = 3.5$ ohm·m).

Using sonic log data, overpressured pore pressure can be obtained by the following equation (Eaton 1975):

$$PP = OBG - (OBG - p_n)\left(\frac{DT_n}{DT}\right)^{3} \tag{3.28}$$

where DT is the interval transit time of shale; DT_n is the interval transit time of shale in normal hydrostatic pressure ($DT_n = 80$ μsec/ft).

A case study involves an offshore Nigeria well for a prediction of in-situ stress and pore pressure prior to drilling. Pore pressure analysis from resistivity and sonic log data is shown in Fig. 3.4 using Drillworks software (Zhang et al. 2006). It can be seen that the formations are overpressured at the studied depths, because the pore pressure is far higher than the normal hydraulic pressure gradient (1.03 sg).

The pore pressure purposed basin modeling is an advanced tool for pore pressure prediction (Matthews and Standifird 2006, Williams and Madatov 2005). This model integrates the effects of pore pressure data in offset wells and formation physical properties (such as pore specific surface area, compaction constant) as well as formation geologic structures. The basin model can also be updated as new data is acquired. This can provide a drilling look-ahead prediction that may be valid for several thousand feet ahead of the bit (Standifird and Matthews 2005). Therefore, the pore pressure prediction from the basin model gives more reasonable results.

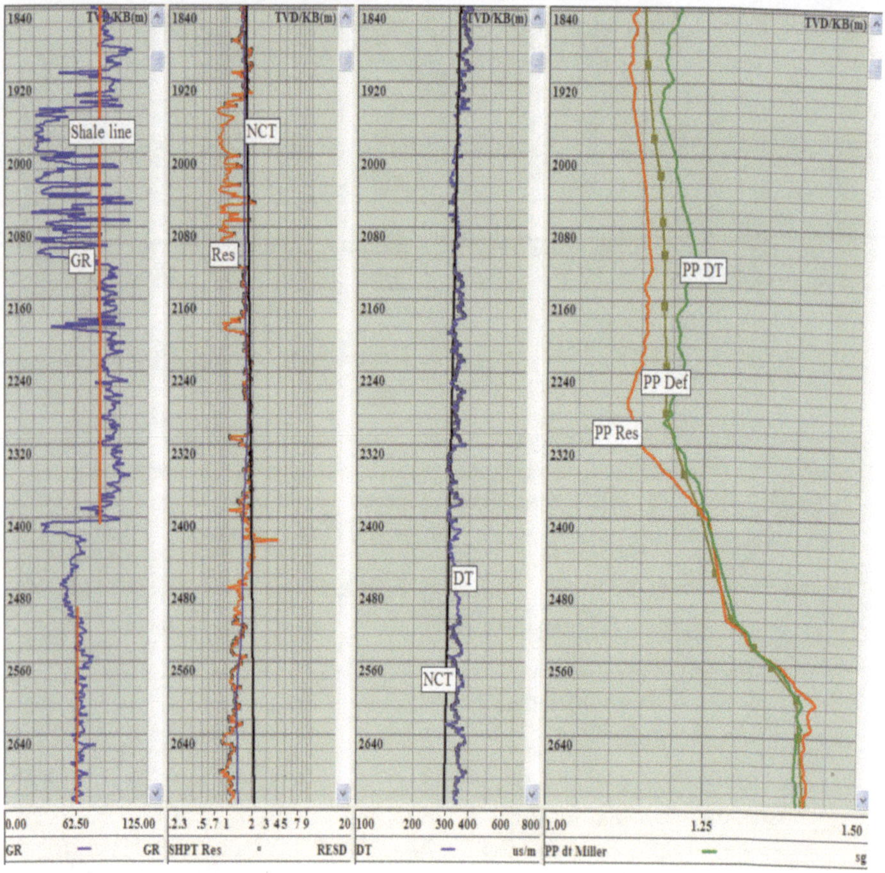

Fig. 3.4. Well log data plots with depth for gamma ray (GR), resistivity (Res), sonic slowness (DT), and pore pressure gradient (PP) predicted by these log data. In the far right track the definitive pore pressure gradient (PP Def) is also plotted according to the well log prediction from resistivity (PP Res), and from sonic (PP DT) and real-time pore pressure measurements. Note that the unit of sg is the specific gravity, and the NCT is the normal compaction trend line (Zhang et al. 2006).

3.6 Numerical modeling of in-situ stress

Research shows that tectonic stress or tectonic residual stress has a significant effect on in-situ stress (Peng 1995). To estimate the distributions of in-situ stress in sedimentary formations it needs to consider the influence of rockmass structure, for instance, lithology and pre-existing geologic structures (Peng 1999). Therefore, understanding sedimentary rockmass geologic structures is important for predicting origin stress state and stress redistributions due to mining and other perturbation (Peng 1998). A case study is examined in coal seam #13-1 and surrounding rock formations in mining face #1221, Pansan colliery, Huainan Coalfield, eastern China (Peng and Meng 2002). The finite element method is applied to model the formation discontinuities and geological structures of sedimentary rocks. Three models are analyzed to investigate the influence of sedimentary rockmass structures on distributions of in-situ stress.

3.6.1 Geological models

Based on the three-dimensional seismic data, the buried depth of coal seam #13-1 in face #1221 is 524 m. The average thickness of the seam is 3.85 m. The immediate roof of coal seam consists of mudstone and sandy mudstone with a thin layer of coal, and the main roof consists of medium-grained sandstone and fine-grained sandstone (Peng and Meng 1999). The thicknesses of both roof varies. There are two high-angle normal faults in the west side of mining panel, the offset of the faults varies from 3 to 8 m, as shown in Fig. 3.5. In the finite element modeling three models are adopted. The length of Model 1 is 208 m, which is located in the east of the mining panel. The lithology and thickness of the coal roof keep basically constant in the lateral direction. Therefore, the rocks belong to continuous media (Fig. 3.6).

Model 2 has a length of 112 m, close to Model 1 as shown in Fig.3.5. In this model both rock lithology and thickness vary in transverse direction, as shown in Fig. 3.7. The length of Model 3 is 132 m, located in the west of the mining panel. The surrounding strata of the coal seam is cut by high-angle normal faults, thus the formations belong to fractured media (Fig. 3.8).

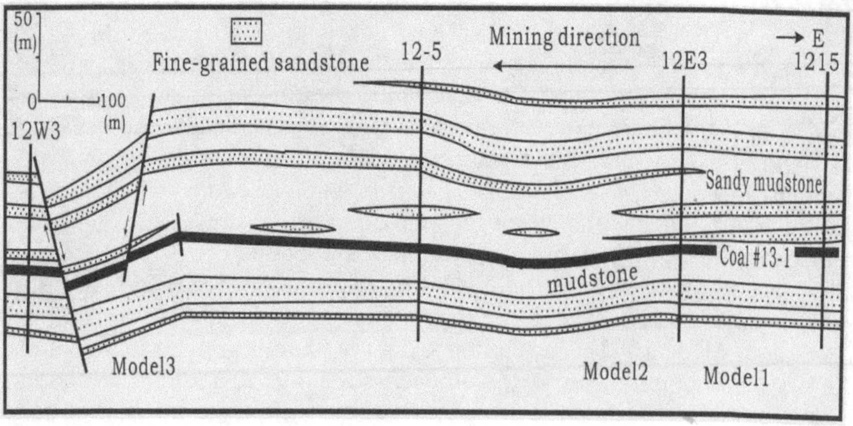

Fig. 3.5. Geologic cross section used for numerical modeling in mining Face #1221.

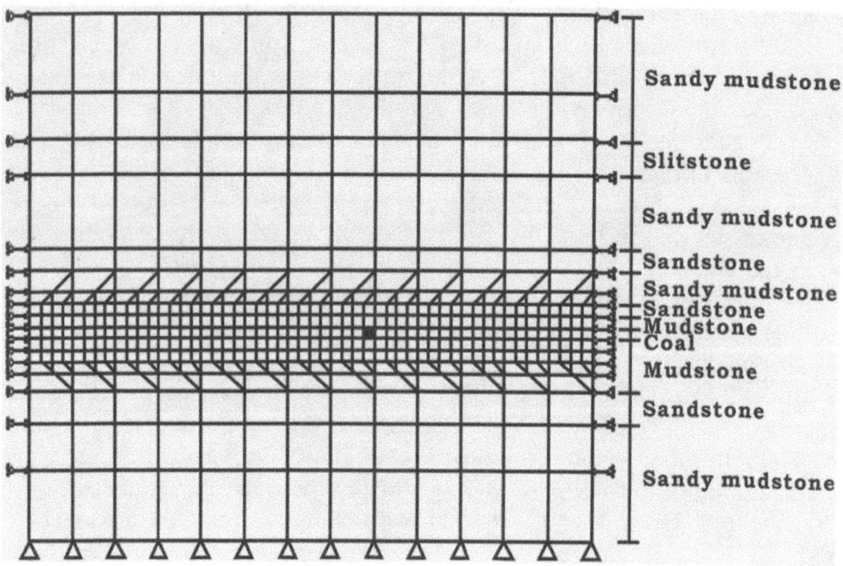

Fig. 3.6. Finite element model for Model 1.

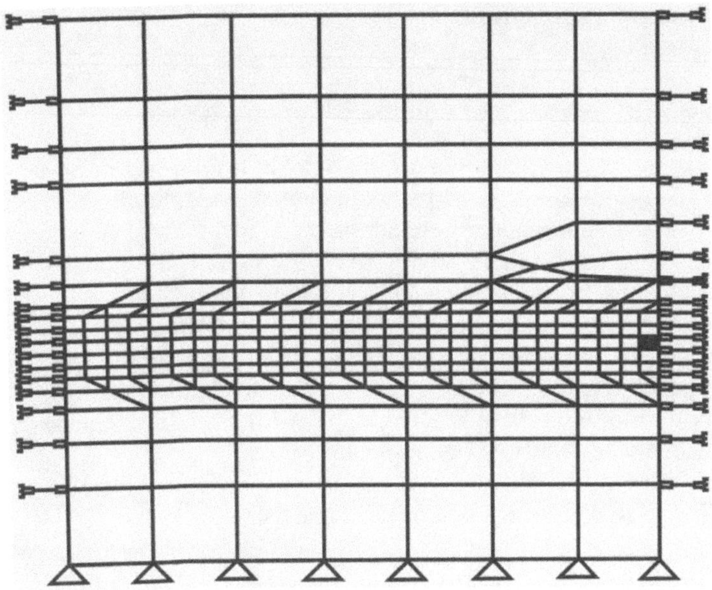

Fig. 3.7. Finite element model for Model 2.

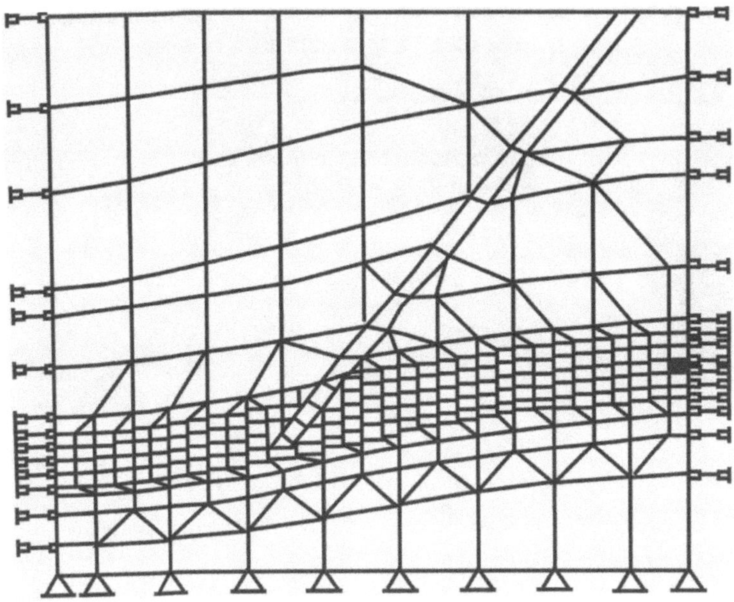

Fig. 3.8. Finite element model for Model 3 with fault effects.

3.6.2 Computational models

Finite element meshes and boundary conditions

According to the geological conditions shown in Fig. 3.5, the meshes presented in Figs 3.6-3.8 are used for finite element calculation. In the modeling the quadrangular elements are adopted. Model 1 has 520 elements and 512 nodes (Fig. 3.6), Model 2 has 284 elements and 286 nodes (Fig. 3.7), and Model 3 has 333 elements and 318 nodes (Fig. 3.8).

In the horizontal direction the uniaxial strain boundary conditions are applied, i.e. the horizontal displacements at left and right boundaries are zero, and only vertical movements at boundaries are allowed.

The bottom of the model is fixed, namely both the horizontal and vertical displacements at bottom boundary are zero.

The vertical stress, calculated from the buried depth of model is 10.35 MPa. This stress is applied on the top boundary.

Mechanical model and properties of the rocks

Elastoplastic model is adopted to describe the mechanical behavior of sedimentary rockmass. The Drucker-Prager failure criterion is applied, and it can be expressed as follows (see Chap. 4 for more details):

$$f = \alpha I_1 + \sqrt{J_2} - k = 0 \tag{3.29}$$

where α and k are rock strength parameters and can be calculated from cohesive strength (c) and internal friction angle (φ), as follows:

$$\alpha = \frac{\sin\varphi}{(9 + 3\sin^2\varphi)^{1/2}} \tag{3.30}$$

$$k = \frac{3c \cdot \cos\varphi}{(9 + 3\sin^2\varphi)^{1/2}} \tag{3.31}$$

Rock physical and mechanical parameters are obtained from lab tests and are listed in Table 3.1.

Table 3.1. Rock physical and mechanical parameters in the models

lithology	Density (g/cm^3)	UCS (MPa)	Tensile strength (MPa)	Friction angle (°)	Cohesion (MPa)	Young's modulus (MPa)	Poisson's ratio
Medium-grained sandstone	2.63	116.68	2.2	36.0	12.11	36104	0.178
Sandy mudstone	2.30	71.40	1.0	35.0	10.2	19566	0.216
mudstone	2.63	66.11	0.80	42.0	1.43	18261	0.268
coal	1.52	22.30	0.88	38.4	3.50	18480	0.303
Sandstone with mudstone	1.52	11.50	0.88	38.4	3.50	18480	0.268
Fine-grained sandstone	2.61	104.12	3.2	40.0	11.6	36410	0.143
Siltstone	2.50	94.54	1.2	37.0	11.0	20871	0.164
Faulted zone	1.32	3.31	0.040	30	0.072	913.10	0.268

Numerical results

In Model 1, where the rocks belong to continuous media, the in-situ stress field is mainly induced by overburden gravitational load. The vertical stress increases as the buried depth increases (Fig. 3.9). However, in Model 2 the thickness of sandstone of the roof become gradually thinner, consequently, the initial stress field is obviously disturbed in the location where sandstone disappeared. The in-situ vertical stress concentration occurs in the layer where the lithology changes, as shown in Fig. 3.10. In Model 3 the in-situ stress field is obviously different from Model 1 because of existence of the fault, as shown in Fig. 3.11. A low vertical stress distribution area appears within the faulted zone. However, high stress zone exists between the normal rocks and the fault. The high stress is 1.5 times higher than the normal in-situ vertical stress. The additional high stress is mainly caused by the discontinuity of formations, which disturbs initial stress field. This will affect stability of underground structures, such as tunnels.

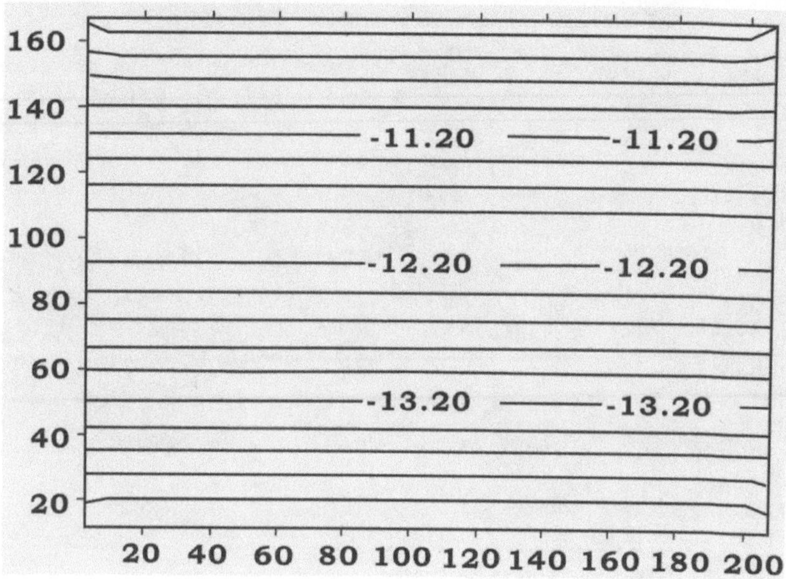

Fig. 3.9. In-situ vertical stress distribution in Model 1. The coordinates are distances in m, and contours are in-situ vertical stress in MPa. Negative value in this calculation means compressive stress.

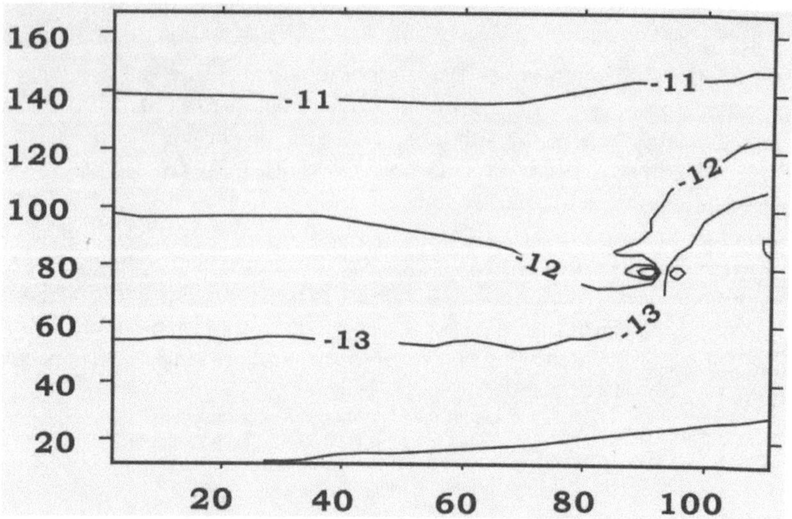

Fig. 3.10. In-situ vertical stress distribution with variable thickness of the roof in Model 2. The coordinates are distances in m, and contours are in-situ vertical stress in MPa. Negative value in this calculation means compressive stress.

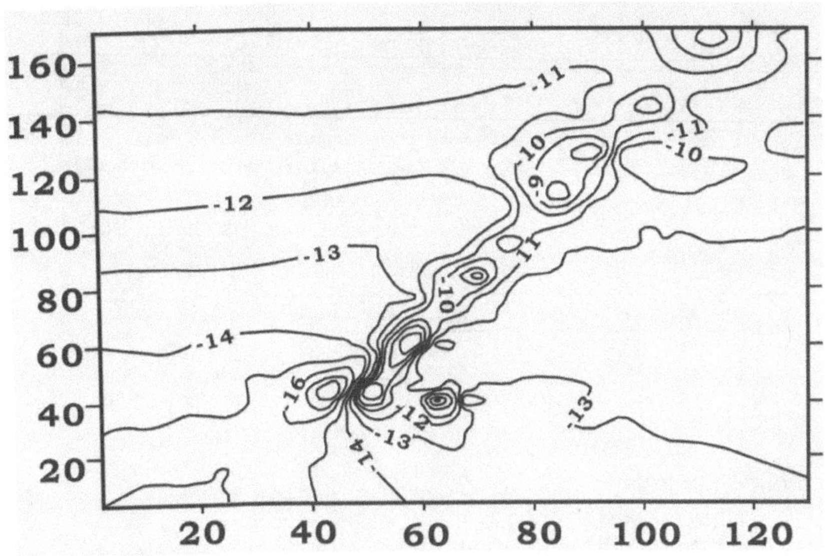

Fig. 3.11. In-situ vertical stress distribution with variable thickness of the roof in Model 3. The coordinates are distances in m, and contours are in-situ vertical stress in MPa. Negative value in this calculation means compressive stress.

3.6.3 Numerical model for calculating horizontal stresses

Calculation model

The studied area is located in the west side in Fangezhuang colliery, where the coal seam is going to be exploited in an area of 2 km^2 . The finite difference method is used to calculate in-situ stress. The seam floor is chosen as the calculation plane. The length of model is 2940 m, and the width is 640 m, as shown in Fig. 3.12. The buried depth of coal seam #12 in the studied area varies from 495.18 to 807.25 m, and the average depth is 659.75 m. The thickness of coal seam is 0.29-5.24m with an average thickness of 2.51. The calculation model is discretized into 4704 quadrangle elements. Stress boundary is applied (Peng and Meng 2002).

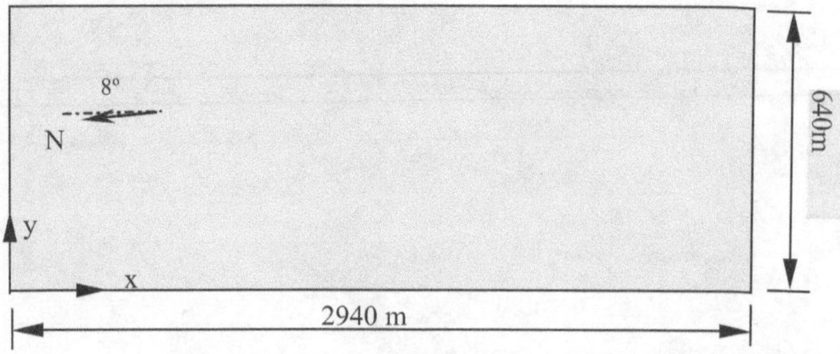

Fig. 3.12. Numerical model to calculate horizontal stresses in Fangezhuang Colliery of Kailuan Coalfield, China.

Calculation results

The vertical stress in this analysis is obtained by using Eq. 3.10, which is

$$\sigma_V = 0.027z = 17.55\,\text{(MPa)}$$

The horizontal stresses are calculated by using the numerical method. Figures 3.13-3.15 present the calculated maximum principal horizontal stress, minimum principal horizontal stress, and maximum shear stress components, respectively. It can be seen from the figures that the maximum horizontal principal stress varies from 54.24 to 5.59 MPa with an average of 24.71 MPa and the minimum principal horizontal stress are 26.89 to 3.80 MPa with an average of 13.53 MPa. The maximum horizontal principal stress is in EW and NW-SE direction.

In-situ stress measurement was conducted by overcoring method in this area. The results show that this region belongs to strike-slip faulting stress regime. The maximum horizontal stress varies from 20.46 to 24.34 MPa, and direction is N103 - N142. The intermediate principal stress varies from 15.31 to 16.01 MPa, whose direction is near vertical with an inclination of 78.67-81.88° from the horizontal direction. The minimum principal stress varies from 7.64 to 13.9 MPa. The calculated results are close to the data from in-situ measurement. The differences are mainly caused by tectonics, geological structures, and lithology. For instances, the faults cause stress concentration, causing changes in both stress magnitudes and directions.

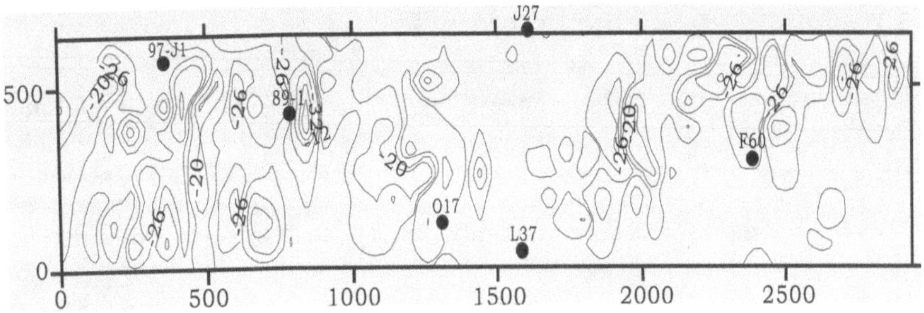

Fig. 3.13. Maximum principal horizontal stress contours calculated by numerical method. The coordinates are distances in m, and contours are maximum horizontal vertical stress changes in MPa. Negative value in this calculation means compressive stress.

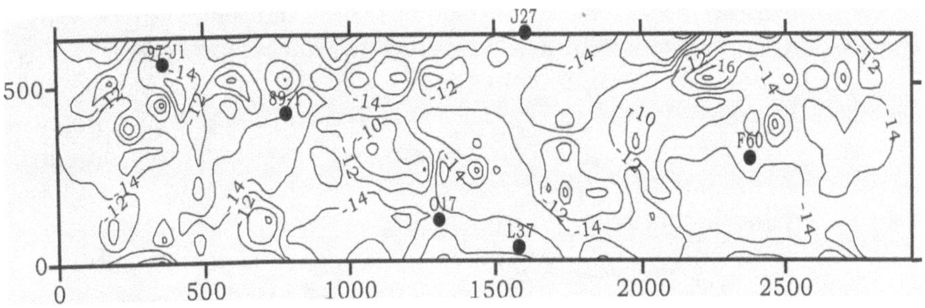

Fig. 3.14. Minimum principal horizontal stress contours calculated by numerical method. The coordinates are distances in m, and contours are minimum horizontal stress changes in MPa. Negative value in this calculation means compressive stress.

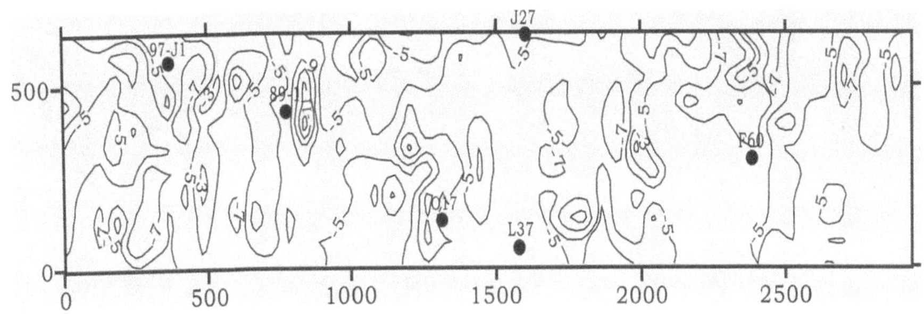

Fig. 3.15. Maximum shear stress contours calculated by numerical method. The coordinates are distances in m, and contours are shear stress changes in MPa. Negative value in this calculation means compressive stress.

3.7 In-situ stress measurements

In-situ stress is a non-uniform stress field and is one of the most important aspects in engineering geology. In-situ stress measurement is one of the most accurate methods to obtain the in-situ stress magnitudes and directions. Therefore, the measurement of in-situ stress magnitudes and directions is important for underground engineering, such as mine exploitation, oil and gas exploration and development, geothermal development, and civil engineering.

In-situ stress measurements can be carried out in deep boreholes from the surface, using hydraulic fracturing or LOT techniques, as described in Sec. 3.3 of this chapter. The in-situ stress measurement can also be conducted from underground access using overcoring methods and acoustic emission method (Peng and Meng 2002).

In-situ stress measurements are to measure the undisturbed three-dimensional stress state in formations. The three-dimensional stress state at any point inside the rock formation can be expressed by six independent stress components (σ_x, σ_y, σ_z, τ_{xy}, τ_{yz}, τ_{zx}) as shown in Fig. 3.16. From the six stress components, the magnitudes and directions of three principal stresses can be calculated. The three principal stresses (σ_V, σ_H, σ_h) are usually called the in-situ stress.

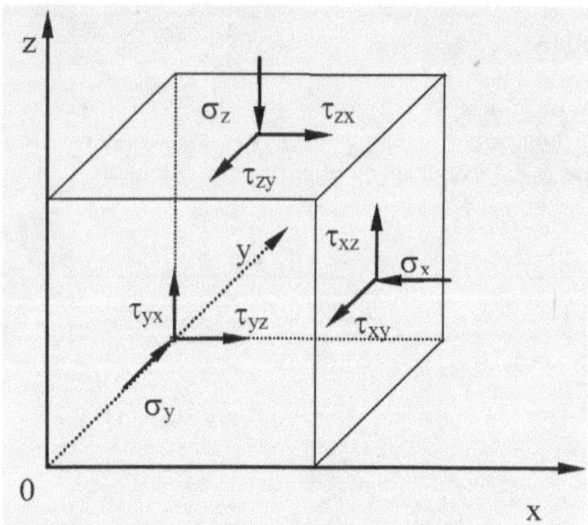

Fig. 3.16. Three-dimensional stress components in any point inside of rock formations.

Any system utilized for estimating the in-situ stress state must involve a minimum of six independent measurements. There are methods of direct stress measurement and methods of estimating the stresses via various indirect or indicator methods (Hudson and Harrison 1997).

The four direct methods recommended by the ISRM (Kim and Franklin 1987) are shown as follows:

1. The flatjack test.
2. The hydraulic fracturing test.
3. The United States Bureau of Mines (USBM) overcoring torpedo.
4. The Commonwealth Scientific and Industrial Research Origination (CSIRO) overcoring gauge.

In this book the overcoring method is introduced.

3.7.1 Overcoring method

Overcoring stress measurement is conducted by drilling a borehole into the formation, then a stress meter or deformation/strain gauge is installed to measure stress or strain. The following procedures are normally used for overcoring measurements.

1. A borehole of the 130-150 mm overcored is drilled into the formation to certain depth, where the rock is not disturbed, as shown in Fig. 3.17 (1).
2. The bottom surface of the borehole is finished by using a tapered bit. A 40-mm-diameter pilot borehole of about 50 cm length is then drilled from the bottom center of this overcored tapered borehole, as shown in Fig. 3.17 (2).
3. The stress meter or strain gauge is inserted into the 40 mm diameter pilot borehole using the insertion rods fitted with insertion device and direction device (Fig. 3.17 (3)).
4. Overcoring operation using double core tube is started after the insertion rod with the insertion device is extruded, as shown in Fig. 3.17 (4).
5. After the core is recovered with the stress meter or strain gauge inside, field experiment to evaluate the elastic properties of the rock is carried out. Then the stress meter or strain gauge is taken out from the core, data recorded by the data-logger in the stress meter or strain gauge are downloaded for in-situ stress calculation.

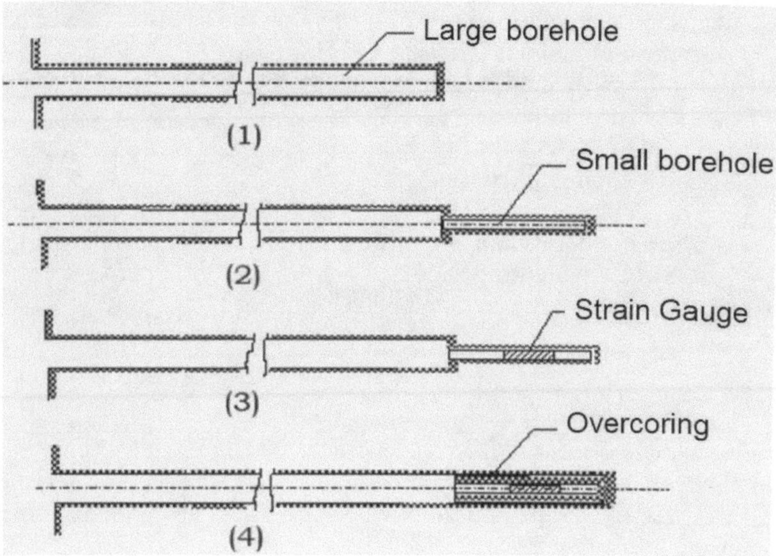

Fig. 3.17. Schematic drawing of the overcoring in-situ stress measurement.

3.7.2 Case application

In-situ stress measurement in Huainan Coalfield

Hydraulic fracturing and overcoring methods were applied to determine in-situ stress in Huainan Coalfield. The measurement results of in-situ stress are listed in Table 3.2 (Peng and Meng 2002). It shows that the ratio of the maximum horizontal principal stress to vertical stress ranges from 0.82-1.82 with an average of 1.23, when the depth of the cover is less than 600 m. When the depth is greater than 600 m, the ratio is 0.51-1.33 with an average of 0.91. Figure 3.18 presents the measurement results of the maximum horizontal principal stress to vertical stress ratio (λ). Field measurements also show that the two horizontal stresses are not equal, and the ratio of two horizontal principal stresses ranges from 0.54 to 0.95 with an average of 0.74.

The maximum in-situ stress in Huainan Coalfield is the vertical stress, thus this area belongs to normal faulting stress regime without the impact of active tectonic activity. The direction of maximum horizontal principal stress is EW, and changes to N49-N55° due to the influence of faults (Peng and Li 1996).

Table 3.2. In-situ stress measurements in Huainan Coalfield, China

Location	Depth (m)	Measurement method	Vertical stress (MPa)	Max. horizontal stress (MPa), direction	Min. horizontal stress (MPa), direction
Panji Mine, Huainan	550	Overcoring	16.47	13.43 N78.94°	7.22 N124.21°
Xinzhuangzi Mine	650	Overcoring	19.50	15.70 N0°	12.80 N101°
Xinji Borehole # 201	350.7	Hydraulic Fracturing	9.47	14.20 N53°	9.50
	373.6		10.09	10.10 N49°	7.72
	460.1		12.42	14.90	10.20
	553.46		14.94	26.63	16.53
	632.63		17.08	8.63	7.93
	666.6		18.0	10.36 N52°	9.86
Xinji Borehole #1202	555.18	Hydraulic Fracturing	14.99	18.78	13.52
	589.74		15.92	16.74	12.84
	652.72		17.62	23.52 N51°	16.52
	718.64		19.40	20.78	15.28
	757.6		20.46	19.37 N55°	14.87
	845.1		22.82	25.97	18.84

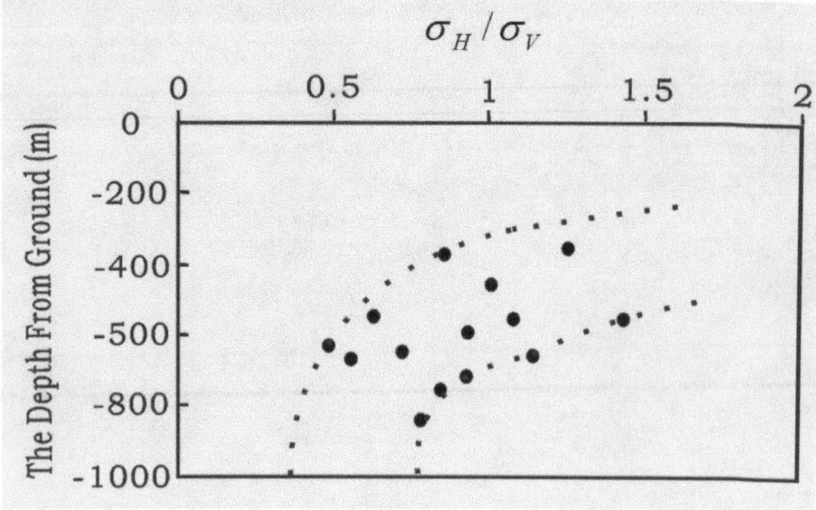

Fig. 3.18. The ratio of maximum horizontal stress to vertical stress versus the depth obtained by in-situ stress measurement in Huainan Coalfield, China.

In-situ stress measurement results in China

According to the in-situ stress measurement in China, the maximum horizontal and minimum horizontal principal stresses have the following linear relationships with depth, when the burial depth is less than 500 m:

$$\sigma_H = (4.5 \pm 2.5) + 0.049\, z \tag{3.32}$$

$$\sigma_h = (1.5 \pm 1.0) + 0.030\, z \tag{3.33}$$

where σ_H and σ_h are the maximum and minimum horizontal stresses respectively, in MPa; z is the depth in m.

The average horizontal stress (σ_a), in MPa, and depth, in m, have the following relationship:

$$\sigma_a = 0.72 + 0.041\, z \tag{3.34}$$

Figure 3.19 shows the average horizontal to vertical stress ratio and depth relationship. The vertical and horizontal stresses have the following relationships:

$$\sigma_H / \sigma_v = 150/z + 1.4 \tag{3.35}$$

$$\sigma_h / \sigma_v = 128/z + 0.5 \tag{3.36}$$

where z is the depth in m.

Figure 3.20 presents the measured maximum stress and depth relation in some coal mines in central and northern China. It shows that the maximum stress increases as the depth increases. The maximum horizontal stress magnitude in those coal mines is small compared to other regions.

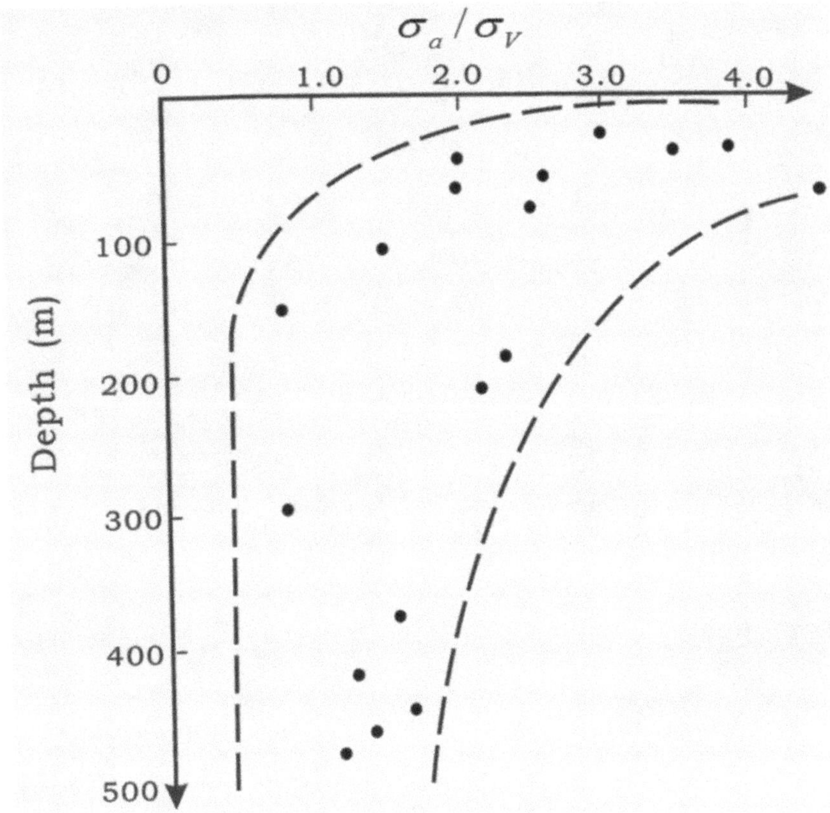

Fig. 3.19. The average horizontal stress to vertical stress ratio versus depth obtained by in-situ stress measurement in China.

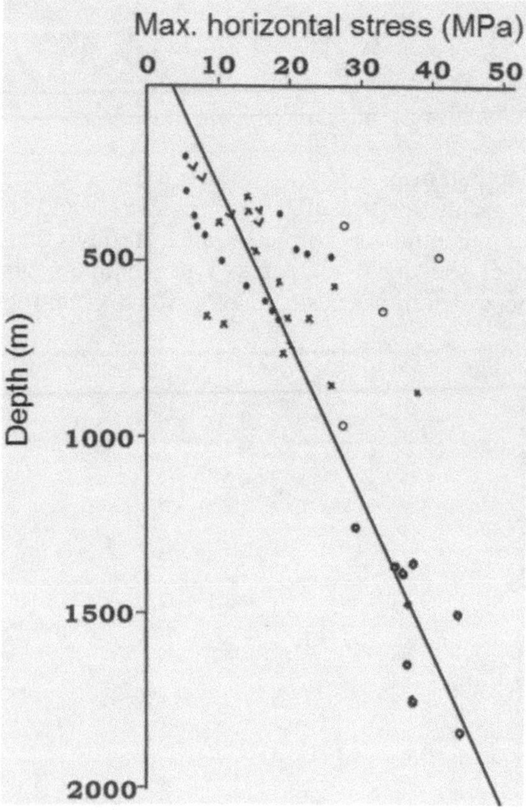

Fig. 3.20. The maximum horizontal stress and depth relationship measured in coal mines in central and northern China.

References

Addis MA, Last N, Yassir N (1996) Estimation of horizontal stresses at depth in faulted regions and their relationship to pore pressure variations. SPE formation evaluation 48 (3):11-18

Barton CA, Zoback MD, Burns KL (1988) In-situ stress orientation and magnitude at the Fenton geothermal site, New Mexico, determined from wellbore breakouts. Geophys Res Lett 15(5):467-470

Breckel IM, von Eekelen HAM (1982) Relationship between horizontal stress and depth in sedimentary basins. J Petro Tech, September, 2191-2199

Bredeoeft JD, Wolf RG, Keys WS, Shutter E (1976) Hydraulic fracturing to determine the regional in situ stress field in the Piceance Basin, Colorado. Geol Soc Am Bull 87:250-258

Eaton BA (1975) The equation for geopressure prediction from well logs. Paper SPE5544

Edwards ST, Bratton TR, Standifird WB (2002) Accidental geomechanics – capturing in-situ stress from mud losses encountered while drilling. SPE/ISRM 78205 presented at the SPE/ISRM rock mechanics conference held in Irving, Texas

Fjær E, Holt RM, Horsrud P, Raaen AM (1992) Petroleum related rock mechanics. Elsevier

Haimson BC, Fairhurst C (1970) In situ stress determination at great depth by means of hydraulic fracturing. Rock Mechanics - Theory and Practice, Somerton WH (ed). Am Inst Mining Eng pp559-584

Hoek E, Brown ET (1980) Underground excavations in rock. Inst Mining and Metallurgy London

Hubbert MK, Willis DG (1957) Mechanics of hydraulic fracturing. Petro Trans AIME 210:153-163

Hudson JA, Harrison JP (1997) Engineering rock mechanics, an introduction to the principles. Pergamon

Ito T, Sato A, Hayashi K (1997) Two methods for hydraulic fracturing stress measurements. Proc Int Sym Rock Mech, Kumamoto, Japan. AA Balkema pp127-132

Kim K, Franklin JA (1987) Suggested methods for rock stress determination. Int J Rock Mech Min Sci Geomech Abstr. 24:53-74

Knowledge Systems (2001) Best practice procedures for predicting pre-drill geopressures in deep water Gulf of Mexico. DEA119 JIP report

Matthews MD, Standifird W (2006) Forward model wellbore pore pressures. Hart's E & P 79(11):77-79

Madge C, Zhang J, Standifird WB (2006) Predict predrill pressure and stress. E&P, May, pp81-83

Peng S (1995) Geological modeling techniques for longwall mining roof stability: a case study. Proc Rock Mechanics and Strata Control in Mining and Geotech Eng, Beijing

Peng S, Li Y (1996) Wedge-shaped sandstone roof strata and roof stability, examples from Seam 13-1 in Huainan Coalfield, China. Proc for Rock Mechanics and Strata Control in mining and Geotech Eng Xi'an

Peng S (1998) Approach on mechanical properties of clastic rocks and rock microstructure. Proc. 5th Ann Conf of Young Scientists, China Coal Society. Coal Industry Press (in Chinese)

Peng S (1999) The state-of-art and the future work of engineering disasters induced by coal mining at deep depth in China. Symp Engineering and Disaster, Dept Material Science and Engineering, China NSF, Mt Jiuhua, Jianxi (in Chinese)

Peng S, Meng Z (1999) Numerical modeling on coal roof and floor stability in longwall mining. J China Univ Min Tech 28(1):41-45 (in Chinese)

Peng S, Meng Z (2002) Theory and practice of mining engineering geology. Geological Press (in Chinese)

Reinecker J, Heidbach O, Tingay M, Connolly P, Muller B (2006) The 2006 release of the world stress map

Rutqvist J, Tsang C-F, Stephansson O (2000) Uncertainty in the maximum principal stress estimated from hydraulic fracturing measurements due to the presence of the induced fracture. Int J Rock Mech Min Sci 37:107-120

Standifird WB, Matthews MD (2005) Real time basin modeling: improving geopressure and earth stress prediction. Paper SPE 96464 presented at Offshore Europe 2005 held in Aberdeen, Scotland, UK, 6–9 September

Standifird WB, Paine K, Matthews MD (2004) Improving drilling success requires better technology and models. World oil 225(10):51-56

Williams K, Madatov AG (2005) Analysis of pore pressure compartments in extensional basin. GCSSEPM 25th Ann Research Conference: Petroleum Systems of Divergent Continental Margin Basins, pp862-909

Zhang J, Standifird WB, Adesina K, Keaney G (2006) Wellbore stability with consideration of pore pressure and drilling fluid interactions. Paper ARMA /USRMS 06-922 presented at 41st US Symp Rock Mechanics held in Golden, Colorado

Zoback MD, Zoback ML (1989) Stress in the earth's lithosphere. Encyclopedia of Earth Sciences Series (Fairbridge RW, Series Ed), Van Nostrand Reinhold Co, New York pp 1221-1232

Zoback MD et al. (2003) Determination of stress orientation and magnitude in deep wells. Int J Rock Mech Min Sci 40:1049-1076

4 Rock strength experiments and failure criteria

4.1 Introduction

There are many types of laboratory tests to obtain rock mechanical properties. Laboratory tests usually consist of simple experiments appropriate to the nature of the rock in which important quantities, often stress and strain, are determined (Jaeger and Cook 1979). Before laboratory tests, rock sample preparation and physical analysis are needed. The ISRM standards gave the suggested methods for both core preparation and testing (Brown 1981, Kovari et al. 1983).

4.2 Uniaxial tensile test

Tensile strength describes the capacity of the rock to resist tensile stress. There are direct and indirect methods for measurement of tensile strength (ISRM 1985). The indirect methods have been dominant in determining tensile strength of rocks in the past due to their ease in sample preparation and testing procedure. The indirect methods include point load test and Brazilian test.

4.2.1 Direct tensile strength test

The direct measurement of tensile strength can be performed by gluing both ends of a cylindrical rock specimen to metal front plates which are fixed to the platens of a load frame (Fjær et al. 1992). The cement has to be more resistant to tensile stress than the specimen.

4.2.2 Point load test

The standardized equipment in the point load test includes a pair of 60° conical point loading platens installed either on a hydraulic hand-pump, for field use, or on a loading frame for laboratory use, as shown in Fig. 4.1. The rock sample can be of regular or irregular shape, and is compressed to failure. The load at failure, P, is recorded for the strength calculation following these three steps (Zeng 2002):

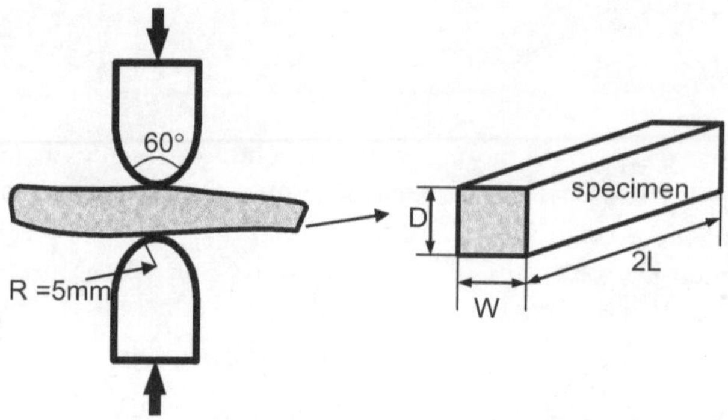

Fig. 4.1. Schematic figure showing point load test apparatus and the specimen.

Step 1: calculate initial index, I_s:

$$I_s = \frac{P}{D_e^2} \tag{4.1}$$

where D_e is the specimen equivalent diameter in mm. According to the sample geometry and loading direction, D_e is calculated as:

For diametral test, $D_e = D$

For axial, block and lump test, $D_e = \sqrt{WD/\pi}$

where D, W are the parameters related to sample size in mm, as shown in Fig. 4.1.

Step 2: calculate standard index, $I_{s(50)}$ for size effect:

$$I_{s(50)} = \left(\frac{D_e}{50}\right)^{0.45} I_s \tag{4.2}$$

Step 3: calculate uniaxial tensile strength, T_0:

$$T_0 = \frac{S_a P}{\left(L - \frac{1.7P}{22I_{s(50)}}\right)^2} \tag{4.3}$$

where L is the sample size in mm, as defined in Fig. 4.1; S_a is the shape factor determined by:

For diametral test $S_a = 0.79$

For other test $S_a = 0.79\,D/L$

The point load test also can be used for determining rock uniaxial compressive strength (UCS). A linear regression between the mean $I_{s(50)}$ and mean UCS values determined for 908 samples in US coal measure rocks yields the following equation (Rusnak and Mark 2000):

$$UCS = 1970 + 17.6 I_{s(50)} \tag{4.4}$$

where UCS and $I_{s(50)}$ are in psi, and 1 MPa is approximately equal to 145 psi.

The zero-intercept regression equation obtained from the entire data set is as follows:

$$UCS = 21 I_{s(50)} \tag{4.5}$$

Early studies (Bieniawski 1975) were conducted on hard, strong rocks, and found that relationship between UCS and the point load strength could be expressed as:

$$UCS = 24 I_{s(50)} \tag{4.6}$$

4.2.3 Brazilian test

The Brazilian test is performed by applying a load by two platens diametrically compressed to a rock cylinder (Fig. 4.2), which is normally shorter than or equal in thickness (t) to its diameter (D). The technique involves loading disc-shaped specimens in compression across their diameters. Such loading generates a tensile stress at the center of the disc in a direction perpendicular to the direction of the applied load. Failure occurs by an extensional fracture in or close to the loaded diametral plane. The tensile strength is given by the ratio of the peak load F_c to the diameter and thickness product (Fjær et al. 1992):

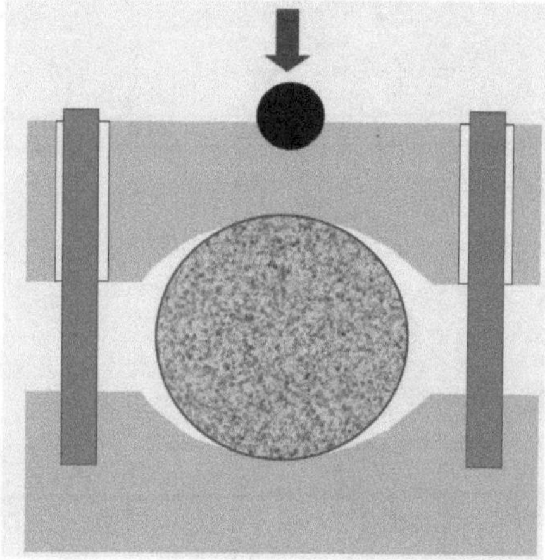

Fig. 4.2. Schematic Brazilian test apparatus and the specimen.

$$T_0 = \frac{2F_c}{\pi D t} \tag{4.7}$$

where D, t are the sample diameter and thickness in mm; F_c is the yield load in Newtons; T_0 is the tensile strength in MPa.

Brazilian method often overestimates the tensile strength of the poorly consolidated rocks, because the line load applied during the test typically widens for soft materials. In such a case, a line load is no longer applied and the force is distributed over a larger area. Consequently, a higher tensile strength is calculated.

4.3 Uniaxial compressive test

4.3.1 Uniaxial compressive test of rock samples

Uniaxial compressive test is used to determine uniaxial compressive strength, Poisson's ratio, and Young's modulus. Uniaxial compressive test is also called unconfined compressive test. In this test, circular cylinders of rock samples are compressed parallel to their longitudinal axis. It is the oldest and simplest test, and continues to be one of the most convenient and useful ways for determining the properties of rocks (Jaeger and Cook 1979).

The ISRM recommends that the diameter (D) and height (H) of cylindrical samples of rocks should have the following relationship:

$$H / D = 2 \tag{4.8}$$

In the uniaxial test, the following equation is used for calculating uniaxial compressive strength (UCS) for ISRM suggested samples.

$$UCS = \frac{P_{\max}}{A} \tag{4.9}$$

where P_{max} is the maximum load exerted in the sample; A is the area of the cross section of the sample.

If the sample size does not meet ISRM requirement in Eq. 4.8, the strength test results need to be corrected. The following equation may be used to correct the UCS:

$$UCS = 0.889\sigma_c[0.778 + 0.222H / D] \tag{4.10}$$

where UCS is the uniaxial compressive strength and σ_c is the test strength of rock sample when its H/D is not equal to 2.

When the stress-strain curve is nonlinear, or the Young's modulus (E) and Poisson's ratio (ν) depend on the axial stress level, it is common to specify the values of E and ν at 50% of the peak stress, i.e.:

$$E = \frac{\sigma_{c(50)}}{\varepsilon_{a(50)}} \tag{4.11}$$

$$\nu = -\frac{\varepsilon_{r(50)}}{\varepsilon_{a(50)}} \tag{4.12}$$

where $\sigma_{c(50)}$ is the stress at 50% of the peak stress; $\varepsilon_{a(50)}$ and $\varepsilon_{r(50)}$ are the axial and radial strains at $\sigma_{c(50)}$, respectively.

Figures 4.3 - 4.5 show the uniaxial compressive test results for sandstones and shales in China's Permian coal measures. Because of the intrinsic relations between the uniaxial compressive strength, Young's modulus and Poisson's ratio, these three parameters are usually determined from a single experiment. Table 4.1 presents the uniaxial compressive test results for different rocks in Eastern China's Permian coal measures (Peng and Wang 2001). It can be seen that sandstones have much greater uniaxial compressive strength and Young's modulus. Figures 4.6 and 4.7 also indicate that sandstones have much higher UCS, E, and tensile strength than the mudstones (Meng et al. 2006).

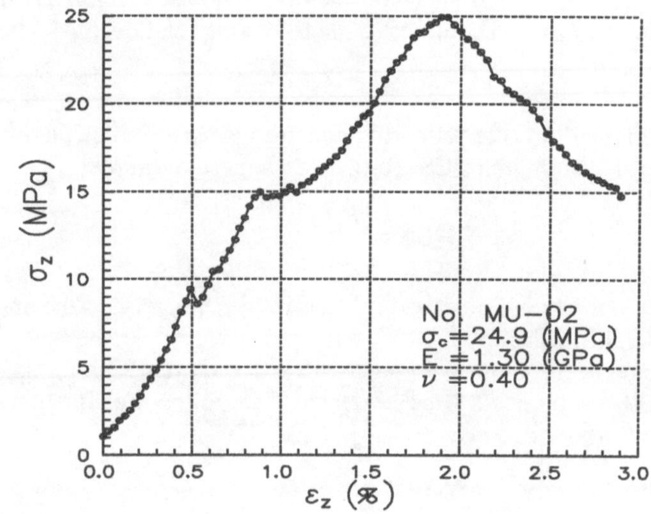

Fig. 4.3. The uniaxial compressive test for mudstone in Eastern China's coal measures. In the figure σ_c equals UCS.

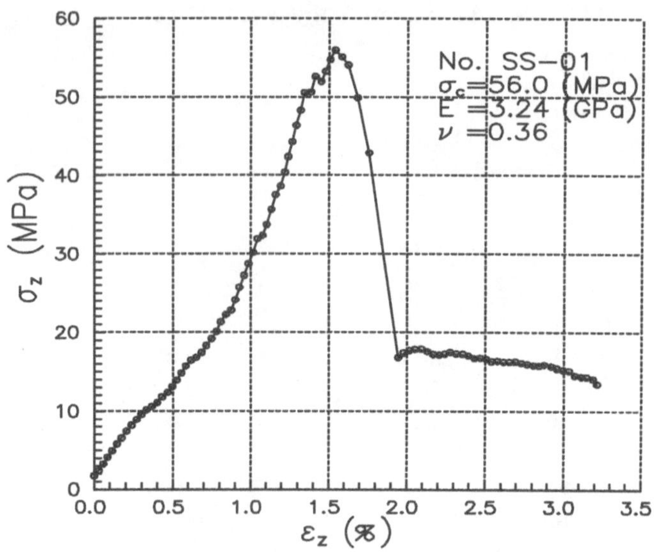

Fig. 4.4. The uniaxial compressive test for sandy shale in Eastern China's coal measures. In the figure σ_c equals UCS.

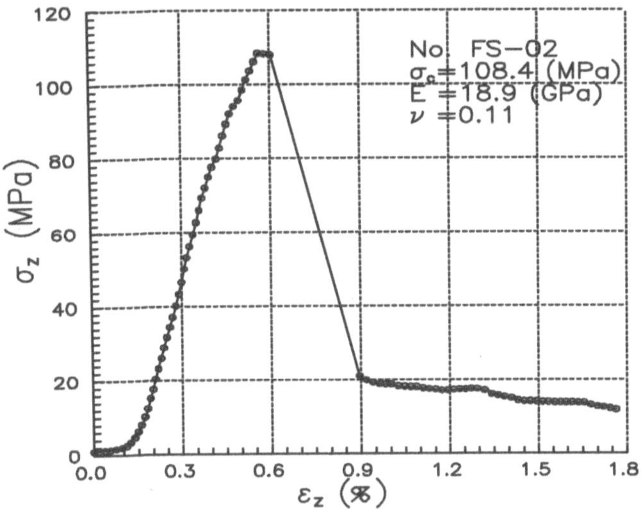

Fig. 4.5. The uniaxial compressive test for fine-grained sandstone in Eastern China's coal measures. In the figure σ_c equals UCS.

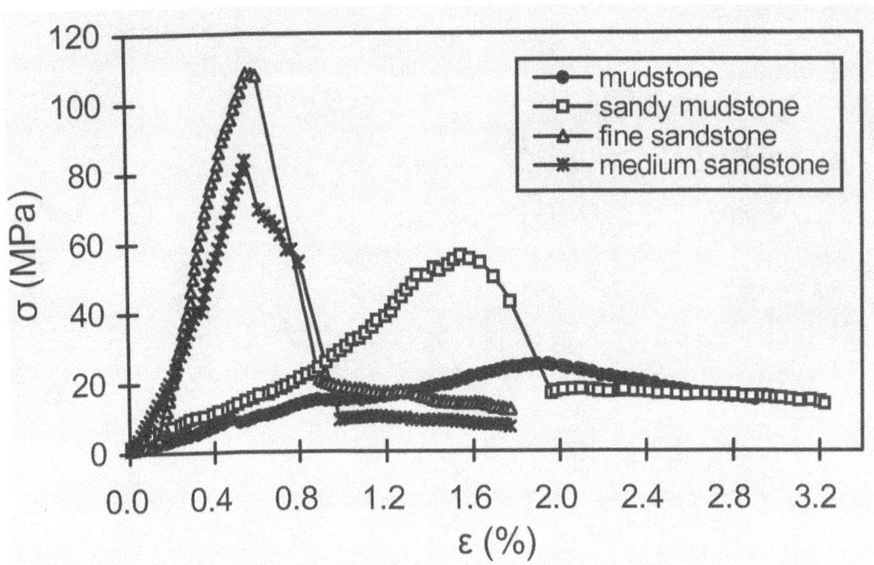

Fig. 4.6. Complete stress-strain curves for sandstones and mudstones in Eastern China's coal measures under the uniaxial compressive test.

Table 4.1. Uniaxial compressive test results in Eastern China's coal measures

Rocks	UCS (MPa)	E (GPa)	ν
Mudstone	32.2	4.26	0.22
Mudstone	27.7	1.30	0.40
Mudstone	42.2	3.23	0.42
Mudstone	22.9	3.01	0.39
Sandy mudstone	34.5	5.68	0.23
Sandy shale	48.8	3.24	0.35
Shale	47.5	4.86	0.38
Shale	43.1	4.71	0.32
Fine-grained sandstone	102	19.4	0.18
Fine-grained sandstone	107.6	18.9	0.11
Fine-grained sandstone	108.4	20.4	0.11
Medium-grained sandstone	83.5	13.6	0.16
Medium-grained sandstone	70.4	11.4	0.10

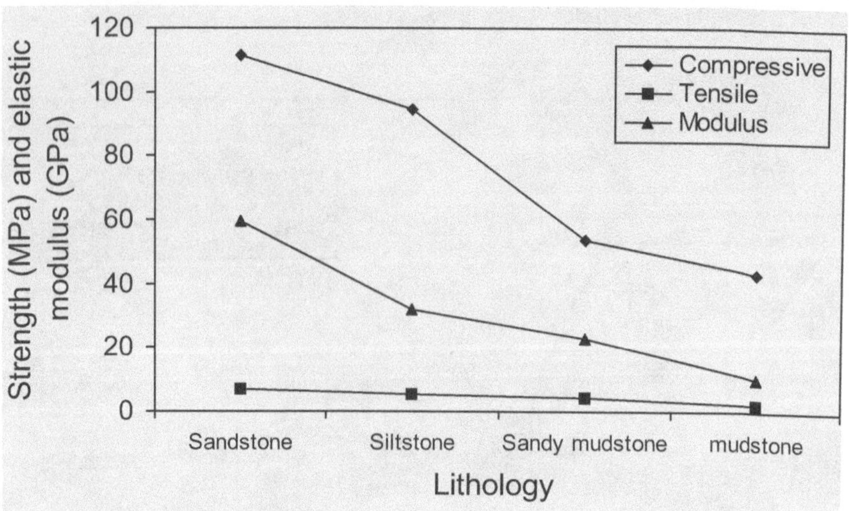

Fig. 4.7. The comparison of the compressive and tensile strength and elastic modulus for different lithologic samples under the uniaxial compressive tests.

It is noticed from Fig. 4.6 that the stress-strain curves for different lithologies have obvious differences, particularly after the peak value. For sandstone, in the stress-strain curve the stress drops sharply down with a very steep slope after the peak strength, and the residual strength is only 1/10 - 1/20 of its peak value. For the mudstone sample, which has the lowest strength, after the peak load the stress-strain curve has a gentle slope, and the sample can still keep its certain strength. The residual strength in the mudstone is about 1/3 of the peak value. The stress-strain curve for sandy mudstone lies in between of sandstone and mudstone, the residual strength is about 1/4 -1/8 of its peak value.

4.3.2 Influence of sample size

The influence of sample size upon rock strength has been widely discussed in geotechnical literature and it is generally assumed that there is a significant reduction in strength with increasing sample size. Based upon an analysis of published data, Hoek and Brown (1980) have suggested that the uniaxial compressive strength σ_{cd} of a rock specimen with a diameter of d mm is related to the uniaxial compressive strength σ_{c50} of a 50 mm diameter sample by the following relationship:

$$\sigma_{cd} = \sigma_{c50}(50/d)^{0.18} \tag{4.13}$$

This relationship, together with the data upon which it was based, is shown in Fig. 4.8 (Hoek and Brown 1980). It shows that the rock compressive strength decreases as the sample size increases.

4.4 Triaxial compressive test

The triaxial compression test has proved to be the most useful test in the study of the mechanical properties of rocks over a wide range of values (Jeager and Cook 1979). In a triaxial compressive test, the major principal stress (σ_1) is applied along the axis of a cylindrical rock specimen and an equal minor principal stress (σ_3) is applied to the curved surfaces of the specimen by fluid-confining pressure. Figure 4.9 shows a typical servo-controlled triaxial compressive test apparatus made by MTS Systems (Peng et al. 2000).

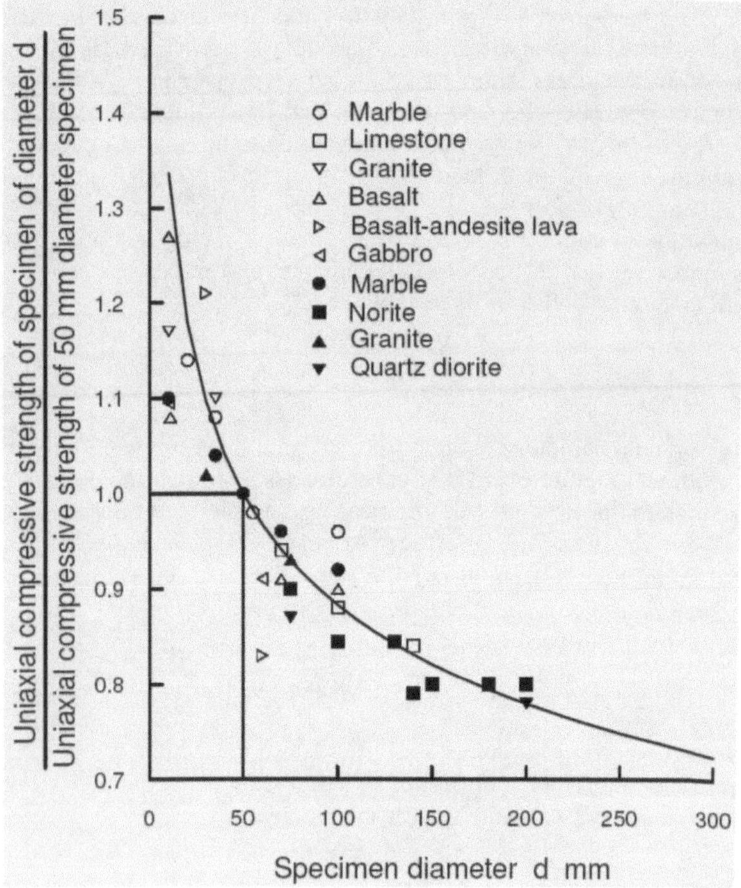

Fig. 4.8. Influence of specimen size on the strength of intact rocks (after Hoek and Brown 1980).

In the triaxial compressive tests both the UCS and failure envelope can be determined, when enough rock samples are tested. Figure 4.10 presents complete stress-strain curves for different confining pressures under the triaxial tests in a medium-grained sandstone in the Huinan coal mines of Eastern China's Permian coal measures (Peng and Meng 2002). It can be seen from this figure that rock strength increases as the confining pressure increases. Therefore, the rock has a higher strength when it is subjected to a three-dimensional stress condition. The triaxial tests also show that the elastic modulus is not a constant, but increases as the confining pressure increases, as shown in Fig. 4.10. It should be noted that the elastic/Young's modulus is directly related to the slope of each stress-strain curve prior to the peak load. The experimental results indicate that the elastic modulus and confining pressure have a non-linear relationship for each lithology (Meng et al. 2006). It can be expressed as follows:

$$E = b_2\sigma_3^2 + b_1\sigma_3 + b_0 \tag{4.14}$$

where E is the rock elastic modulus in GPa; σ_3 is the confining pressure in MPa; b_0, b_1, and b_2 are the parameters depended on lithology. For different lithologies, the parameters in Eq. 4.14 differ (Table 4.2). Therefore, the elastic modulus and confining pressure in the rocks in the Huainan coal-field in eastern China have a pronounced power relation, which best fits the following equation proposed by Brown et al. (1989), with the exponent of c = 2:

$$E = (a\sigma_3 + b)^c \tag{4.15}$$

where a, b, and c are the constants.

Fig. 4.9. Servo-controlled triaxial compressive test apparatus of MTS Systems in China University of Mining and Technology.

Table 4.2. The parameters for different lithologies in Eq. 4.14

Lithology	b_2	b_1	b_0	σ_3 (MPa)
Medium-, and fine-grained sandstone	0.016	-0.121	38.113	≤50
Sandy mudstone	0.040	-0.678	26.909	≤40
Mudstone	0.006	0.254	28.876	≤50

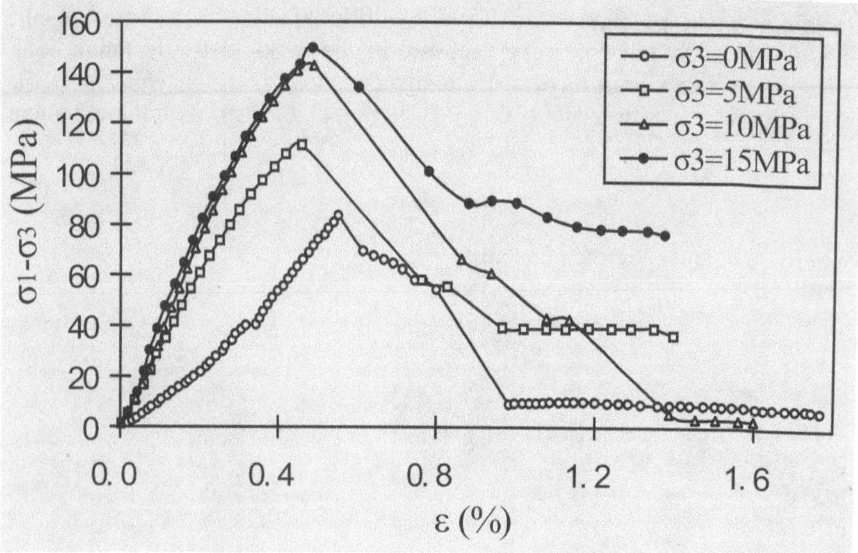

Fig. 4.10. Complete stress-strain curves for different confining pressures under the triaxial tests in a medium-grained sandstone.

Applying Mohr-Coulomb strength criterion in the triaxial test results (refer to Chap. 4, Sec. 4.6), the Mohr circle and rock failure envelope can be obtained. The Mohr-Coulomb strength criterion indicates that the normal stress (σ), shear stress (τ), and principal stresses (σ_1 and σ_3) have the following relationship for dry rocks:

$$\left(\sigma - \frac{\sigma_1 + \sigma_3}{2}\right)^2 + \tau^2 = \left(\frac{\sigma_1 - \sigma_3}{2}\right)^2 \tag{4.16}$$

Obviously, it is a circle in the (σ, τ) plane with the center at $(\frac{\sigma_1 + \sigma_3}{2}, 0)$ and the radius of $\frac{\sigma_1 - \sigma_3}{2}$. This circle is called Mohr circle as shown in Fig. 4.11 and can be used to determine rock failure envelope, cohesive strength (c) and internal friction angle (φ) (refer to Sec. 4.6.3 for more details).

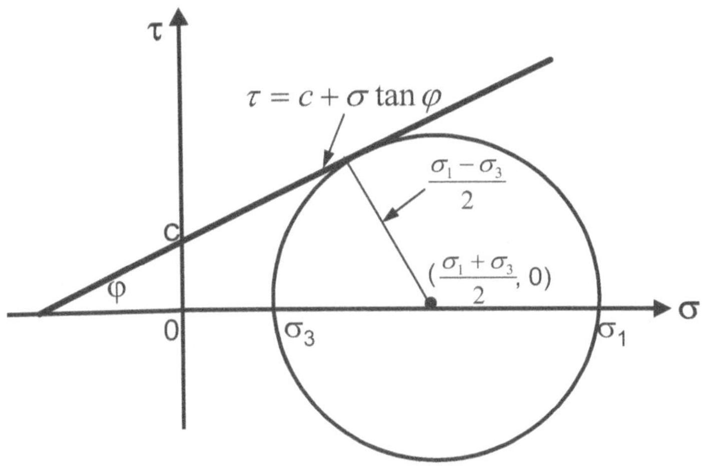

Fig. 4.11. Mohr circle and strength envelope for dry materials.

Figures 4.12 - 4.15 present the Mohr circles and failure envelopes from series triaxial compressive tests for different rocks sampled from Eastern China's Permian coal measures (Peng and Wang 2001). From the tests rock cohesion and internal angle of friction can be obtained from the Mohr circles and strength envelopes. Theses experimental results show that the mudstone and sandy shale have the cohesions ranged from 10.7 to 11.6 MPa and the internal friction angle varied from 30.7 to 40.3°. The cohesions in sandstones ranges from 16.5 to 22.1 MPa; the internal friction angle ranges from 40.6 to 46.23°. It can be seen that mudstone and shale have much lower cohesion than that in sandstones. Therefore, mudstones are normally most likely to have compressive/shear failure than sandstones.

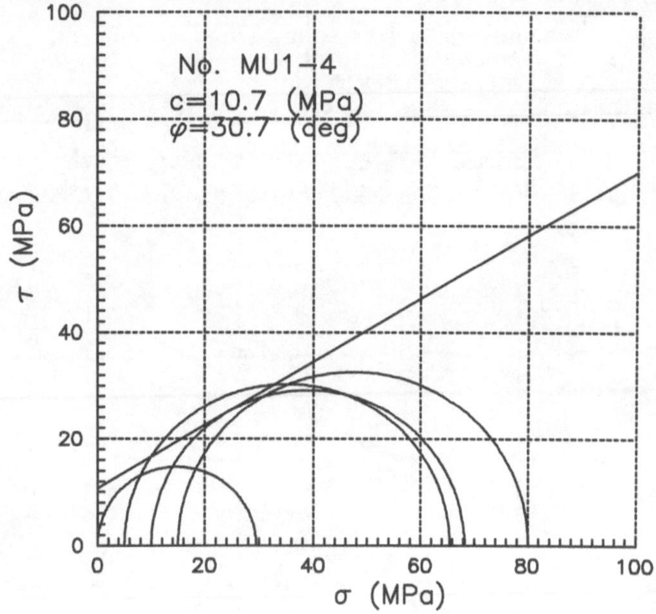

Fig. 4.12. Mohr circles and strength envelope from triaxial tests in mudstone.

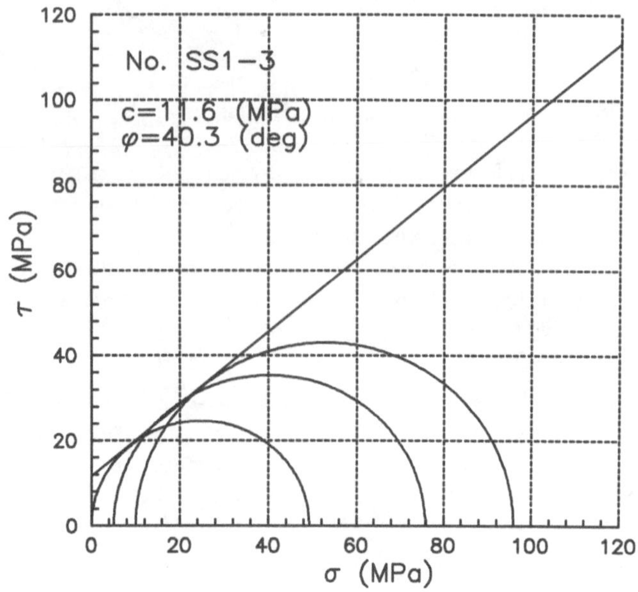

Fig. 4.13. Mohr circles and strength envelope from triaxial tests in sandy shale.

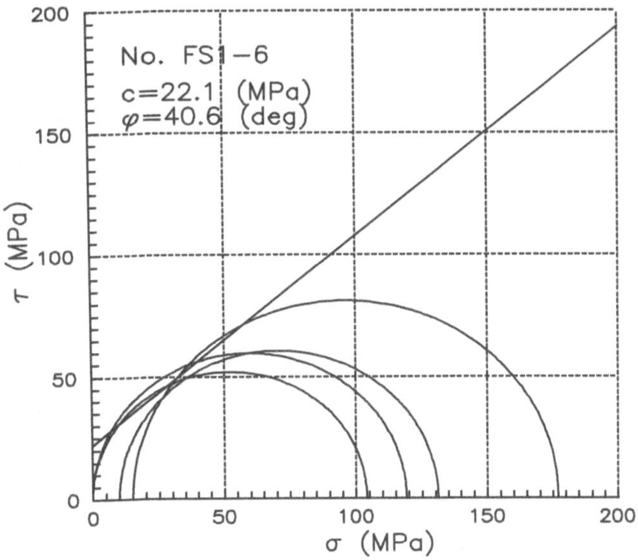

Fig. 4.14. Mohr circles and strength envelope from triaxial tests in fine-grained sandstone.

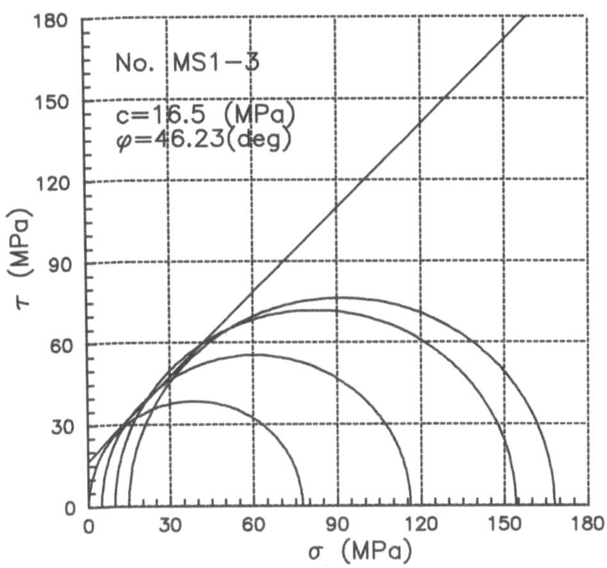

Fig. 4.15. Mohr circles and strength envelope from triaxial tests in medium-grained sandstone.

4.5 Polyaxial compressive test

Mogi (1971) constructed perhaps the first apparatus that enabled the application of three independent and mutually perpendicular uniform loads to the faces of a rectangular prismatic specimen with minimum friction. He subjected Dunham dolomite and other rocks to different intermediate principal compressive stresses for the same least principal stress, and then raised the maximum principal stress to failure.

The polyaxial cells built by Mogi (1971) and Spetzler et al. (1981) were mainly suited for testing deformational and strength characteristics of weak to medium strength rocks. The University of Wisconsin designed and fabricated a true triaxial testing system suitable for testing both weak and strong rocks (Haimson and Chang 2000). The true triaxial tests can better describe the stress-strain relationship for rocks situated in a three dimensional stress domain.

Polyaxial compressive tests demonstrate experimentally that rock strength is a function of the major principal stress (σ_1) and the minor principal stress (σ_3) as well as the intermediate stress (σ_2). Therefore, rock failure characteristic depends on the effects of all three principal stresses.

4.6 Rock failure criteria

4.6.1 Rock failure types

Rocks fail when the surrounding stress exceeds the tensile, the compressive, or the shear strengths of the rock formation, whichever is reached first. There are several failure types depended on rock lithology, rock microstructures, and applied stresses. Jaeger and Cook (1979) described rock failures at various confining pressures, as shown in Fig. 4.16.

In unconfined compression (Fig. 4.16a), irregular longitudinal splitting is observed. With a moderate amount of confining pressures, the rock failure is characterized by a single plane of fracture, inclined at an angle of less than 45° to the direction of σ_1, as shown in Fig. 4.16b and Fig. 4.17 (Peng and Meng 2002). This is a typical shear failure under compressive stresses, and it generates a shear displacement along the surface of the shear fracture. If the confining pressure is increased so that the material becomes fully ductile (Jaeger and Cook 1979), a network of shear fractures accompanied by plastic deformation appears, as shown in Fig. 4.16c. The second basic type of failure is tensile failure, which appears typically in uniaxial tension. Its characteristic feature is a clean separation with no offset between the surfaces (Fig. 4.16d). If a slab is compressed between line loads as shown in Fig.4.16e, a tensile fracture appears between the loads.

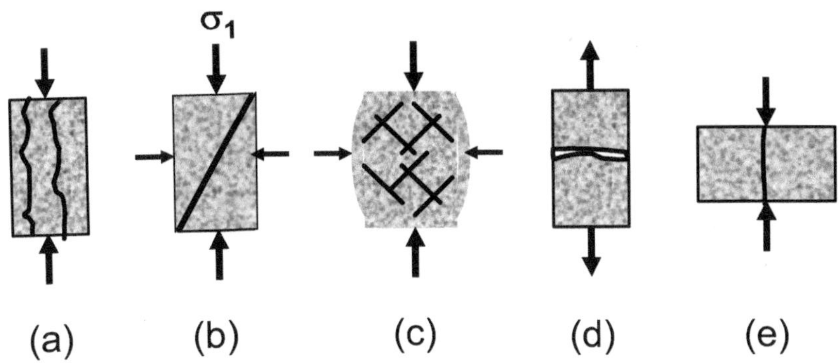

Fig. 4.16. Rock failure types. **a.** splitting; **b.** shear failure; **c.** multiple shear fractures; **d.** tensile failure; **e.** tensile failure induced by point loads.

Fig. 4.17. Shear failure induced by the axial compressive loading in limestone samples.

There are various failure/strength criteria applied to compare rock stresses and strength to describe rock failures. For the porous media the effective stress concentrations near underground engineering due to the far-field stresses need to be considered for failure initiation. It is commonly accepted that rock failure is controlled by Terzaghi's effective stress concept, because Biot's effective stress coefficient approaches 1.0 when rock failure is approached. Terzaghi's effective stress can be expressed as follows:

$$\sigma'_{ij} = \sigma_{ij} - \delta_{ij} p \tag{4.17}$$

where σ'_{ij} is the effective stress tensor, p is the pore pressure and δ is the Kronecker delta, and $\delta = 1$ when $i = j$; $\delta = 0$ when $i \neq j$.

Various failure criteria, such as tensile, Mohr-Coulomb, and Drucker-Prager, were presented for different types of underground engineering.

4.6.2 Tensile failure and Griffith criterion

Tensile failure will occur when the effective stress becomes tensile and equals or exceeds the formation tensile strength, that is:

$$\sigma'_3 = -T_0 \tag{4.18}$$

where T_0 is the tensile strength; σ'_3 is the minimum effective principal tensile stresses (note that σ'_3 is negative).

Griffith (1921) proposed that fractures are caused by stress concentration at the tips of minute Griffith cracks supposed to pervade the material. A fracture is initiated when the maximum stress near the tip of most favorably oriented crack reaches a value characteristic of the material. The Griffith failure takes place if the following equation exists for a dry material:

If $\sigma_1 + 3\sigma_3 > 0$,

$$(\sigma_1 - \sigma_3)^2 = 8T_0(\sigma_1 + \sigma_3) \tag{4.19}$$

If $\sigma_1 + 3\sigma_3 < 0$,

$$\sigma_3 = -T_0 \tag{4.20}$$

When $\sigma_3 = 0$, it is uniaxial compression, so that the uniaxial compressive strength (UCS) predicted by Eq. 4.19 (UCS = σ_1) is:

$$UCS = 8T_0 \tag{4.21}$$

It should be noted that for some rocks, Eq. 4.21 underestimates the uniaxial compressive strength. Modified Griffith theory (McClintock and Walsh 1962) suggested that the uniaxial compressive strength and tensile strength have the following relationship:

$$UCS = \frac{4(1+\sin\varphi)}{1-\sin\varphi} T_0 \tag{4.22}$$

where φ is the internal friction angle of the material.

Another extension of the Griffith criterion gives that the uniaxial compressive strength is 12 times of the tensile strength (Murrell 1963):

$$UCS = 12T_0 \tag{4.23}$$

4.6.3 Mohr-Coulomb criterion

Shear failure occurs when the shear strength of the formation is exceeded. The Mohr-Coulomb failure criterion can be used to determine such a failure mechanism for a porous medium; i.e.:

$$\tau = c + \sigma' \tan\varphi \tag{4.24}$$

where σ' is the normal effective stress; τ is the shear stress; φ is the angle of internal friction and c is the cohesion and also called inherent shear strength of the material.

In the principal space, $(\sigma'_1, \sigma'_2, \sigma'_3)$, the Mohr-Coulomb failure criterion can be expressed as:

$$\sigma'_1 = UCS + q\sigma'_3 \tag{4.25}$$

where σ'_1 , σ'_3 are the maximum and minimum effective principal stresses, respectively, and

$$q = (1+\sin\varphi)/(1-\sin\varphi) \tag{4.26}$$

$$UCS = 2c\cos\varphi/(1-\sin\varphi) \tag{4.27}$$

Figure 4.18 shows the Mohr circle and Mohr-Coulomb strength envelope. The failure will not take place if the values of σ' and τ lie below the strength envelope. In the Mohr-Coulomb criterion, two conclusions should be noted: (i) the intermediate principal stress σ'_2 does not affect failure and this may overestimate failure; (ii) the plane of shear fracture passes through the direction of the intermediate stress. This conclusion can be applied to determine the direction of the in-situ intermediate stress, when the stress regime is known.

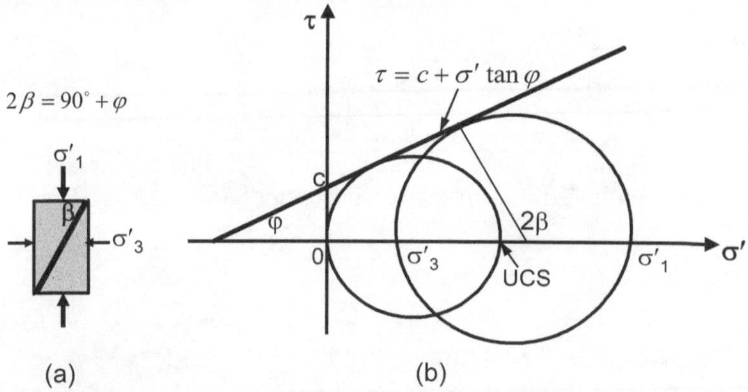

Fig.4.18. Mohr-Coulomb strength in effective stress domain. **a.** shear failure plane under triaxial compression, and in the shear plane, $\tan 2\beta = -1/\tan\varphi$; **b.** Mohr circles and strength criterion.

4.6.4 Drüker-Prager criterion

The effect of the intermediate principal stress is not considered in any of the failure criteria mentioned above. Laboratory data have shown that the intermediate principal stress plays an important role in the failure of rocks and can be conveniently represented by the Drüker-Prager yield condition (Drüker and Prager 1952); that is:

$$\sqrt{J_2} = \alpha I_1' + \kappa \tag{4.28}$$

where α and κ are material constants; and, I'_1 and J_2 are the stress invariants, which can be expressed as:

$$I_1' = \sigma_x' + \sigma_y' + \sigma_z' \tag{4.29}$$

$$J_2 = \frac{1}{6}[(\sigma_x - \sigma_y)^2 + (\sigma_y - \sigma_z)^2 + (\sigma_z - \sigma_x)^2] + \tau_{xy}^2 + \tau_{yz}^2 + \tau_{zx}^2 \tag{4.30}$$

The Drüker-Prager strength criterion is an extended form of the Mohr-Coulomb failure criterion. In the principal stress space, a section in the deviatoric plane shows that it is possible in various ways to relate the two failure criteria (Fig. 4.19). For plane-strain conditions it is recommended that the inside cone tangent to the pyramid is used (Kovari 1977). In this case the conversion formulae for the material parameters from the Mohr-Coulomb criterion to the Drüker-Prager criterion are (Zhang 2002):

$$\alpha = \frac{\tan\phi}{\sqrt{9+12\tan^2\phi}} \tag{4.31}$$

$$\kappa = \frac{3c}{\sqrt{9+12\tan^2\phi}} \tag{4.32}$$

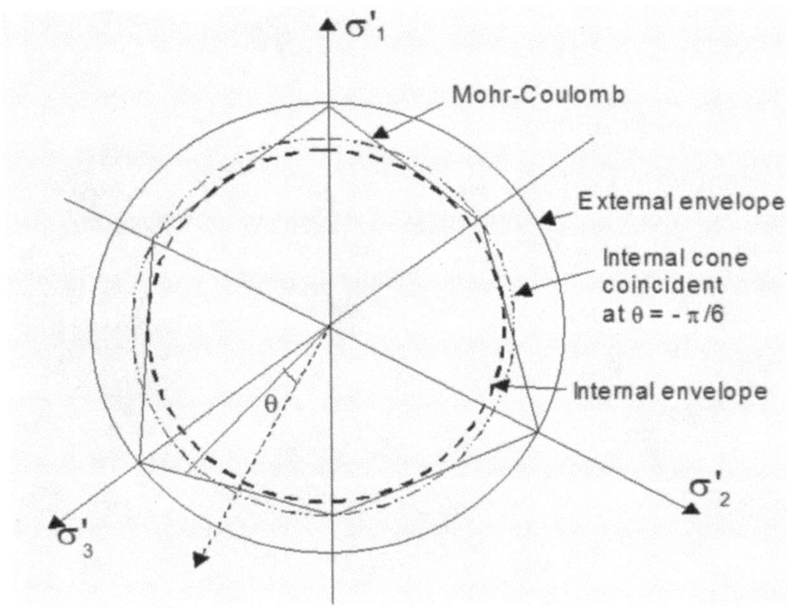

Fig.4.19. The Mohr-Coulomb yield surface and different possibilities for corresponding Drücker-Prager yield conditions.

When the external pyramid envelope is used, the conversion formulae for the material parameters from the Mohr-Coulomb criterion to the Drüker-Prager criterion are as follows:

$$\alpha = \frac{2\sin\phi}{\sqrt{3}(3-\sin\phi)} \tag{4.33}$$

$$\kappa = \frac{6c\cos\phi}{\sqrt{3}(3-\sin\phi)} \tag{4.34}$$

When the internal cone envelope coincident at $\theta = \pi/6$ is used, the conversion formulae for the material parameters from the Mhor-Coulomb criterion to the Drüker-Prager criterion become:

$$\alpha = \frac{2\sin\phi}{\sqrt{3}(3+\sin\phi)} \tag{4.35}$$

$$\kappa = \frac{6c\cos\phi}{\sqrt{3}(3+\sin\phi)} \tag{4.36}$$

4.6.5 Modified Lade criterion

The modified Lade criterion was proposed by Ewy (1998) based on Lade strength criterion (Lade 1984). In this criterion the three effective principal stresses were considered, and this criterion finds applications in petroleum engineering for wellbore stability and sand production. The modified Lade can be expressed as follows:

$$\frac{(I_1^{''})^3}{I_3^{''}} = 27 + \eta \tag{4.37}$$

where

$$I_1^{''} = (\sigma_1'+S) + (\sigma_2'+S) + (\sigma_3'+S)$$

$$I_3^{''} = (\sigma_1'+S)(\sigma_2'+S)(\sigma_3'+S)$$

$$S = \frac{c}{\tan\varphi}$$

$$\eta = \frac{4\tan^2\varphi(9-7\sin\varphi)}{1-\sin\varphi}$$

Modified Lade criterion considers the influence of intermediate principal stress on rock failures. The failure predicted by this criterion is less failure/conservative than the Mohr-Coulomb criterion because the latter ignores the influence of intermediate principal stress (σ_2). The modified Lade predicts more failure/aggressive than the Drüker-Prager criterion, because the latter predicts a greater strengthening effect from σ_2.

4.6.6 Hoek-Brown criterion

Through study of the experimental results of a wide variety of rocks, Hoek and Brown (1980) presented the following empirical rock failure criterion for jointed rock masses:

$$\sigma_1' = \sigma_3' + UCS(m\sigma_3'/UCS + s)^a \tag{4.38}$$

where σ_1', σ_3' are the maximum and minimum effective principal stresses, respectively; m and s are the constants depending on the properties of the rock and on the extent to which it has been broken before being subject to the stresses; a is the constant depended upon the rock mass characteristics; m ranges from 0.001 (extremely week rock) to 25 (extremely strong rock) and $s = 1$ for intact rock; $s < 1$ for previously broken rock.

For intact rocks that make up the rock mass, Eq. 4.38 simplifies to:

$$\sigma_1' = \sigma_3' + \sqrt{mUCS\sigma_3' + sUCS^2} \tag{4.39}$$

For clastic sediments, Hoek and Brown (1997) suggested using the following values for m:

$m = 22$ for conglomerate; $m = 19$ for sandstone; $m = 9$ for siltstone; and $m = 4$ for claystone.

The uniaxial compressive strength of the rock mass (σ_c) is obtained by setting $\sigma_3' = 0$ in Eq. 4.38 (Hoek 2007):

$$\sigma_c = s^a UCS \tag{4.40}$$

The tensile strength of the rock mass (σ_t) can be obtained from the following equation:

$$\sigma_t = -sUCS/m \tag{4.41}$$

The constant a, s, and m can be expressed as the follows:

$$a = \frac{1}{2} + \frac{1}{6}(e^{-GSI/15} - e^{-20/3}) \tag{4.42}$$

$$m = m_i \exp\left(\frac{GSI - 100}{28 - 14D}\right) \tag{4.43}$$

$$s = \exp\left(\frac{GSI - 100}{9 - 3D}\right) \tag{4.44}$$

where m is the material constant for the intact rock; D is the disturbance factor depended on the degree of disturbance to which the rock mass has been subjected by blast damage and stress relaxation. It varies from 0 for undisturbed in-situ rock masses to 1 for very disturbed rock masses; GSI is the Geological Strength Index introduced by Hoek (1994). GSI varies from 0-100 and can be estimated from Fig. 4.20 (Hoek 2007).

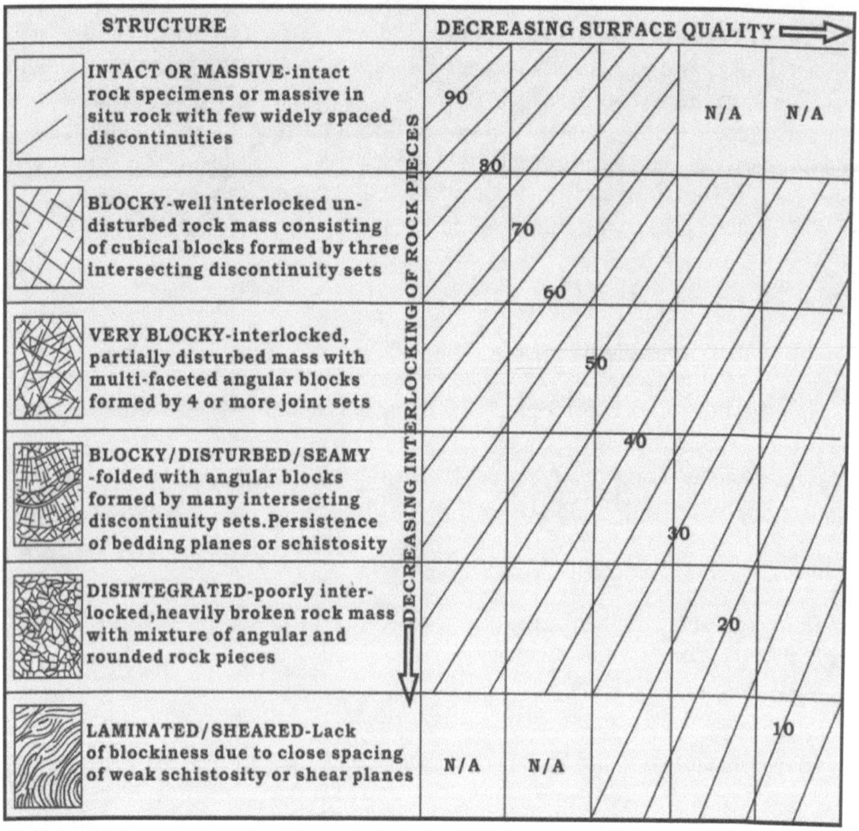

Fig. 4.20. Estimate of Geological Strength Index (GSI) based on geological descriptions (after Hoek 2007).

References

Brown ET (ed) (1981) Rock characterization, testing and monitoring: ISRM suggested methods. Pergamon Press, Oxford

Bieniawski ZT (1975) The point load test in geotechnical practice. Eng Geol, Sept, pp 1-11

Brown ET, Bray JW, Santarelli FJ (1989) Influence of stress-dependent elastic moduli on stresses and strains around axisymmetric boreholes. Rock Mech Rock Engng 22:189-203

Ewy RT (1998) Wellbore stability prediction using a modified Lade criterion. Paper SPE 47251 presented at SPE/ISRM Eurock'98 held in Trondheim, Norway

Drüker DC, Prager W (1952) Soil mechanics and plastic analysis or limit design. Q Appl Math 10:157-165

Fjær E, Holt RM, Horsrud P, Raaen AM (1992) Petroleum related rock mechanics. Elsevier

Griffith AA (1921) The phenomena of rupture and flow in solids. Phil Trans Roy Soc London A221:163-198

Haimson B, Chang C 2000 A new true triaxial cell for testing mechanical properties of rock and its use to determine rock strength and deformability of Westerly granite. Int J of Rock Mech Min Sci 37:285-296

Hoek E, Brown ET (1980) Underground excavations in rock. Inst Min Metall London

Hoek E (1994) Strength of rock and rock masses. ISRM News J 2(2):4-16

Hoek E, Brown ET (1997) Practical estimates of rock mass. Int J Rock Mech Min Sci 34 (8);1165-1186

Hoek E (2007) Practical Rock Engineering. Rocscience.

ISRM (International Society of Rock Mechanics) (1985) Suggested methods for determining point load strength. Int J Rock Mech Min Sci Geomech Abstr 22:71-96

Jaeger JC, Cook NGW (1979) Fundamentals of rock mechanics, 3rd edition. Chapman & Hall

Kovari K (1977) The elasto-plastic analysis in the design practice of underground openings. Finite Elements in Geomechanics Gudehus G (ed), John Wiley & Sons pp377-412

Kovari K, Tisa A, Einstein HH, Franklin JA (1983) Suggested methods for determining the strength of rock materials in triaxial compression: revised version. Int J Rock Mech Min Sci Geomech Abstr 20, 283-290

Lade PV (1984) Failure criterion for frictional materials. Mechanics of Engineering Materials Desai CS, Gallager RH (ed) John Wiley & Sons

McClintock FA, Walsh JB (1962) Friction on Griffith cracks under pressure. 4th US National Congress of Appl Mech Proc p1015-1021

Meng Z, Zhang J, Peng S (2006) Influence of sedimentary environments on mechanical properties of clastic rocks. Environ Geol 51:113-120

Meng Z, Peng S, Li G (2003) Influence of lithological characters of coal bearing formation on the coal roof stability. J of China Univ of Min Tech 13(1):1-6 (in Chinese)

Meng Z, Peng S, Ling B (2000) characteristics of strength and deformation for sedimentary rocks under different confining stresses. J of China Coal Soceity 25(1):15-18 (in Chinese)

Mogi K (1971) Fracture and flow of rocks under high triaxial compression. J Geophys Res 76:1255-1269

Murrell SAF (1963) A criterion for brittle fracture of rocks and concrete under triaxial stress and the effect of pore pressure on the criterion. Proc 5th Rock Mech Symp in Rock Mechanics Fairhurst C (ed), Pergamon p563-577

Peng S (1998) Approach on mechanical properties of clastic rocks and rock microstructure. Proc. 5th Ann Conf of Young Scientists, China Coal Society. Coal Industry Press (in Chinese)

Peng S, Meng Z (1999) Classification and its application of roof stability using lithology and geological structure. Symp 46th Young scientists, China Sci Tech Soc, China Sci Tech Press Beijing pp35-42 (in Chinese)

Peng S, Qu H, Luo L, Wang L, Duan Y (2000) Complete stress-strain and permeability experiments. J China Coal Society 25(2):113-116 (in Chinese)

Peng S, Meng Z (2002) Theory and practice of mining engineering geology. Geological Press (in Chinese)

Peng S, Wang J (2001) Safe mining over confining aquifers. Coal Industry Press (in Chinese)

Rusnak J, Mark C (2000) Using the point load test to determine the uniaxial compressive strength of coal measure rock. Proc 19th Int Conf on Ground Control in Mining. West Virginia University, pp 362-371

Spetzler HA, Sobolev GA, Sondergeld CH, Salov BG, Getting IC, Koltsov A (1981) Surface deformation, crack formation, and acoustic velocity changes in pyrophyllite under polyaxial loading. J Geophys Res 86:1070-1080

Zeng Z (2002) Imaging hydraulic fractures using microseismicity. PhD dissertation, Univ of Oklahoma

Zhang J (2002) Dual-porosity approach to wellbore stability in naturally fractured reservoirs. PhD dissertation, Univ of Oklahoma

5 Sedimentary rock masses and discontinuities

5.1 Rock mass and discontinuities

5.1.1 Introduction

One of the most prominent features of the earth's upper crust is the presence of joints and fractures at all scales. The fractures and joints usually are named to be discontinuities. A rock mass consists of both rock blocks and discontinuities. A rock block/matrix and a rock mass have significant differences in geomechanical behaviors. Nearly every physical property of sedimentary rock (such as mechanical properties; hydraulic, thermal, and electrical conductivities; and acoustic properties) is determined to some extent on the fractures and the fluids they contain. The success of many applications such as efficient recovery from fractured reservoirs, hazardous waste disposal, and geothermal energy extraction depends on a thorough understanding of fracture behavior. Many petroleum reservoirs and coal seams are situated in fractured porous formations. Thus, prior to any rock engineering it is necessary to determine the geometrical and mechanical properties of the fractures and rock mass from lab and the field. For instance, the geometry, locations, orientation, aperture variations and fluid-mechanical properties of fractures and rock mass are needed to be determined.

The discontinuities can be the single most important factor governing the deformability, strength, and permeability of the rock mass. Moreover, a particularly large and persistent discontinuity could critically affect the stability of any underground excavation (Hudson and Harrison 1997). For these reasons, it is necessary to develop a thorough understanding of the geomechanical properties of discontinuities and rock mass and understand the way in which discontinuities affect geological engineering.

5.1.2 Discontinuities

Bedding

There are different kinds of discontinuities in rock formations. Bedding is one of the most common geologic phenomena in sedimentary rocks. The bedding plane or interface between two layers in the sedimentary rock mass represents geologically the transitive interface from one sedimentary environment to another, and it also represents an interruption of sedimentary planes. A bedding plane is a weak plane, usually consisting of plant detritus, mica, and other weak minerals. Therefore, it is usually treated mechanically as a weak plane.

Unconformity

Unconformity plane is a surface that represents a break in the geologic record, with the rock unit immediately above it being considerably younger than the rock beneath. There are three major aspects to consider:

1. Time. An unconformity develops during a period of time in which no sediment is deposited.
2. Deposition. Any interruption of deposition or breaks in sedimentation is an unconformity.
3. Structure. Structurally, unconformity may be regarded as planar structures separating older rocks below from younger rocks above.

A plane of unconformity may be a surface of weathering, erosion, or denudation, or a surface of non-deposition, or possibly some combination of these factors. It may be parallel to the upper strata, make an angle with the upper strata, or be irregular. Subsequent Earth movements may have folded or faulted it. Therefore, rock mechanical behaviors are very different cross the unconformity plane.

Weak layer

A weak layer in the sedimentary rock mass is the layer with very low strength. Rock construction is most likely to fail in this layer. Therefore, the weak plane plays a key role in stability of rock engineering.

Tectonic structural plane

Tectonic structural plane includes faults, fractures, and other discontinuities induced by tectonic activities.

5.2 Mechanical characteristics of clastic rocks

The mechanical properties of clastic rocks are not only affected by the external factors, such as the state of stress and surrounding geological environment, but also strongly dependent on the depositional and sedimentary environments, including components, textures, and structures of the rocks. It is of importance to study the mechanical properties of clastic rocks for energy industry, because oil and gas, as well as coal, are situated in these rocks (Peng and Meng 2002). Until now, most research on the microstructure of the rock and its deformation and strength behaviors mainly focuses on granites and other hard rocks. Therefore, relations between the sedimentary characteristics and mechanical properties of clastic rocks need to be addressed, particularly those of soft rocks such as mudstone and shale (Peng 1998).

The formations of coal measures are primarily composed of clastic rocks and coal seam. The rocks in coal measures mainly consist of sandstones, siltstones, and mudstones. For example, coal measures of the Permian period in the Huainan coalfield in China were formed from a delta plain environment (Peng and Meng 2002), which contains 33.66% of sandstones, 8.7% of siltstones, 50.38% of mudstones, and 7.23% of coal seams. Sedimentary rocks in coal measures originated in shallow crust, and their formation and existing environments are distinctly different from those of igneous or metamorphic rocks. Because of their lithological characteristics, complex components, and textures, clastic rocks differ considerably from other rocks in mechanical properties. Therefore, it is important to establish the qualitative and quantitative relations between sedimentary parameters and mechanical properties in clastic rocks for recognizing the mechanical properties in different sedimentary rocks, revealing deformation and failure mechanisms of clastic rocks, and solving the related engineering problems (Meng et al. 2006). The following sections present lab experiments to determine clastic rocks mechanical properties.

5.2.1 Sample preparation and test methods

The rock samples were obtained from the roof and floor strata of the primary coal seams in the Permian coal measures in the Xinji coal mine, the Huainan coalfield in China. In order to make the testing results conform as much as possible to the natural status, the rock samples were promptly packaged and fully waxed after they were picked up. The lab experiments are conducted to test the mechanical properties, observe the microstructures of rock slices by using optical microscope, and analyze the mineral and chemical components for 55 samples. The quantitatively statistical analyses are conducted for these experimental results.

The mechanical properties of rocks are tested in the servo-controlled testing system (MTS 815.02). Using the optical microscope, the sedimentary parameters such as the component, grain size, intersertal material, cementation type, and structural support type in all samples of clastic rocks are observed and analyzed statistically. In the process of quantitative analyses of the granularity of the rock slices, the object glass desk of the microscope is moved in both horizontal and vertical directions to determine the size of detritus grain (lattice number) of the sample slice lying in the cross-line in the microscope. In each rock slice 500 points are measured; then the grain diameter can be obtained by multiplying the measured lattice numbers by the value of the actual size (in mm). The value of the actual size for each lattice is measured using the standard micro-ruler. The rock grain size is measured by the Phi (ϕ) scale as a log transformation of the grain diameter in millimeters in order to simplify computation of statistical parameters. Because the ϕ scale is now used almost exclusively for computation and is gradually replacing the millimeter scale, it is used in the ongoing analyses. The following expression is used to calculate the standard grain-size scale:

$$\phi = -\log_2 D \tag{5.1}$$

where: ϕ is the grain-size scale; D is the diameter of the grain in mm. It can be seen that the larger diameter of the grain size has a smaller ϕ scale.

The average grain size in the ϕ scale can be calculated by the following equation:

$$\Phi = \frac{(\phi_{16} + \phi_{50} + \phi_{84})}{3} \tag{5.2}$$

where: Φ is the average grain-size ϕ scale, which reflects the hydrodynamic performance during deposition of terrigenous clasts; ϕ_{16}、ϕ_{50} and ϕ_{84} are the corresponding grain sizes of the cumulative contents of 16%, 50%, and 84% in the cumulative curve of the clastic grains, respectively.

5.2.2 Clastic rock contents versus rock mechanical properties

Statistical analyses of experimental results show that the mechanical properties are highly related to the content of the clastic grains and quartz in clastic rocks. In this research, the clasts whose grain diameters are more than 0.03 mm (ϕ scale < 5) are counted, which are called contents of detritus. Table 5.1 lists the mechanical properties of sedimentary rocks with different contents of detritus. It shows that the mechanical properties of clastic rocks vary in a wide range. However, the general trend is that average uniaxial compressive strength and elastic modulus increase as the contents of detritus increase.

Table 5.1. The mechanical properties of sedimentary rocks with different detritus contents

Detritus content (%)	Compressive strength (MPa)	Elastic modulus (GPa)
< 25	$\frac{35.5-56.47}{51.45}$	$\frac{24-27}{25.5}$
25~50	$\frac{28.81-77.01}{49.25}$	$\frac{18-31}{24.5}$
50~75	$\frac{54.59-110.93}{92.64}$	$\frac{9.58-54}{35.82}$
> 75	$\frac{30.5-176}{107.67}$	$\frac{12.97-86.44}{41.1}$

Note, $\frac{Min-Max}{Ave}$

Figures 5.1-5.4 show the uniaxial compressive strength and elastic modulus of rocks versus the contents of detritus and quartz (Peng and Meng 2002, Meng et al. 2006). It can be seen that the contents of detritus and quartz have profound effects on the strength and elastic modulus. In each figure, the maximal uniaxial compressive strength or elastic modulus in various contents of detritus or quartz is connected to form an envelope of the maximum strength or modulus. It can be found from Figs. 5.1-5.4 that there are two obvious turning points in each envelope showing two jumps in the compressive strength or elastic modulus of clastic rocks.

The first jump takes place at the content of detritus of 50%, or the content of quartz of 40%. In this interval (area I in Figs. 5.1-5.4) the uniaxial compressive strength and elastic modulus of clastic rocks are small. As the contents increase, the enveloping lines slowly climb up; i.e. the uniaxial compressive strength increases from 54 MPa to 70 MPa in Fig. 5.1, and the elastic modulus from 24 GPa to 35 GPa in Fig. 5.3. The dispersion in the uniaxial compressive strength or elastic modulus in this interval is small. The cause is that clastic rocks in this interval are mainly fine-grained siltstones, sandy mudstones, and mudstones with smaller grain sizes (grain scale $\phi > 5$). Because the content of detritus is small and the mutual connection of the grains of detritus is quite weak, the change in the content of detritus is not large enough to cause the apparent changes of the mechanical properties. The mechanical strength in this case is mainly dependent on the cementation of clay minerals. The rocks behave elastoplastic, plastic, or visco-elastoplastic deformations.

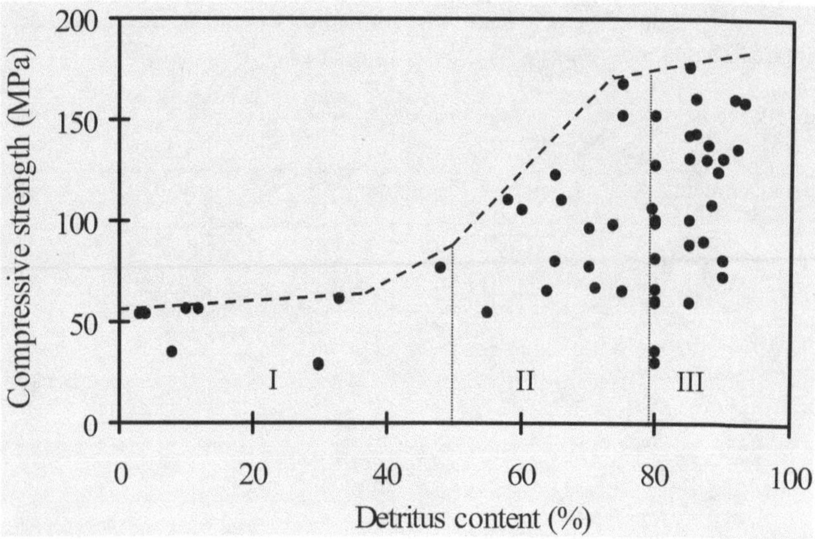

Fig. 5.1. Experimental results of the content of detritus and the uniaxial compressive strength.

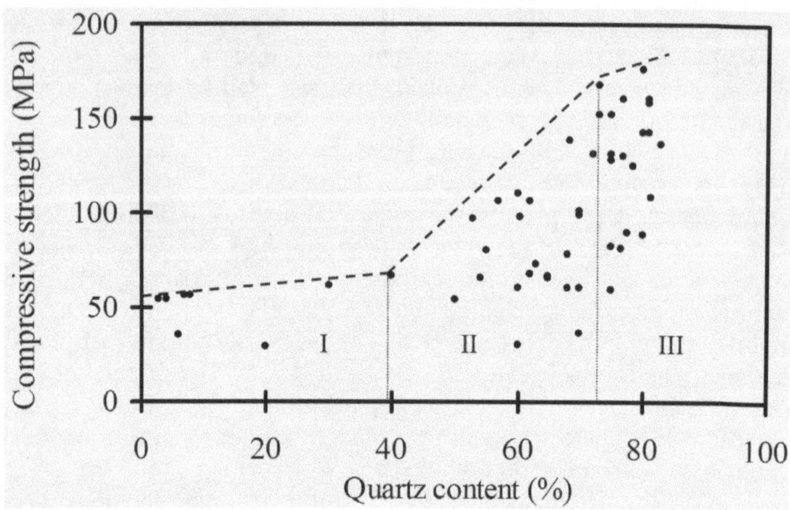

Fig. 5.2. Experimental results of the content of quartz and the uniaxial compressive strength.

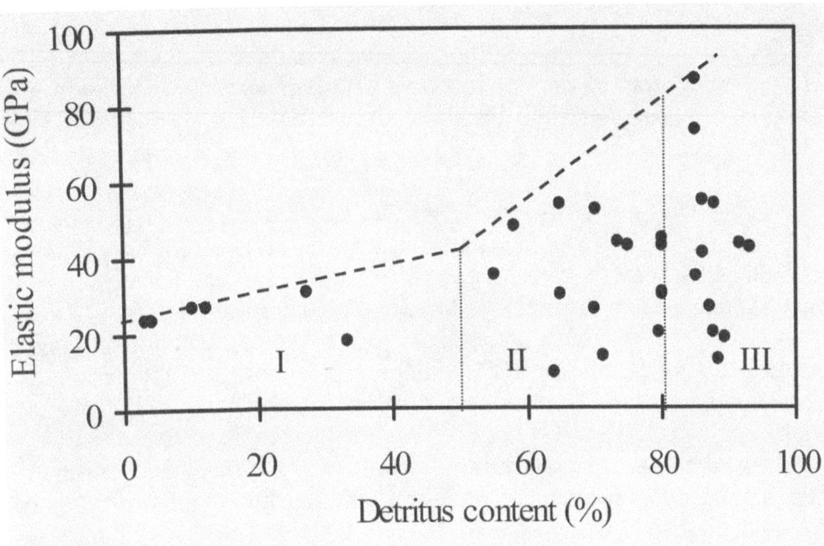

Fig. 5.3. Experimental results of the content of detritus and the elastic modulus.

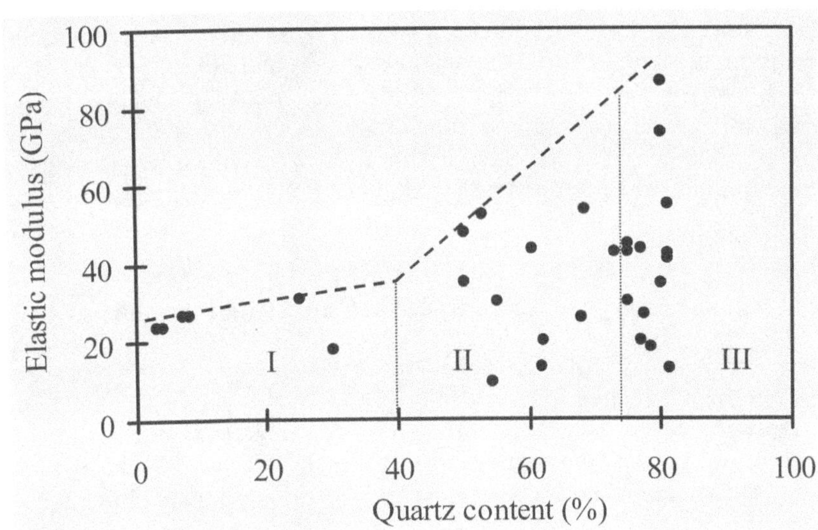

Fig. 5.4. Experimental results of the content of quartz and the elastic modulus.

In area II in Figs. 5.1-5.4, each envelope has a steep slope, showing that the uniaxial compressive strength or the elastic modulus increases significantly as the content of the detritus or quartz increases. For example, the uniaxial compressive strength increases from 70 MPa to 176 MPa in Fig. 5.1, and the elastic modulus from 35 GPa to 86 GPa in Fig. 5.3. The dispersion of the uniaxial compressive strength or elastic modulus also increases. These are primarily caused by the fact that the large size detritus are the main constitutional parts in clastic rocks. These rocks are sandstones and siltstones of various grades. In this case the contact between grains gradually becomes rigid contact such as the point contact, line contact, and convex and concave contact. The internal cementation of clastic rocks transitions from the basal cementation to the porous cementation, osculant cementation, and inserted cementation. Furthermore, the dispersion of the uniaxial compressive strength obviously increases, because of the influence of the grain size, cementation, and support type of the grains.

The second jump happens at about 80% of detritus content or 75% of quartz content, as shown in area III in Figs. 5.1-5.4. The slope of each envelope in the compressive strength or elastic modulus of clastic rocks increases slightly as the contents of detritus or quartz increases. The rocks have very high compressive strength and elastic modulus. This is because of the fact that these rocks are mainly quartzy sandstones, feldspathic-quartzy sandstones, and debris-quartzy sandstones, when the content of detritus is greater than 80% or the content of quartz more than 75%. For clastic sedimentary rocks, particularly sandstones, the main composition of clastic grains is quartz, which accounts for up to 80-90% of the contents of the clast. Quartz is a rigid mineral with a high strength, hence the clastic rock with high content of quartz has high compressive strength and elastic modulus.

5.2.3 Clastic rock grain size versus rock mechanical properties

The experimental results show that the mechanical properties of clastic rocks are highly related to their textures. The texture parameters that considerably affect the mechanical properties are the grain size, cementation between grains, and cementation type in the clast. Table 5.2 lists the experimental results of rock mechanical properties for various clastic rocks with different grain sizes. It can be seen that the dispersion is large between the mechanical properties and grain sizes, which illustrates that the grain size is not the only factor to control the mechanical properties.

Table 5.2. The influence of the grain sizes on rock mechanical properties

Lithology	Grain size (ϕ)	Compression strength (MPa)	Elastic modulus (GPa)
Coarse- grained sandstone	0-1.0	$\frac{97.55-136.05}{116.8}$	$\frac{32-32}{32}$
Medium- grained sandstone	1.0-2.0	$\frac{54.59-160.26}{124.7}$	$\frac{35.1-55}{45.68}$
Fine- grained sandstone	2.0-3.3	$\frac{59.44-176.14}{116.99}$	$\frac{18.69-79.89}{40.68}$
Siltstone	3.3-6.67	$\frac{60.02-130.56}{84.36}$	$\frac{9.58-52.47}{23.87}$
Mudstone	> 6.67	$\frac{28.81-56.47}{45.1}$	$\frac{24-27}{25.5}$

Note, $\frac{Min-Max}{Ave}$

Figures 5.5 and 5.6 plot the uniaxial compressive strength and elastic modulus versus the Φ scale (i.e. the average grain-size ϕ scale obtained from Eq. 5.1). It can be seen from Figs. 5.5 and 5.6 that the envelope lines for the maximal uniaxial compressive strength and elastic modulus are closely related to the grain sizes of the rocks (Peng and Meng 2002, Meng et al. 2006).

It can be seen that the envelopes of the mechanical properties versus the grain size of clasts are non-monotonic, which can be explained as follows. The grain sizes in clastic rocks are determined by the internal components and structures of clasts. As the Φ scale increases, the internal components of clastic rocks change gradually from the detritus, polycrystalline quartz, feldspar, single-crystal quartz into clay and mica. In general, the strength and rigidity of single-crystal quartz are greater than those of polycrystalline quartz.

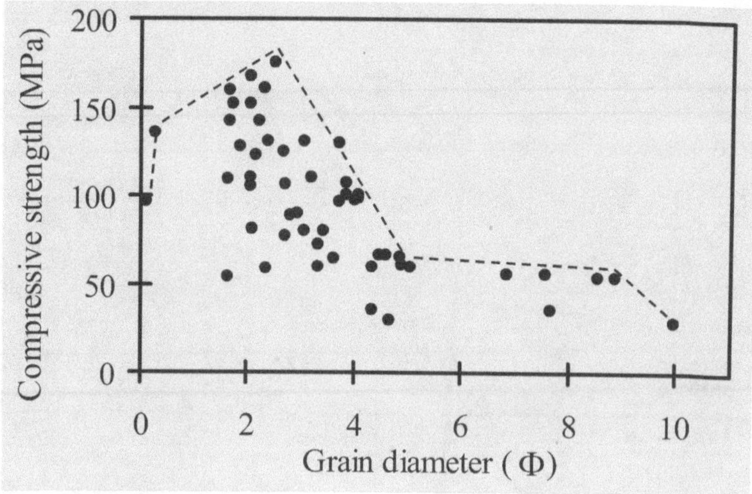

Fig. 5.5. Experimental results of the scale of grain diameter and the compressive strength.

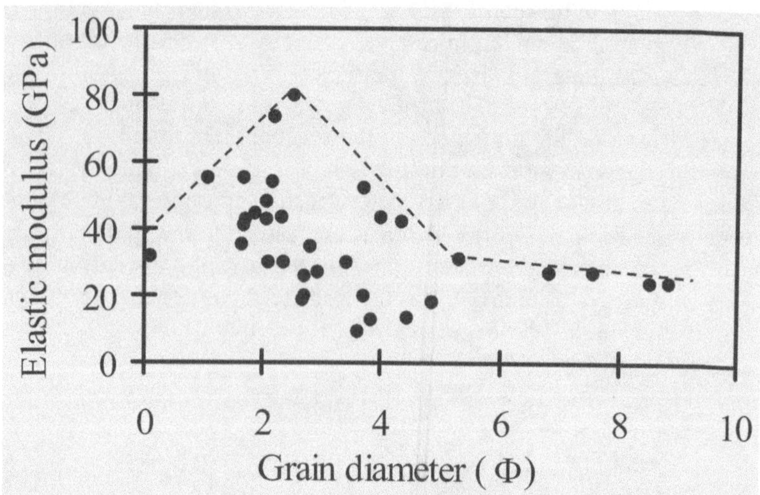

Fig. 5.6. Experimental results of the scale of grain diameter and the elastic modulus.

From the sedimentological theory, the distribution of the grain sizes has a certain regularity. The detritus and polycrystalline quartz mostly exist in the conglomerate and coarse-grained sandstone, and the single-crystal quartz mainly exists in the fine-grained sandstone. Therefore, the mechanical properties for the largest grain size are not the highest. However, the mechanical properties in the fine-grained sandstone ($\Phi = 2.5$) are the highest, as shown in the peak values in Figs. 5.5 and 5.6. As the Φ scale increases, namely, as the grains in clasts become fine, the contents of clay and mica minerals increase, and the structures of clastic rocks change. Therefore, the mechanical properties of the rocks decrease. However, the mechanical properties of clastic rocks are also influenced by other factors, such as the cementation and its type, because the clasts were bonded by cementation to form the consolidated rocks.

5.2.4 Clastic rock lithology versus rock mechanical properties

Under the uniaxial compression test

The uniaxial compression results (Table 5.3) show that the mechanical properties in different clastic rocks (sandstone, siltstone, sandy mudstone and mudstone) are quite different (Peng and Meng 2002, Meng et al.2006). The average values of the compressive and tensile strength are the largest in sandstone, being 111.5 MPa and 6.66 MPa, respectively; and the values in mudstone are the least, being 42.75 MPa and 1.91 MPa, respectively (refer to Fig. 4.7). These indicate that the strength and other mechanical properties of the rocks strongly depend upon lithology (Peng and Meng 1999).

Table 5.3. Lab test results of the mechanical properties in clastic rocks in China

Properties	Sandstone	Siltstone	Sandy mudstone	Mudstone
Density (g/cm^3)	$\frac{2.47\text{-}3.47}{2.76}$	$\frac{2.43\text{-}2.63}{2.56}$	$\frac{2.64\text{-}2.98}{2.72}$	$\frac{2.05\text{-}2.97}{2.68}$
Compressive strength (MPa)	$\frac{50.60\text{-}281.3}{111.50}$	$\frac{67.2828\text{-}130.1}{94.54}$	$\frac{13.50\text{-}112.1}{53.46}$	$\frac{9.81\text{-}81.5}{42.75}$
Tensile strength (MPa)	$\frac{1.77\text{-}10.67}{6.66}$	$\frac{1.20\text{-}9.2}{5.20}$	$\frac{0.70\text{-}8.70}{4.39}$	$\frac{0.30\text{-}7.29}{1.91}$
Cohesion (MPa)	$\frac{1.91\text{-}13.07}{6.30}$	$\frac{1.25\text{-}2.40}{2.33}$	$\frac{4.00\text{-}11.9}{6.22}$	$\frac{0.14\text{-}8.4}{3.95}$
Internal friction angle (°)	$\frac{33.41\text{-}39.15}{36.49}$	$\frac{39.00\text{-}40.03}{39.52}$	$\frac{31.90\text{-}38.4}{34.14}$	$\frac{31.8\text{-}41.52}{36.72}$

Note, $\frac{Min - Max}{Ave}$

Under the triaxial compression test

Triaxial compression tests demonstrate that the deformations and strength of clastic rocks are closely related to confining pressures (Meng et al. 2000), as shown in Figs. 5.7-5.9. The slopes of the stress-strain curves obviously become steep, and the strength increases as confining pressures increase for all different rocks. This implies that the confining stress is of importance for rock engineering stability.

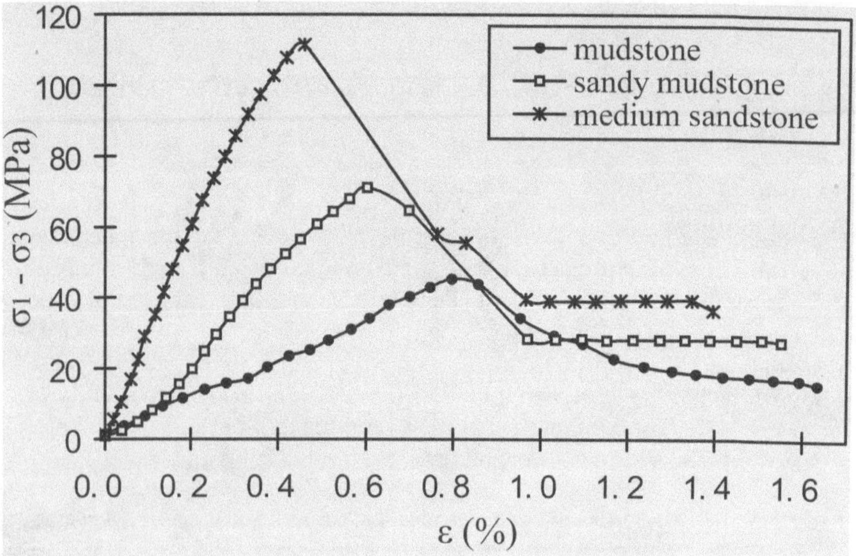

Fig. 5.7. Complete stress-strain curves for different rock samples under the triaxial tests with confining pressure of 5 MPa.

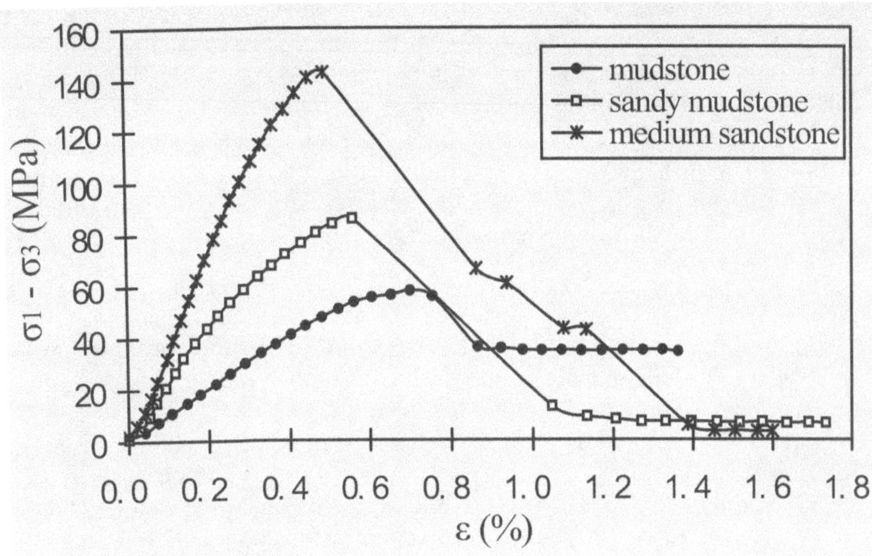

Fig. 5.8. Complete stress-strain curves for different rock samples under the triaxial tests with confining pressure of 10 MPa.

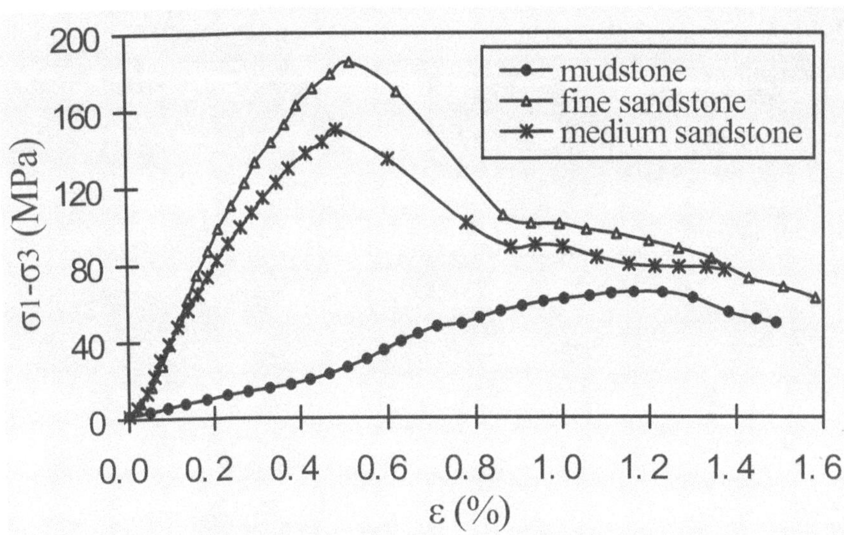

Fig. 5.9. Complete stress-strain curves for different rock samples under the triaxial tests with confining pressure of 15 MPa.

Triaxial compressive tests for sedimentary rock samples in Huinan coal mines of Eastern China's Permian coal measures show that the compressive strength increases as the confining pressure increases (Peng et al. 2000). However, variations are controlled by lithology. Figure 5.10 plots the rock compressive test results in the maximum principal stress (σ_1) versus minimum principal stress (σ_3) domain (Meng et al. 2006). The following relationship between the principal stresses and uniaxial compressive strength is obtained from the experimental results:

$$\sigma_1 = \sigma_c + k\sigma_3 \tag{5.3}$$

where σ_1 is the maximum principal stress (MPa); σ_3 is the confining pressure (MPa); σ_c is the uniaxial compressive strength (MPa); k is the parameter dependent upon lithology (refer to Table 5.4).

Equation 5.3 is another form of expression of Mohr-Coulomb strength criterion (refer to Chap. 4, Sec. 4.6). Based on Eq. 5.3 and the data in Table 5.4, sedimentary rocks have the linear best-fit for Mohr-Coulomb' failure criterion. For the soft formations, such as mudstone and coal, there is perfect fit to linear Mohr-Coulomb strength criterion (Fig. 5.10). This is important for choosing a correct strength criterion in modeling of rock engineering in such rocks.

Table 5.4. The parameters in different lithologies in Eq. 5.3

Lithology	σ_c (MPa)	k	σ_3 (MPa)
Medium-, fine-grained sandstone	127.83	5.89	≤40
Sandy mudstone	73.85	5.72	≤40
Mudstone	40.15	2.66	≤50
Coal	11.23	1.99	≤30

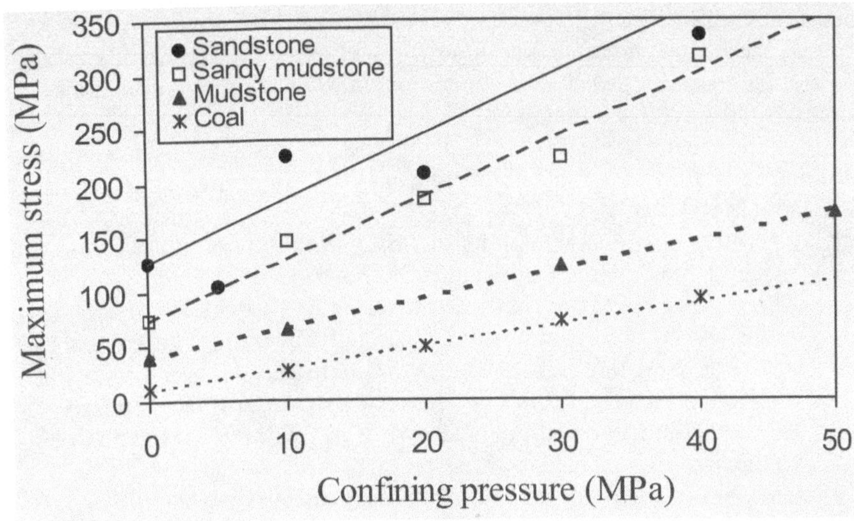

Fig. 5.10. Experimental results showing that clastic rocks fit the linear Mohr-Coulomb strength criterion in terms of the maximum and minimum principal stresses for fine- and medium-grained sandstone, sandy mudstone, mudstone, and coal obtained from the triaxial compression tests. Note that the symbols are the experimental results and the lines represent linear regressions for the different rocks.

5.3 Chemical components and mechanical properties of mudstones

5.3.1 Mineral and chemical components

Since many problems in rock engineering are caused by shales or mudstones, the experiments of mineral composition, chemical components and mechanical properties of mudstone in mudstones are particularly important. Excavation stability in soft rocks such as mudstone and shale has been plaguing the coal mines in the Huainan coalfield, China. The mudstone samples taken in the roof and floor strata of coal seams in the Huainan coalfield are analyzed and tested. The analyzed results using X-ray diffraction show that the main mineral components of mudstone in this studied area are clay minerals. The secondary minerals are quartz and siderite. There are also some potassium feldspars in sandy mudstones. In the mudstone, clay minerals occupy 50-90%, with an average of 69.3%. Among these clay minerals, kaolinite is the essential component, while illite and chlorite are the secondary components. Quartz in the mudstone occupies 10-30%, with an average of 21.5%.

From the quantitative analyses using the fluorescence spectrum, the chemical constitution of the mudstone has the following characteristics (Peng and Meng 2002):

1. The main chemical constitution is SiO_2, and the secondary is Al_2O_3. In sandy mudstone the content of SiO_2 is up to 65.03%.
2. The contents of ferric oxide (especially Fe_2O_3) in granophyric mudstone and light gray mudstone are relatively high. The contents of Fe_2O_3 in dark and carbonaceous mudstone and dark gray mudstone are low. Hence the former two mudstones were formed in the oxidation environments, and the latter two were in the deoxidization environments.
3. The ignition loss in the immediate roof of coal seam # 13 is relatively high. The main ignited substances include carbon, sulfur, and other organic substances.

5.3.2 Chemical component effect on mechanical properties

There is a close relationship between the mechanical properties of mudstone and its chemical composition, particularly the content of SiO_2 and ignition loss. Figures 5.11 and 5.12 present the uniaxial compressive strength and elastic modulus and the content of SiO_2 relationships in a mudstone sample (Peng and Meng 2002).

It can be seen from Figs. 5.11 and 5.12 that the uniaxial compressive strength and elastic modulus of mudstone increase as the content of SiO_2 increases. When the content of SiO_2 is less than 40%, the compressive strength and elastic modulus of mudstone increase slightly. However, when the content of SiO_2 is greater than 40%, the mechanical strength and elastic modulus increase significantly as the content of SiO_2 increases. This is because SiO_2 has a very high uniaxial compressive strength and elastic modulus. The mechanical properties and the contents of SiO_2 have the following exponential relations:

$$UCS = 6.85e^{0.025S} \tag{5.4}$$

$$E = 0.873e^{0.048S} \tag{5.5}$$

where UCS is the uniaxial compressive strength (MPa); S is the content of SiO_2 (%); E is the elastic modulus (GPa).

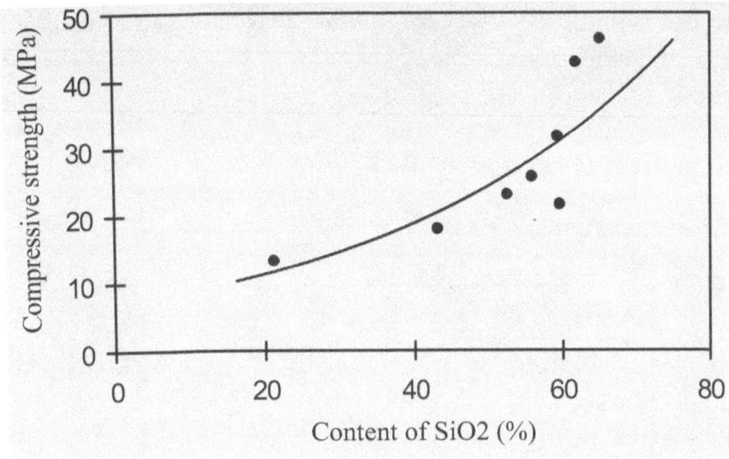

Fig. 5.11. Uniaxial compressive strength versus the content of SiO_2 in the mudstone.

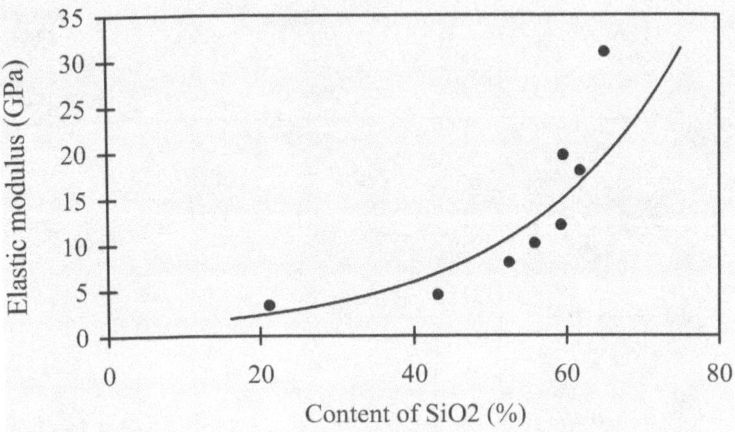

Fig. 5.12. Elastic modulus versus the content of SiO_2 in the mudstone.

Figures 5.13 and 5.14 plot the uniaxial compressive strength and elastic modulus versus the ignition loss obtained by calculating the loss of mass after heating the mudstone to 1000 °C. It can be found that the compressive strength and elastic modulus decrease as the ignition loss increases. When the ignition loss is greater than 10%, the compressive strength or elastic modulus of mudstone decreases sharply and then tends to have a very low value. This indicates that the mudstone with a high content of carbon, such as carbonaceous mudstone, has much lower strength and elastic modulus.

The mechanical properties and the ignition loss have the following negative power relations:

$$UCS = 180.18I^{-0.73} \tag{5.6}$$

$$E = 491.65I^{-1.45} \tag{5.7}$$

where I is the ignition loss (%).

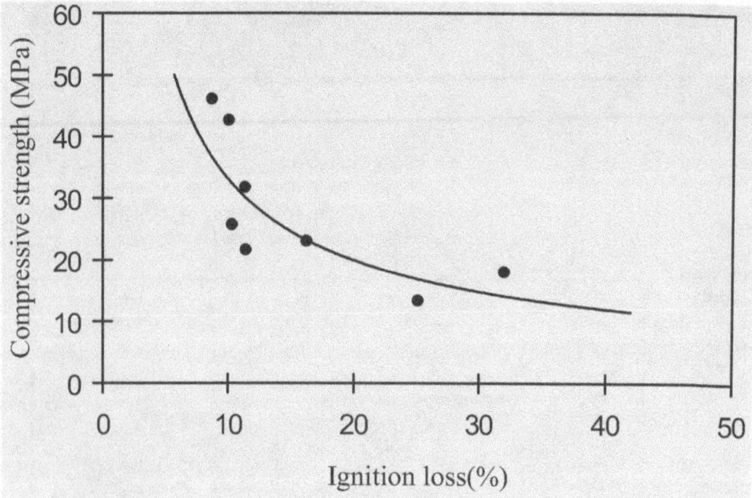

Fig. 5.13. Uniaxial compressive strength versus the ignition loss in the mudstone.

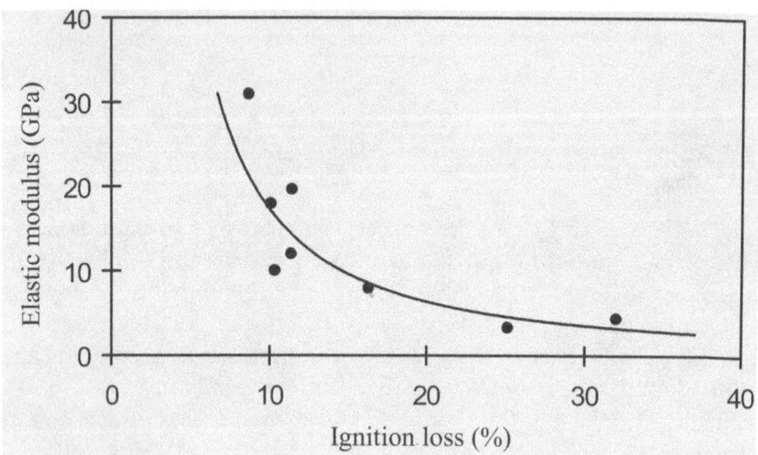

Fig. 5.14. Elastic modulus of mudstone versus the ignition loss in the mudstone.

5.4 Mechanical behaviors of discontinuities

Bedding is one of the most common geologic phenomena in sedimentary rocks. Generally a bedding plane is a weak plane, compared to rock blocks, it has prominently lower strength and higher compressibility (Peng 1998). Therefore, the bedding plane has much lower compression resistance, shear resistance, and tension resistance. In rock engineering, the rock is likely to fail in a bedding plane.

There are two basic parameters to describe rock bedding/discontinuity mechanical behaviors, normal and shear stiffnesses.

The normal stiffness of a rock is defined as the slope in the normal stress and deformation curve when a discontinuity is loaded under a normal stress perpendicular to the direction of the discontinuity, i.e.:

$$k_n = \frac{\partial \sigma_n}{\partial u} \tag{5.8}$$

where k_n is the normal stiffness of the discontinuity; σ_n is the normal stress applied perpendicularly to the discontinuity; u is the normal displacement of the discontinuity.

When a normal compressive stress exerts on a discontinuity, the normal stress and discontinuity displacement have a hyperbolic relation (Goodman 1976, Hudson and Harrison 1997):

$$u = \frac{\sigma_n}{c + d\sigma_n} \tag{5.9}$$

where u is the normal displacement of the discontinuity; σ_n is the normal compressive stress; and c and d are the experimental constants.

Brandis et al. (1983, 1985) proposed an empirical model for an interlocked joint:

$$\Delta\sigma = \frac{\Delta V_j}{c - b\Delta V_j} \tag{5.10}$$

where c, b are constants, and $c = 1/k_{ni}$, $c/d = V_m$; k_{ni} is the initial stiffness; V_m is the maximum closure of the joint; ΔV_j is the joint closure; $\Delta\sigma$ is the normal compressive stress; and

$$V_m = A + B(JRC) + C\left(\frac{JCS}{a_{t0}}\right)^D \tag{5.11}$$

$$k_{ni} = 0.02(\frac{JCS}{a_{t0}}) + 1.75JRC - 7.15 \tag{5.12}$$

where *A*, *B*, *C*, and *D* are constants determined from cyclic loading tests; *JRC* is the joint roughness coefficient; *JCS* is the joint wall compression strength, and a_{t0} is the initial mechanical aperture or initial true aperture.

For a dislocated joint, the normal stress and displacement has the following relationship:

$$\log\sigma_n = k_{ni} + M\Delta V_j \tag{5.13}$$

where *M* is a constant.

Zhang et al. (1999) proposed the following exponential relationship for normal stress and fracture displacement based on the concrete block tests:

$$u = b_0(1 - e^{-a\sigma_n}) \tag{5.14}$$

where b_0 is the initial fracture aperture; *a* is the experimental constant.

The shear stress and displacement characteristic of a rough discontinuity depend on the normal stress and the discontinuity surface characteristics, because the shear deformation involves dilation and fracture asperities. A hyperbolic function is frequently used to describe the shear-displacement relationship for a discontinuity in the pre-peak stress regime:

$$\tau = \frac{v}{a + bv} \tag{5.15}$$

where v is the shear displacement of the discontinuity; τ is the shear stress; and *c* and *d* are the constants.

The shear stiffness is the slope in the shear stress and shear displacement curve and can be expressed as:

$$k_s = \frac{\partial\tau}{\partial v} \tag{5.16}$$

Table 5.5 presents the mechanical properties of some interfaces of sedimentary rocks in China (Yu 1983). It shows that the shear stiffness in the discontinuity is much smaller than the normal stiffness. In other words, the discontinuity is much weaker to resist shear stress.

Table 5.5. Mechanical properties of the interfaces in some rocks and coal seam

Rocks above and beneath the interface	Shear stiffness k_s (MPa/m)	Normal stiffness k_n (MPa/m)	Cohesion c (MPa)	Internal friction angle ϕ (°)
Limestone and coal seam	24.5	981	0.2	30
Coal seam and shale	14.7	588	0.1	30
Shale and sandstone	14.7	588	0.1	30

5.5 Mechanical behaviors of rock masses

Research shows that the deformation behaviors of a rock matrix and a rock mass are different, For a simple case, if a set of structure planes (such as fractures or bedding planes) are parallel and equally spaced in the rock mass, the Young's moduli of rock mass and rock matrix can be expressed as:

$$\frac{1}{E_m} = \frac{1}{E_r} + \frac{1}{k_n s} \tag{5.17}$$

where E_m is the Young's modulus of the rock mass; E_r is the Young's modulus of the rock matrix; k_n is the normal stiffness of the structure plane; and s is the spacing of structure plane.

The existence of structure planes in a rock mass makes its compressive strength far less than that of the rock matrix. The statistical results from 16 worldwide rock engineering projects show that for the intact rock mass (RQD $\geq$ 90%) the ratio of the uniaxial compressive strength of the rock mass to rock matrix ranges from 0.16 to 0.82, and the average ratio is 0.47. For blocky and fissured rocks, the ratio of uniaxial compressive strength of the rock mass to rock matrix is lower, ranging from 0.16 to 0.48 with an average of 0.32.

The existence of bedding or other joints also makes the rock mass anisotropic in both mechanical properties and behaviors, for example rock mass strength and deformation are different in the parallel and perpendicular directions of bedding. Lab results show that the bedded rock specimens have different mechanical properties when they are loaded in different directions. Figures 5.15 and 5.16 show that the uniaxial compressive strength (UCS) and cohesion are higher when the loading direction is perpendicular to the bedding plane.

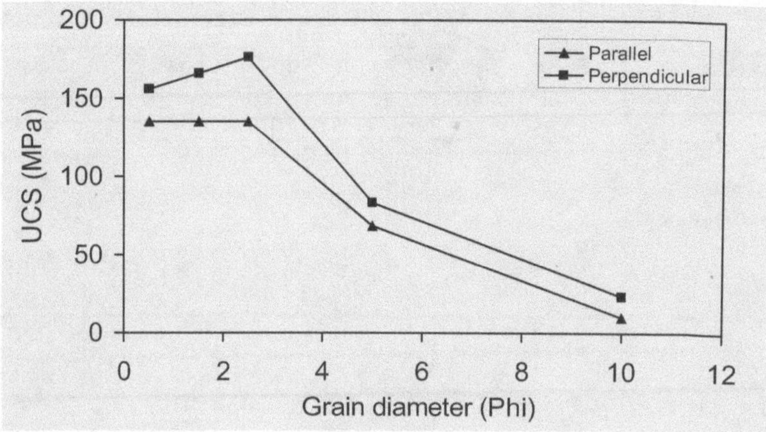

Fig. 5.15. Uniaxial compressive strength versus grain size (phi) for loading parallel and perpendicular to the bedding plane.

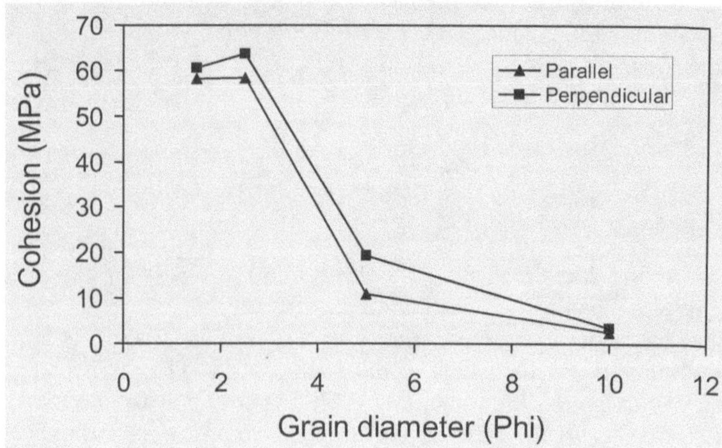

Fig. 5.16. Cohesion versus grain size (phi) for loading parallel and perpendicular to the bedding plane.

The Young's modulus, however, is lower when the loading direction is perpendicular to the bedding plane (Fig. 5.17). This is because the rock mass has a larger displacement in the bedding plane. Table 5.6 presents some experimental results of rock uniaxial compressive strength, cohesion, and Young's modulus for loading parallel and perpendicular to the bedding plane in China's eastern coal measures. It shows that the rock compressive strength is much lower when loading parallel to bedding direction, and that the ratio varies from 0.42 to 0.87.

Table 5.7 lists experimental results of bedding effects on rock properties in China's coal measures. It shows that bedding plane and loading direction have impacts on rock tensile strength, sonic compressional velocity, and dynamic Young's modulus.

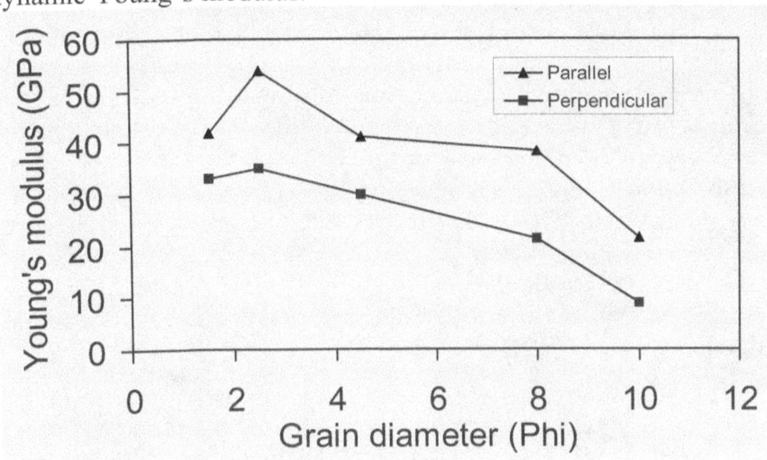

Fig. 5.17. Young's modulus versus grain size (phi) for loading parallel and perpendicular to the bedding plane.

Table 5.6. Ratios of rock strength and Young's modulus for loading parallel and perpendicular to the bedding plane

Properties	Ratio of parallel to and perpendicular to bedding plane
UCS	0.42 – 0.87
Cohesion	0.57 – 0.96
Young's modulus	1.27 – 2.45

Table 5.7. Rock properties and loading directions of parallel and perpendicular to the bedding plane

Rocks	Tensile strength (MPa)		Compressional velocity (m/s)		Dynamic Young's modulus (GPa)	
	Parallel	Perpendicular	Parallel	Perpendicular	Parallel	Perpendicular
Medium-grained sandstone	7.55	5.1	4834	4197		
Fine-grained sandstone	10.67	6.66	5116	4487		
Mudstone		0.86-2.1	2167	1696	10.90	6.68
Coal seam			2151	1790	6.35	4.21

5.6 Water impact on sedimentary rock strength

Natural rocks have experienced many tectonic activities and have been extensively stressed and deformed. Consequently, they are naturally fractured porous media. Most of oil reservoirs and some aquifers fall into this category. These rocks usually contain water or other fluid in the rock pores and fractures. When rock engineering is constructed in such media, the water or fluid effect needs to be considered. Figure 5.18 illustrates the importance of water pressure on rock strength and rock failure. The figure shows that if the Mohr-Coulomb strength envelope is assumed to stay unchanged, then without water effects, the rock is stable. However, the Mohr circle moves toward the opposite direction of the x-axis due to water pressure. As a result, the circle exceeds the strength envelope, causing rock failure. Therefore, in rock engineering design and practice the fluid effect is not negligible.

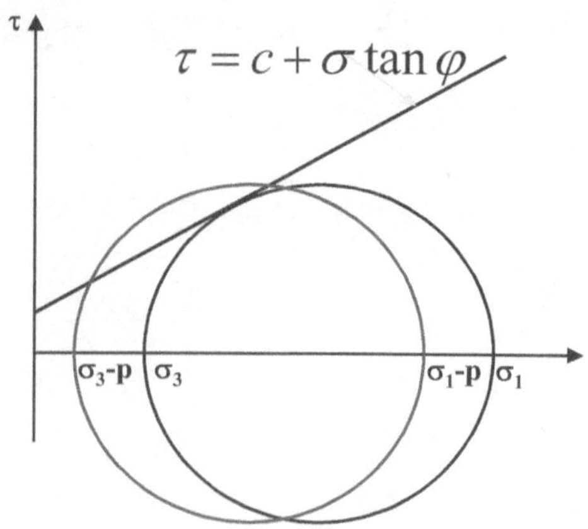

Fig. 5.18. Mohr circles with and without water pressure effects, assuming Biot's constant of 1.

Zhang et al. (1997) proposed that rock strength reduces due to water pressure, and rock strengths in dry rock and in rock with a pore pressure of p have the following relationship (Zhang et al. 1997):

$$UCS_w = UCS - \frac{2\alpha p \sin\varphi}{1 - \sin\varphi} \tag{5.18}$$

where UCS_w is the uniaxial compressive strength in the rock with pore pressure of p; α is the Biot's constant; φ is the internal friction angle.

The cohesion in the rock with pore pressure effect may be expressed as (Zhang et al. 1997):

$$C_w = C - \alpha p \tan \varphi \tag{5.19}$$

where C and C_w are the cohesions without and with the effect of a pore pressure of p.

Rock strength also depends on the water content in the rock. Table 5.8 presents the experimental results of water content's influences on rock strength, Young's modulus, and Poisson's ratio in China's coal measures. The results show that rock compressive strength decreases significantly as the water content increases.

Table 5.8. Experimental results of water influence on rock mechanical properties

Lithology	Water content W (%)	Compressive strength UCS (MPa)	Young's modulus E (GPa)	Poisson's ratio ν
Fine-grained quartz sandstone	0	142.5	73.33	0.330
	0.23	82	48.30	0.340
	0.35	44.56	23.28	0.286
Limestone	0	110.52	41.04	0.032
	0.06	89.92	41.69	0.161
	0.11	58.08	42.35	0.189
Muddy siltstone	0	119.4		
	0.56	88.5		
	0.78	51.6		
	0.84	68.3		

Lab tests show that rock compressive strength decreases as the water content increases and can be expresses as follows (Zhang et al. 1997):

$$UCS_w = UCS - bW \tag{5.20}$$

where W is the water content (%); b is experimental constant.

For the rocks in China's coal measures, the following experimental correlations were given to describe the water effect on rock uniaxial strength (Peng and Meng 2002):

For fine-grained quartz sandstone and water content of less than 0.35%:

$$UCS_w = 143.4 - 277.6W$$

For muddy siltstone and water content of less than 0.84%:

$$UCS_w = 121 - 71.6W$$

For mudstone and water content of less than 5.13%:

$$UCS_w = 30.4 - 4.93W$$

For limestone and water content of less than 0.11%:

$$UCS_w = 112.9 - 472.33W$$

References

Bandis SC, Barton NR, Christianson M (1985) Application of a new numerical model of joint behavior to rock mechanics problems. Proc Int Symp on Fundamentals of Rock Joints, Bjorkliden

Bandis SC, Lumsden AC, Barton NR (1983) Fundamentals of rock joint deformation. Int J Rock Mech Min Sci Geomech Abstr 20:249-268

Goodman RE (1976) Method of geological engineering in discontinuous rocks. West Publishing

Hudson JA, Harrison JP (1997) Engineering rock mechanics, an introduction to the principles. Pergamon

Meng Z, Zhang J, Peng S (2006) Influence of sedimentary environments on mechanical properties of clastic rocks. Environ Geol 51:113-120

Meng Z, Peng S, Qu H (2000) Relationships between rock mechnaical properties and rock contents and structure in coal roof and floor formations. Chinese J Rock Mech Eng 19(2):136-139 (in Chinese)

Peng S (1998) Approach on mechanical properties of clastic rocks and rock microstructure. Proc. 5th Ann Conf of Young Scientists, China Coal Society. Coal Industry Press (in Chinese)

Peng S, Meng Z (1999) Classification and its application of roof stability using lithology and geological structure. Symp 46th Young scientists, China Sci Tech Soc, China Sci Tech Press Beijing pp35-42 (in Chinese)

Peng S, Qu H, Luo L, Wang L, Duan Y (2000) Complete stress-strain and permeability experiments. J China Coal Society 25(2):113-116 (in Chinese)

Peng S, Meng Z (2002) Theory and practice of mining engineering geology. Geological Press (in Chinese)

Yu X (1983) Analysis of underground engineering stability. Coal Industry Press (in Chinese)

Zhang J, Bai M, Roegiers J-C, Liu T (1999) Determining stress-dependent permeability in the laboratory. In: Amadei B, Kranz RL, Scott GA, Smeallie PH (ed) Proc 37th US Rock Mech Symp Colorado. Rotterdam, Balkema pp 341-347

Zhang J, Zhang Y, Liu T (1997) Rock mass permeability and coal mine water inrush. Geological Publ House Beijing (in Chinese)

6 Double porosity poroelasticity and its finite element solution

6.1 Introduction

Field observations have revealed a need for a better and more comprehensive method for geomechanics modeling. For instance, in oil and gas industry, the exploration and production of hydrocarbons now occur in more difficult geological settings; such as in naturally fractured media, in overpressured shale formations, in rubble zones of sub-salt formations, and at great depths. In fractured porous formations, borehole instability has been of major concern due to potential rock movements along fractures at the borehole wall. In the case of shale formations, not only does the state of stress and pressure play a role, but also the properties and interactions between the shale and the drilling fluid.

Most petroleum reservoirs are situated in fractured porous formations. In fact, in order to drain such reservoirs, inclined and horizontal wells must be drilled in such formations. Therefore, traditional borehole modeling methods cannot completely satisfy this requirement. The double porosity approach can be used to model boreholes drilled in the fractured porous medium.

6.2 Double porosity poroelastic model

In the double porosity approach, a naturally fractured reservoir is classified as a system containing two different physical domains. The primary rock matrix contains a large volume of fluid but has a rather low permeability. While the fractures constitute a small volume but have the ability to transmit a large portion of the total flow through the reservoir. In early approaches (Barenblatt et al. 1960, Warren and Root 1963, Kazemi 1969), the reservoir was considered as two overlapping continua: matrix and fractures. Flow between the matrix and the fractures was accounted for by introducing source functions.

A typical dual-porosity formulation is usually written as (Barenblatt et al. 1960):

$$\begin{cases} \dfrac{k_{ma}}{\mu}\nabla^2 p_{ma} = c_{ma} n_{ma} \dfrac{\partial p_{ma}}{\partial t} + \Gamma(p_{fr} - p_{ma}) \\ \dfrac{k_{fr}}{\mu}\nabla^2 p_{fr} = c_{fr} n_{fr} \dfrac{\partial p_{fr}}{\partial t} - \Gamma(p_{fr} - p_{ma}) \end{cases} \tag{6.1}$$

where the subscripts '*ma*' and '*fr*' represent matrix and fractures, respectively; p_{ma} and p_{fr} are the fluid pressures for the matrix and fractures, respectively; k_{ma} and k_{fr} are the permeabilities for the matrix and fractures, respectively; μ is the fluid dynamic viscosity, n_{ma} and n_{fr} are the porosities for the matrix and fracture, respectively; c_{ma}, c_{fr} are the compressibilities for the matrix and fractures, respectively; t is the time, Γ is the geometric leakage factor. It can be seen that no solid deformation is considered in this formulation.

Alternatively, Barenblatt et al. (1990) proposed a more complete dual-porosity formulation than that expressed by Eq. 6.1 by considering the cross-phase storativity interaction:

$$\begin{cases} \dfrac{k_{ma}}{\mu}\nabla^2 p_{ma} = c_{ma} n_{ma} \dfrac{\partial p_{ma}}{\partial t} - c_{mf} n_{ma} \dfrac{\partial p_{fr}}{\partial t} + \Gamma(p_{fr} - p_{ma}) \\ \dfrac{k_{fr}}{\mu}\nabla^2 p_{fr} = c_{fr} n_{fr} \dfrac{\partial p_{fr}}{\partial t} - c_{fm} n_{fr} \dfrac{\partial p_{ma}}{\partial t} - \Gamma(p_{fr} - p_{ma}) \end{cases} \tag{6.2}$$

where c_{mf} and c_{fm} are the cross-coefficients.

However, effects of stresses and deformations on both the matrix and fractures were not considered in Eqs. 6.1 and 6.2. Then, a coupled flow-deformation approach within double porosity poroelastic media was presented (Aifantis 1997, Wilson and Aifantis 1982), and a constitutive model to define response of a fissured medium was given (Elsworth and Bai 1992). Later, a series of papers were published to study the fluid flow and solute transport in multiple porosity media (Bai et al. 1993, Bai et al. 1999a, and Zhang and Roegiers 2005, etc.).

The naturally fractured reservoir can be assumed to be an ideal double porosity and double permeability model as shown in Fig. 6.1 (Zhang and Roegiers 2005). In this model the system, including matrix blocks and fractures, is considered as a continuous medium.

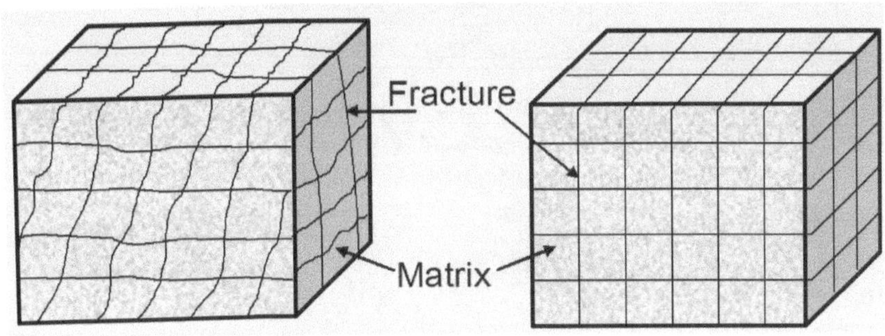

Fig. 6.1. Naturally fractured and idealized double porosity reservoirs.

The relationship between total stress (σ_{ij}) and effective stress (σ_{ij}') was given by Terzaghi (1943) and Biot (1941). For double porosity media, the effective stresses can be expressed as:

$$\sigma_{maij} = \sigma'_{maij} + \alpha_{ma} p_{ma} \delta_{ij} \qquad (6.3)$$

$$\sigma_{frij} = \sigma'_{frij} + \alpha_{fr} p_{fr} \delta_{ij}$$

where p is the pressure; α is the effective stress coefficient; and, δ_{ij} is the Kronecker delta.

For separate but overlapping porous media, the linear elastic constitutive relationships between the effective stresses and strains are defined as (Elsworth and Bai 1992, Bai et al. 1999b):

$$\varepsilon_{makl} = C_{maijkl} \sigma'_{maij} \qquad (6.4)$$

$$\varepsilon_{frkl} = C_{frijkl} \sigma'_{frij}$$

where C_{maijkl} and C_{frijkl} are the compliance tensors for the rock matrix (subscript *ma*) and fracture systems (subscript *fr*), respectively. The detailed expressions for the tensors are listed in Appendix 6.1.

The total strain due to the elastic deformation in each of the systems is given by:

$$\varepsilon_{kl} = \varepsilon_{makl} + \varepsilon_{frkl} \qquad (6.5)$$

Substituting the double effective law, i.e., Eq. 6.3 into Eq. 6.4 while combining with Eq. 6.5 and noting $\sigma_{maij} = \sigma_{frij} = \sigma_{ij}$, the following equation can be obtained:

$$\sigma_{ij} = D_{mfijkl}(\varepsilon_{kl} + C_{maijkl}\alpha_{ma} p_{ma}\delta_{ij} + C_{frijkl}\alpha_{fr} p_{fr}\delta_{ij}) \tag{6.6}$$

The combined elasticity matrix D_{mfijkl} is defined explicitly in a three-dimensional geometry for an isotropic medium as:

$$D_{mfijkl} = (C_{maijkl} + C_{frijkl})^{-1} \tag{6.7}$$

where the detailed expression for D_{mfijkl} is given in Appendix 6.1.

In general, the stress-strain behavior of rocks is non-linear, especially in the case where external loads exceed the elastic strength of material and the material becomes plastic. The elastoplastic stress-strain can be described by:

$$\Delta\boldsymbol{\sigma} = D^{ep}_{mfijkl}\,\Delta\boldsymbol{\varepsilon} \tag{6.8}$$

where $D^{ep}_{mfijkl} = D_{mfijkl} - D^{p}_{mfijkl}$, D^{p}_{mfijkl} is the plastic modulus tensor.

The total strain increment is assumed to be the sum of the elastic and plastic components; i.e.

$$\Delta\boldsymbol{\varepsilon} = \Delta\boldsymbol{\varepsilon}^{e} + \Delta\boldsymbol{\varepsilon}^{p} \tag{6.9}$$

where $\Delta\boldsymbol{\varepsilon}$, $\Delta\boldsymbol{\varepsilon}^{e}$ and $\Delta\boldsymbol{\varepsilon}^{p}$ are the total, elastic and plastic strain increments, respectively. For more details about perfect plasticity solution, refer to Zhang (2002).

For the separate but overlapping model, the double effective stress law (Eq. 6.3) needs to be considered, and the combined elasticity matrix D_{mfijkl} as well as other related elastic constants need to be introduced. Then, the governing equations for solid deformation and fluid phase in the dual-porosity poromechanical formation can be written as (Bai et al. 1999a, Zhang 2002):

$$\begin{aligned}
&G_{mf}u_{i,jj} + (\lambda_{mf} + G_{mf})u_{k,ki} + \alpha_{ma}D_{mfijkl}C_{maijkl}p_{ma,i} \\
&\quad + \alpha_{fr}D_{mfijkl}C_{frijkl}p_{fr,i} = 0 \\
&\frac{k_{ma}}{\mu}p_{ma,kk} = \alpha_{ma}D_{mfijkl}C_{maijkl}\frac{\partial\varepsilon_{kk}}{\partial t} - \beta_{ma}\frac{\partial p_{ma}}{\partial t} + \omega(p_{fr} - p_{ma}) + q_{ma} \\
&\frac{k_{fr}}{\mu}p_{fr,kk} = \alpha_{fr}D_{mfijkl}C_{frijkl}\frac{\partial\varepsilon_{kk}}{\partial t} - \beta_{fr}\frac{\partial p_{fr}}{\partial t} - \omega(p_{fr} - p_{ma}) + q_{fr}
\end{aligned} \tag{6.10}$$

where β is the relative compressibility representing the lumped deformability of the fluid and the solid; u is the solid displacement; ε_{kk} is the total body strain; ω is the transfer coefficient (Warren and Root 1963); s is the fracture spacing; q_{ma} and q_{fr} are the applied fluid flux; λ_{mf} is Lamé's constant for the combined double porosity medium; and, $\lambda_{mf} = \nu D_{mfijkl} / (1+\nu)(1-2\nu)$.

6.3 Finite element discretization of the poroelastic solution

The first step in solving the coupled problem of fluid flow and solid deformations is to discretize the problem domain by replacing it with a collection of nodes and elements, referred to as the finite element mesh. The values of the material properties are usually assumed to be constant within each element but are allowed to vary from one element to the next; making it possible to simulate non-homogeneous problems. The second step in the finite element method is to derive an integral formulation for the governing equations. This leads to a system of algebraic equations that can be solved for values of the field variable at each node in each mesh. The method of weighted residuals can be used for the fluid flow and solid deformation modeling. The Galerkin method is then used whereby the weighting function for a node is identical to the shape function used to define the approximate solution.

6.3.1 Shape function

Interpolation or shape functions are used to map the element displacements and fluid pressures at the nodal points (Peng and Meng 1999).

For the fluid pressure approximation at phase *i*, one has:

$$\mathbf{p}_i^* = \mathbf{N}\mathbf{p}_i \tag{6.11}$$

where **N** is the shape function for the fluid pressure and solid displacement.

At the nodal level, for four-point two-dimensional elements:

$$p_i^* = \sum_{j=1}^{4} N_j p_{ij} \tag{6.12}$$

For the eight-point three-dimensional elements:

$$p_i^* = \sum_{j=1}^{8} N_j p_{ij} \tag{6.13}$$

For eight-point three-dimensional elements, the shape function vector for pressure can be given in the following forms:

$$\begin{aligned}
N_1 &= \frac{1}{8}(1-\xi)(1-\eta)(1-\zeta) \\
N_2 &= \frac{1}{8}(1+\xi)(1-\eta)(1-\zeta) \\
N_3 &= \frac{1}{8}(1+\xi)(1-\eta)(1+\zeta) \\
N_4 &= \frac{1}{8}(1-\xi)(1-\eta)(1+\zeta) \\
N_5 &= \frac{1}{8}(1-\xi)(1+\eta)(1-\zeta) \\
N_6 &= \frac{1}{8}(1+\xi)(1+\eta)(1-\zeta) \\
N_7 &= \frac{1}{8}(1+\xi)(1+\eta)(1+\zeta) \\
N_8 &= \frac{1}{8}(1-\xi)(1+\eta)(1+\zeta)
\end{aligned} \tag{6.14}$$

where ξ, η, and ζ represent local coordinates; and $-1 \le \xi \le 1$, $-1 \le \eta \le 1$ and $-1 \le \zeta \le 1$.

A similar expression for the approximation in mapping nodal displacements can be described as:

$$\mathbf{u}^* = \mathbf{N}\mathbf{u} \tag{6.15}$$

At the nodal level, for four-point two-dimensional elements:

$$u_x^* = \sum_{j=1}^{4} N_j u_{xj}, \qquad u_y^* = \sum_{j=1}^{4} N_j u_{yj} \tag{6.16}$$

For eight-point three-dimensional elements:

$$u_x^* = \sum_{j=1}^{8} N_j u_{xj}, \qquad u_y^* = \sum_{j=1}^{8} N_j u_{yj}, \qquad u_z^* = \sum_{j=1}^{8} N_j u_{zj} \tag{6.17}$$

where **u** is the nodal displacement vector, For simplicity, the superscript '*' indicates the finite element approximations that are omitted in the subsequent description.

Strains within a single element are related to nodal displacements through the derivatives of the shape functions as:

$$\boldsymbol{\varepsilon} = \mathbf{B}\mathbf{u} \tag{6.18}$$

where **B** is the strain-displacement matrix.

The generalized plane strain solutions maintain compatibility with the primary unknown terms equivalent to the three-dimensional formulation, but geometrically they are not related to the z-coordinate, similar to the two-dimensional case. With reference to the finite element formulation, the major differences among the generalized plane strain, the plane strain and the three-dimensional situation are exhibited in the strain-displacement matrix **B**. In a three-dimensional geometry the matrix **B** can be expressed as:

$$\mathbf{B} = \begin{bmatrix} \frac{\partial}{\partial x} & 0 & 0 \\ 0 & \frac{\partial}{\partial y} & 0 \\ 0 & 0 & \frac{\partial}{\partial z} \\ \frac{\partial}{\partial y} & \frac{\partial}{\partial x} & 0 \\ 0 & \frac{\partial}{\partial z} & \frac{\partial}{\partial y} \\ \frac{\partial}{\partial z} & 0 & \frac{\partial}{\partial x} \end{bmatrix} \mathbf{N} \tag{6.19}$$

It is well-known in plane strain problems, i.e., in an *x*-*y* plane, that the displacement and the shear stresses are restricted along the *z*-direction. In generalized plane strain scenarios, however, these restrictions are removed. As a result, the number of tensor components for stresses and strains are identical to that of a three-dimensional setting. In a generalized plane strain formulation, it is assumed that boundary conditions in the form of surface tractions, pore pressure, displacements, and normal flux, do not change along the z-direction. As a result, the displacements, stresses, strains, pore pressure, and fracture pressure are only functions of x, y and time t (Bai et al. 1999a). For the generalized plane strain formulation, **B** can be written as:

$$\mathbf{B} = \begin{bmatrix} \frac{\partial}{\partial x} & 0 & 0 \\ 0 & \frac{\partial}{\partial y} & 0 \\ 0 & 0 & 0 \\ \frac{\partial}{\partial y} & \frac{\partial}{\partial x} & 0 \\ 0 & 0 & \frac{\partial}{\partial y} \\ 0 & 0 & \frac{\partial}{\partial x} \end{bmatrix} \mathbf{N} \tag{6.20}$$

6.3.2 Conservation equations

The general force equilibrium equation, in terms of nodal variables for a double porosity medium in generalized plane strain domain, is given by (Zhang 2002, Zhang and Roegiers 2005):

$$\int_{\Omega} \mathbf{B}^{\mathrm{T}} \mathbf{D}_{\mathbf{mf}} \boldsymbol{\varepsilon} d\Omega + \alpha_{ma} \int_{\Omega} \mathbf{B}^{\mathrm{T}} \mathbf{D}_{\mathbf{mf}} \mathbf{C}_{\mathbf{ma}} \mathbf{m} \mathbf{N} \mathbf{p}_{\mathbf{ma}} d\Omega + \alpha_{fr} \int_{\Omega} \mathbf{B}^{\mathrm{T}} \mathbf{D}_{\mathbf{mf}} \mathbf{C}_{\mathbf{fr}} \mathbf{m} \mathbf{N} \mathbf{p}_{\mathbf{fr}} d\Omega = \int_{S} \mathbf{N} \mathbf{f} dS \tag{6.21}$$

where Ω represents the integral domain; **C** is the compliance matrix defined in Appendix 6.1 with $\mathbf{C}_{\mathbf{ma}} = C_{maijkl}$, $\mathbf{C}_{\mathbf{fr}} = C_{frijkl}$; $\mathbf{D}_{\mathbf{mf}}$ is the matrix expression of D_{mfijkl}; **f** is the vector of applied boundary tractions; S is the domain surface on which the surface traction **f** is applied.

Substituting Eq. 6.18 and dividing through by Δt, the momentum balance equation (Eq. 6.21) in the finite element form can be expressed as:

$$\int_{\Omega} \mathbf{B}^{\mathrm{T}}\mathbf{D}_{\mathbf{mf}}\mathbf{B}\frac{\partial \mathbf{u}}{\partial \mathbf{t}}d\Omega + \alpha_{ma}\int_{\Omega} \mathbf{B}^{\mathrm{T}}\mathbf{D}_{\mathbf{mf}}\mathbf{C}_{\mathbf{ma}}\mathbf{m}\mathbf{N}\frac{\partial \mathbf{p}_{\mathbf{ma}}}{\partial \mathbf{t}}d\Omega + \alpha_{fr}\int_{\Omega} \mathbf{B}^{\mathrm{T}}\mathbf{D}_{\mathbf{mf}}\mathbf{C}_{\mathbf{fr}}\mathbf{m}\mathbf{N}\frac{\partial \mathbf{p}_{\mathbf{fr}}}{\partial \mathbf{t}}d\Omega = \int_{S}\mathbf{N}\frac{\partial \mathbf{f}}{\partial \mathbf{t}}dS \tag{6.22}$$

or,

$$\mathbf{K}\frac{\partial \mathbf{u}}{\partial \mathbf{t}} + \mathbf{R}_1\frac{\partial \mathbf{p}_{\mathbf{ma}}}{\partial \mathbf{t}} + \mathbf{R}_2\frac{\partial \mathbf{p}_{\mathbf{fr}}}{\partial \mathbf{t}} = \frac{\partial \mathbf{F}}{\partial \mathbf{t}} \tag{6.23}$$

where detailed expressions of the coefficients are given in Appendix 6.2.

Using the Gaussian quadrature method (Zienkiewicz 1977), the double porosity mass balance equations (last two equations in Eq. 6.10) in the finite element forms can be given for each system.

For the rock matrix system:

$$\frac{1}{\mu}\int_{\Omega}\mathbf{N}^{\mathrm{T}}\mathbf{k}_{\mathbf{ma}}\nabla\mathbf{N}d\Omega\mathbf{p}_{\mathbf{ma}} = \alpha_{ma}\int_{\Omega}\mathbf{N}^{\mathrm{T}}\mathbf{m}\mathbf{D}_{\mathbf{mf}}\mathbf{C}_{\mathbf{ma}}\mathbf{B}d\Omega\frac{\partial \mathbf{u}}{\partial \mathbf{t}} - \beta_{ma}\int_{\Omega}\mathbf{N}^{\mathrm{T}}\mathbf{N}d\Omega\frac{\partial \mathbf{p}_{\mathbf{ma}}}{\partial \mathbf{t}} + \omega\int_{\Omega}\mathbf{N}^{\mathrm{T}}\mathbf{N}d\Omega\Delta\mathbf{p} + \int_{\Gamma}\mathbf{N}^{\mathrm{T}}\mathbf{N}d\Gamma\mathbf{q}_{\mathbf{ma}} \tag{6.24}$$

For the fracture system:

$$\frac{1}{\mu}\int_{\Omega}\mathbf{N}^{\mathrm{T}}\mathbf{k}_{\mathbf{fr}}\nabla\mathbf{N}d\Omega\mathbf{p}_{\mathbf{fr}} = \alpha_{fr}\int_{\Omega}\mathbf{N}^{\mathrm{T}}\mathbf{m}\mathbf{D}_{\mathbf{mf}}\mathbf{C}_{\mathbf{fr}}\mathbf{B}d\Omega\frac{\partial \mathbf{u}}{\partial \mathbf{t}} - \beta_{fr}\int_{\Omega}\mathbf{N}^{\mathrm{T}}\mathbf{N}d\Omega\frac{\partial \mathbf{p}_{\mathbf{fr}}}{\partial \mathbf{t}} - \omega\int_{\Omega}\mathbf{N}^{\mathrm{T}}\mathbf{N}d\Omega\Delta\mathbf{p} + \int_{\Gamma}\mathbf{N}^{\mathrm{T}}\mathbf{N}d\Gamma\mathbf{q}_{\mathbf{fr}} \tag{6.25}$$

where $\mathbf{\Delta p} = \mathbf{p}_{\mathbf{fr}} - \mathbf{p}_{\mathbf{ma}}$, $\mathbf{m}^{\mathrm{T}} = (1\ 1\ 1\ 0\ 0\ 0)$; Γ is the domain surface on which the fluid flux $\mathbf{q}$ is applied; and Biot's effective stress coefficients, α, can be evaluated as:

$$\alpha_{ma} = 1 - \frac{K_{sk}}{K_s}$$
$$\alpha_{fr} = 1 - \frac{K_{sk}^{*}}{K_{fr}} \tag{6.26}$$

where K_{sk} and K^*_{sk} are the bulk moduli of the skeleton for the matrix blocks and the fractures, respectively; K_s and K_{fr} are the bulk moduli of the solid grains and fractures, respectively; and the relative compressibilities, β, can be written as:

$$\beta_{ma} = \frac{n_{ma}}{K_f} + \frac{\alpha_{ma} - n_{ma}}{K_s} \tag{6.27}$$

$$\beta_{fr} = \frac{n_{fr}}{sK_f} + \frac{\alpha_{fr} - n_{fr}}{sK_n}$$

where K_f and K_n are the bulk modulus of the fluid and the normal stiffness of the fractures, respectively; n is the porosity; and s is the fracture spacing.

Equations 6.24 and 6.25 can be written as the following finite element forms:

$$\mathbf{M}_1 \frac{\partial \mathbf{u}}{\partial \mathbf{t}} + (\mathbf{Q} - \mathbf{L}_1)\mathbf{p}_{\mathbf{ma}} - \mathbf{Q}\mathbf{p}_{\mathbf{fr}} - \mathbf{N}_1 \frac{\partial \mathbf{p}_{\mathbf{ma}}}{\partial \mathbf{t}} = -\mathbf{Q}_{\mathbf{ma}} \tag{6.28}$$

$$\mathbf{M}_2 \frac{\partial \mathbf{u}}{\partial \mathbf{t}} - \mathbf{Q}\mathbf{p}_{\mathbf{ma}} + (\mathbf{Q} - \mathbf{L}_2)\mathbf{p}_{\mathbf{fr}} - \mathbf{N}_2 \frac{\partial \mathbf{p}_{\mathbf{fr}}}{\partial \mathbf{t}} = -\mathbf{Q}_{\mathbf{fr}} \tag{6.29}$$

where detailed expressions of the above coefficients are listed in Appendix 6.2.

Equations 6.23, 6.28 and 6.29 represent a set of differential equations in time and can be expressed in matrix form as follows:

$$\begin{bmatrix} \mathbf{0} & \mathbf{0} & \mathbf{0} \\ \mathbf{0} & \mathbf{Q} - \mathbf{L}_1 & -\mathbf{Q} \\ \mathbf{0} & -\mathbf{Q} & \mathbf{Q} - \mathbf{L}_2 \end{bmatrix} \begin{bmatrix} \mathbf{u} \\ \mathbf{p}_{\mathbf{ma}} \\ \mathbf{p}_{\mathbf{fr}} \end{bmatrix} + \begin{bmatrix} \mathbf{K} & \mathbf{R}_1 & \mathbf{R}_2 \\ \mathbf{M}_1 & -\mathbf{N}_1 & \mathbf{0} \\ \mathbf{M}_2 & \mathbf{0} & -\mathbf{N}_2 \end{bmatrix} \frac{d}{dt} \begin{bmatrix} \mathbf{u} \\ \mathbf{p}_{\mathbf{ma}} \\ \mathbf{p}_{\mathbf{fr}} \end{bmatrix} = \begin{bmatrix} \frac{\mathbf{dF}}{\mathbf{dt}} \\ -\mathbf{Q}_{\mathbf{ma}} \\ -\mathbf{Q}_{\mathbf{fr}} \end{bmatrix} \tag{6.30}$$

The discretization in space has been completed, and Equation 6.30 now represents a set of differential equations in time.

6.3.3 Finite element discretization in time

Using a fully implicit finite difference scheme in the time discretization domain, such that:

$$\begin{cases} \dfrac{d\mathbf{u}^{t+\Delta t}}{dt} = \dfrac{1}{\Delta t}(\mathbf{u}^{t+\Delta t} - \mathbf{u}^{t}) \\ \dfrac{d\mathbf{p}_{ma}^{t+\Delta t}}{dt} = \dfrac{1}{\Delta t}(\mathbf{p}_{ma}^{t+\Delta t} - \mathbf{p}_{ma}^{t}) \\ \dfrac{d\mathbf{p}_{fr}^{t+\Delta t}}{dt} = \dfrac{1}{\Delta t}(\mathbf{p}_{fr}^{t+\Delta t} - \mathbf{p}_{fr}^{t}) \end{cases} \tag{6.31}$$

and substituting Eq. 6.31 into Eq. 6.30, the finite element equations in the matrix form for a double porosity poroelastic medium can be expressed as follows:

$$\frac{1}{\Delta t}\begin{bmatrix} \mathbf{K} & \mathbf{R}_1 & \mathbf{R}_2 \\ \mathbf{M}_1 & (\mathbf{Q}-\mathbf{L}_1)\Delta t - \mathbf{N}_1 & -\mathbf{Q}\Delta t \\ \mathbf{M}_2 & -\mathbf{Q}\Delta t & (\mathbf{Q}-\mathbf{L}_2)\Delta t - \mathbf{N}_2 \end{bmatrix}\begin{bmatrix} \mathbf{u} \\ \mathbf{p}_{ma} \\ \mathbf{p}_{fr} \end{bmatrix}^{t+\Delta t} \tag{6.32}$$

$$= \frac{1}{\Delta t}\begin{bmatrix} \mathbf{K} & \mathbf{R}_1 & \mathbf{R}_2 \\ \mathbf{M}_1 & -\mathbf{N}_1 & 0 \\ \mathbf{M}_2 & 0 & -\mathbf{N}_2 \end{bmatrix}\begin{bmatrix} \mathbf{u} \\ \mathbf{p}_{ma} \\ \mathbf{p}_{fr} \end{bmatrix}^{t} + \begin{bmatrix} \dfrac{\mathbf{F}}{\Delta t} \\ \mathbf{Q}_{ma} \\ \mathbf{Q}_{fr} \end{bmatrix}^{t+\Delta t} - \begin{bmatrix} \dfrac{\mathbf{F}}{\Delta t} \\ 0 \\ 0 \end{bmatrix}^{t}$$

There are 5 unknowns (u_x , u_y , u_z , p_{ma} , p_{fr}) and 5 equations per node; therefore, displacements and pressures can be solved. In addition, strains and stresses can be obtained through the following equation and Eq. 6.6.

$$\varepsilon_{ij} = \frac{1}{2}(u_{i,j} + u_{j,i}) \tag{6.33}$$

6.3.4 Stress conversion for an inclined borehole

For an inclined borehole with its axis inclined with respect to the principal axes of the far-field stresses (see Fig. 6.2), the following equations can be used to convert the global coordinate (far-field stress coordinate, x', y', z') into the local coordinate (borehole coordinate, x, y, z) system.

$$\begin{Bmatrix} S_x \\ S_y \\ S_z \\ S_{xy} \\ S_{yz} \\ S_{xz} \end{Bmatrix} = \begin{bmatrix} l_{xx'}^2 & l_{xy'}^2 & l_{xz'}^2 \\ l_{yx'}^2 & l_{yy'}^2 & l_{yz'}^2 \\ l_{zx'}^2 & l_{zy'}^2 & l_{zz'}^2 \\ l_{xx'}l_{yx'} & l_{xy'}l_{yy'} & l_{xz'}l_{yz'} \\ l_{yx'}l_{zx'} & l_{zy'}l_{yy'} & l_{zz'}l_{yz'} \\ l_{zx'}l_{xx'} & l_{zy'}l_{xy'} & l_{zz'}l_{xz'} \end{bmatrix} \begin{Bmatrix} S_{x'} \\ S_{y'} \\ S_{z'} \end{Bmatrix} \tag{6.34}$$

where,

$$\begin{Bmatrix} l_{xx'} & l_{xy'} & l_{xz'} \\ l_{yx'} & l_{yy'} & l_{yz'} \\ l_{zx'} & l_{zx'} & l_{zz'} \end{Bmatrix} = \begin{bmatrix} \cos\varphi_x \cos\varphi_z & \sin\varphi_x \cos\varphi_x & -\sin\varphi_z \\ -\sin\varphi_x & \cos\varphi_x & 0 \\ \cos\varphi_x \sin\varphi_z & \sin\varphi_x \sin\varphi_z & \cos\varphi_z \end{bmatrix} \tag{6.35}$$

$S_{x'}$, $S_{y'}$, and $S_{z'}$ are the far-field stresses; S_x , S_y, S_z , S_{xy}, S_{yz} and S_{zx} are the local borehole coordinate stresses; φ_x is the angle between the global and local coordinates (Fig. 6.2); and φ_z is the borehole deviation.

After the conversion, the finite element analysis can be worked out in the local coordinate system, i.e., the section perpendicular to the borehole axial direction (Fig. 6.3).

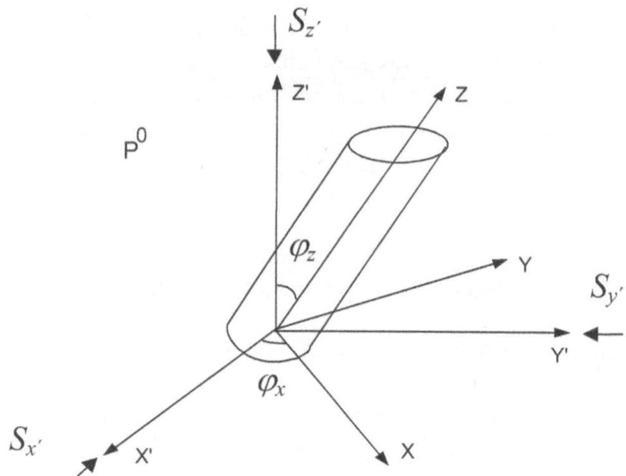

Fig. 6.2. Local and global coordinate systems for an inclined borehole.

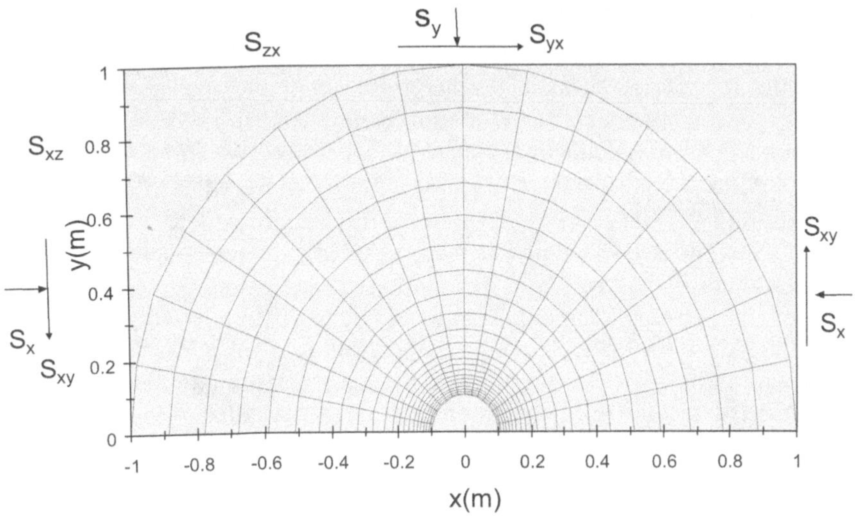

Fig. 6.3. Finite element mesh in the local coordinate system.

6.4 Model validation

The formulations presented in the foregoing section are coded in the pseudo-three-dimensional and time domain using four-node rectangular elements. Some analytical problems, such as elastic and poroelastic analytical solutions for inclined borehole problems, have been examined to validate the computer program.

The geometric loading for an inclined borehole problem is depicted in Fig. 6.2. The Cartesian coordinate system ($x'y'z'$) is chosen to coincide with the principal axes of the in-situ compressive stresses, respectively, designated as S_x, S_y and S_z. The initial formation pore pressure is denoted by p^0. The local coordinate system (Fig. 6.2) is formed by a rotation of the azimuth angle, φ_x, about the x'-axis, and then by an inclination of the zenith angle, φ_z, from the z'-axis toward the z-axis.

Then, at the local coordinate system the boundary conditions at the far-field ($r \to \infty$) are characterized by the normal stresses:

$$\sigma_x^0 = S_x, \qquad \sigma_y^0 = S_y, \qquad \sigma_z^0 = S_z$$

and the shear stresses:

$$\tau_{xy}^0 = S_{xy}, \qquad \tau_{yz}^0 = S_{yz}, \qquad \tau_{xz}^0 = S_{xz}$$

as well as the matrix and fracture pore pressures at the far-field:

$$p_{ma} = p^0_{ma}, \qquad p_{fr} = p^0_{fr}$$

where the superscript '0' indicates the virgin state.

In the following analysis, the in-situ stresses and initial pore pressures are obtained from Woodland (1990) and Cui et al. (1997a): $S_{x'} = 29$ MPa, $S_{y'} = 20$ MPa, $S_{z'} = 25$ MPa, and $p^0 = p_{ma}^0 = p_{fr}^0 = 10$ MPa. The wellbore inclination is $\varphi_x = 0°$ and $\varphi_z = 70°$. The wellbore radius is R = 0.1 m. The load at the wellbore is assumed as being applied instantaneously. In the local coordinate system (after 70° inclination), these values become (Zhang and Roegiers 2002, Zhang et al. 2003): $S_x = 25.5$ MPa, $S_y = 20$ MPa, $S_z = 28.5$ MPa, $S_{xz} = 1.3$ MPa, $S_{xy} = S_{yz} = 0$ MPa, and $p^0 = p_{ma}^0 = p_{fr}^0 = 10$ MPa, as shown in Fig. 6.4. The formation materials are assumed to be isotropic, characterized by the following properties: Biot modulus, M =15.8 GPa; Biot's effective stress coefficient, $\alpha_{ma} = 0.771$ and $\alpha_{fr} = 0.91$; permeability, $k_{ma} = 1 \times 10^{-7}$ Darcy; and fluid dynamic viscosity, μ = 0.001 Pa·s. The analytical solution for this particular generalized plane strain poroelastic problem was provided by Cui et al. (1997a). The corresponding equivalent parameters for the dual-porosity poroelastic model are listed in Table 6.1, in which the selection of an exceptionally large fracture spacing, *s*, denotes the approximation of a homogeneous single-porosity medium.

Table 6.1. Parameters for inclined borehole analysis

Parameter	Unit	Magnitude
Elastic modulus (E)	GPa	20.6
Poisson's ratio (ν)	-	0.189
Fracture stiffness (K_n, K_{sh})	MPa/m	4.821×10^5
Fluid bulk modulus (K_f)	MPa	419.17
Grain bulk modulus (K_s)	GPa	48.21
Matrix porosity (n_{ma})	-	0.02
Fracture porosity (n_{fr})	-	0.002
Matrix mobility (k_{ma}/μ)	m^2/MPa·s	10^{-10}
Fracture mobility (k_{fr}/μ)	m^2/MPa·s	10^{-9}
Fracture spacing (s)*	m	10^7

* The use of extremely large fracture spacing is only for the validation with the single porosity method.

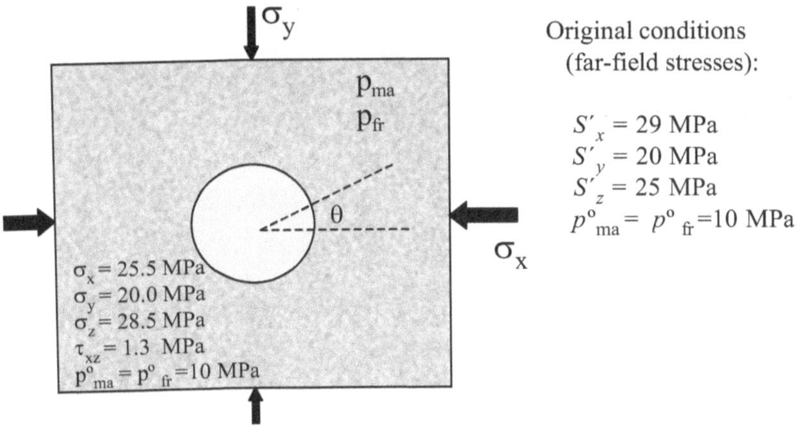

Fig. 6.4. State of stress in the borehole local coordinate (the FEM section) for 70° of borehole inclination.

Figure 6.5 represents the pore pressure variations into the rock formation. The comparative results between the analytical solution and the numerical dual-porosity solution (for large s) along the radial direction θ = 5.7° are shown at two different times; the numerical results appear to agree well with the analytical solution. It should be noted that the angle of θ is defined in Fig. 6.4.

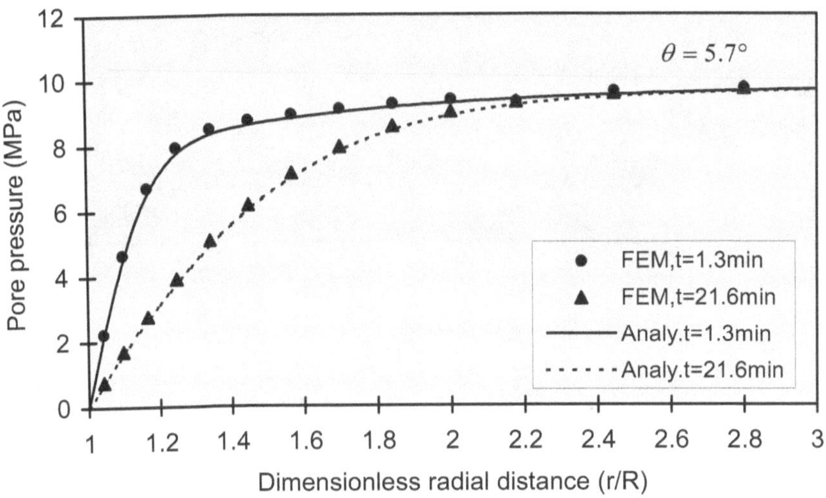

Fig. 6.5. Comparison between the finite element and analytical solutions for pore pressure (θ = 5.7°).

For the same data set, the Terzaghi's effective radial stresses, defined as the difference between the total radial stress and the pore pressure, are plotted in Fig. 6.6 for $\theta = 5.7°$. Except for a slight difference in the near-wellbore region, the analytical and finite element solutions match well. The tensile region developed at early time is due to the non-monotonic pore pressure distribution, which is consistent with the case reported by Cui et al. (1997b).

Figure 6.7 presents the total tangential stresses for two different radial directions ($\theta = 5.7$ and $84.4°$) and two different times ($t = 1.3$ and 21.6 min), and excellent agreement can be seen between the finite element and analytical solutions for a larger time ($t = 21.6$ min). The small differences in the near wellbore region at a small time are induced by initial conditions, time step and boundary effects.

Also, good matches (refer to Fig. 6.8) are obtained for the total axial stresses for two different radial directions ($\theta = 0$ and $90°$) and two different times ($t = 1.3$ and 21.6 min).

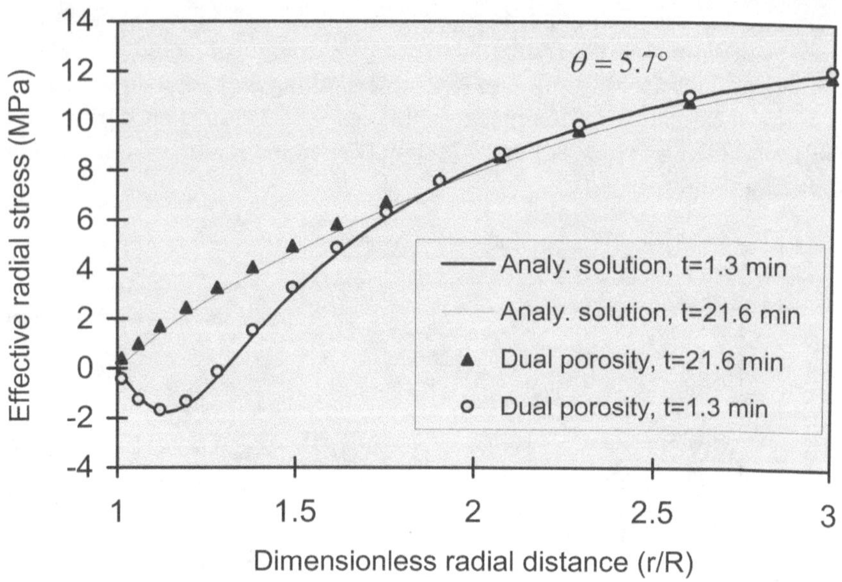

Fig. 6.6. Comparison between the finite element and analytical solutions for Terzaghi's effective radial stress ($\theta = 5.7°$).

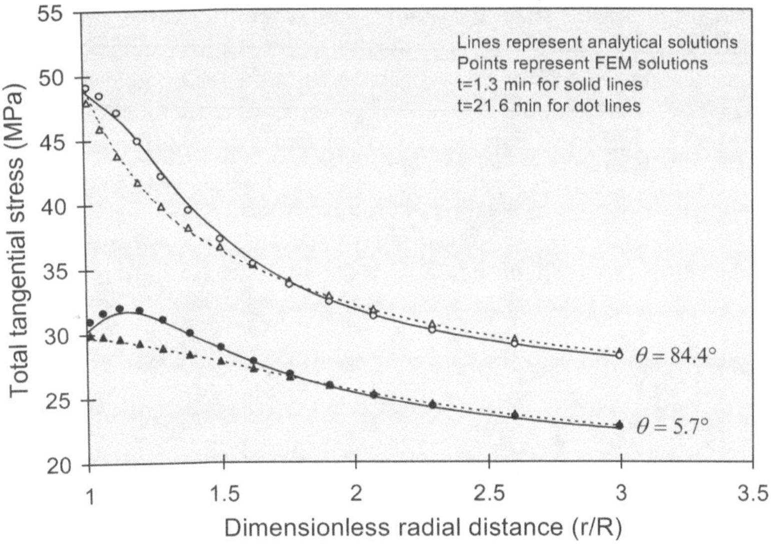

Fig. 6.7. Comparison between the finite element and analytical solutions for total tangential stress along different radial sections ($\theta = 5.7°$, $\theta = 84.4°$) and times ($t = 1.3$ min, $t = 21.6$ min).

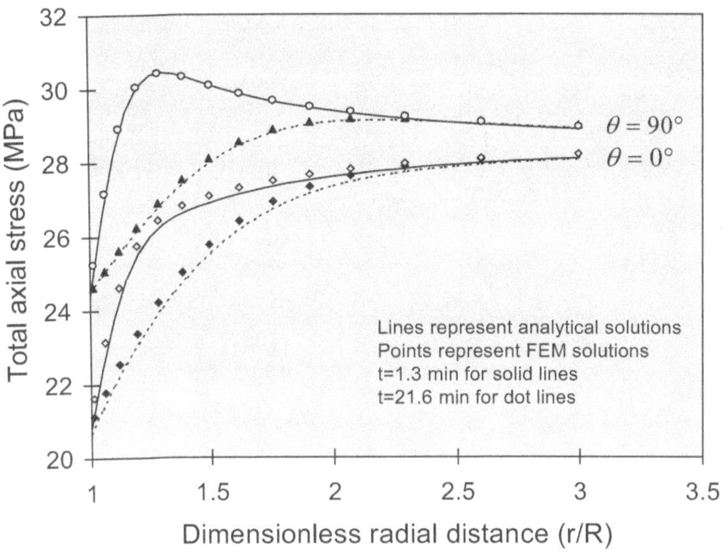

Fig. 6.8. Comparison between the finite element and analytical solutions for total axial stress along different radial sections ($\theta = 0°$, $\theta = 90°$) and times ($t = 1.3$ min, $t = 21.6$ min).

For the impermeable case, the double porosity finite element model with extremely large fracture spacing and zero pore and fracture pressures presents the approximation to the elastic solution (analytical solution in Bradley 1979). The parameters used for the finite element model are the same as in the previous calculation except that a mud pressure (p_w = 10 MPa) is applied along the borehole wall. Along the radial section (θ = 30°), the numerical and analytical solutions for radial and tangential stresses are compared in Fig. 6.9. It shows that the proposed finite element solutions return an excellent match with the analytical solutions except for small discrepancies near the borehole wall.

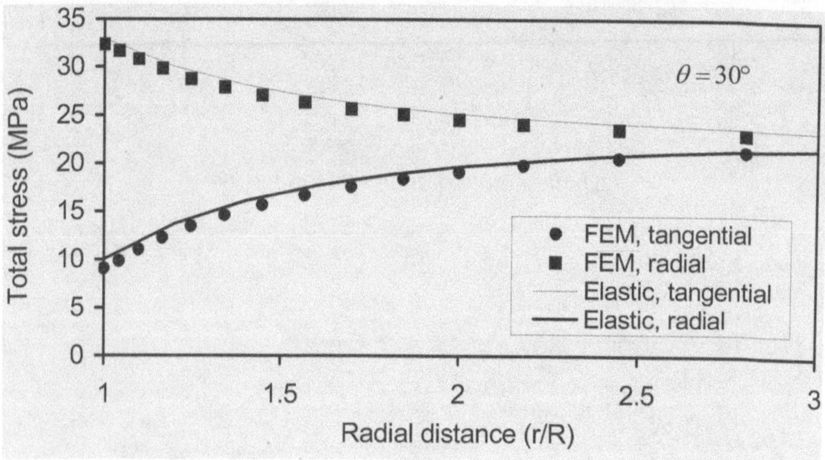

Fig. 6.9. Comparison between the finite element and analytical solutions for total radial and tangential stresses along the radial sections (θ = 30°) for the impermeable model.

6.5 Parametric analyses and application for borehole stability

In this section, examples are given to demonstrate how certain parameters in a double porosity medium affect the pore pressure and stress results. The in-situ stresses, initial pore pressures, geometry and material properties used in the ongoing analyses are identical to those listed Table 6.1, except for the fracture spacing s = 0.1 m. All the results are presented at a borehole inclination angle of φ_z = 70°. Figure 6.4 shows the far-field and local stress states. In this case the x-direction ($\theta = 0°$) is the local far-field maximum stress direction and the y-direction ($\theta = 90°$) is the local far-field minimum one.

In the following analyses, double-porosity, time, fracture, mud weight and inclination effects are considered for inclined borehole problems. For each analysis, only one specific parameter is allowed to be varied.

6.5.1 Double-porosity effects

The pore pressure at t = 100 s and $\theta = 0°$ for the single-porosity model as well as the one for the dual-porosity model are compared in Fig. 6.10. The difference is evidenced by the increase in pore pressure in the matrix for the dual-porosity media due to the associated large fracture compliance. A similar comparison is made for Terzaghi's effective radial stresses (Fig. 6.11). Although the pore pressure induced by dual-porosity effect increases, the tensile stress does not increase, but decreases (Fig. 6.11). This is because the total radial stress has a larger increment than the pore pressure for the dual-porosity medium (Zhang 2002). Therefore, the effective compressive stress, which equals to the difference of the total stress and the pore pressure, increases and the effective tensile stress decreases, which reduces the potential for borehole spalling (tensile failure).

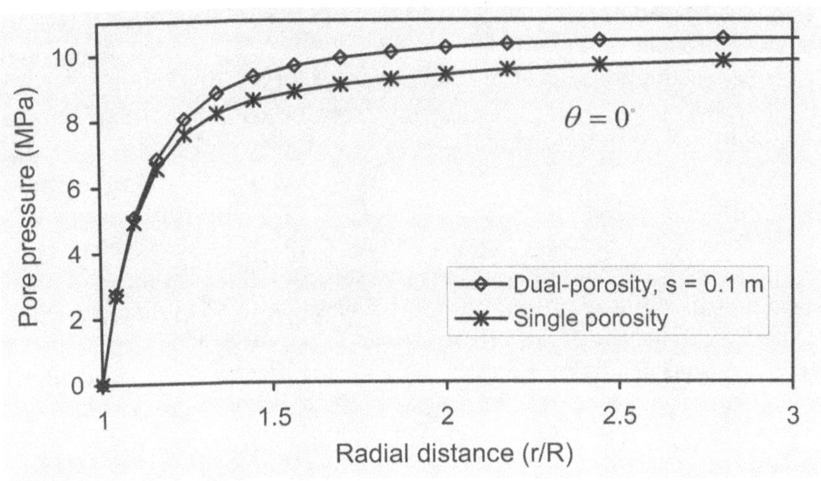

Fig. 6.10. Comparison of pore pressure for single- and double-porosity models in the local far-field maximum stress direction at t = 100 s.

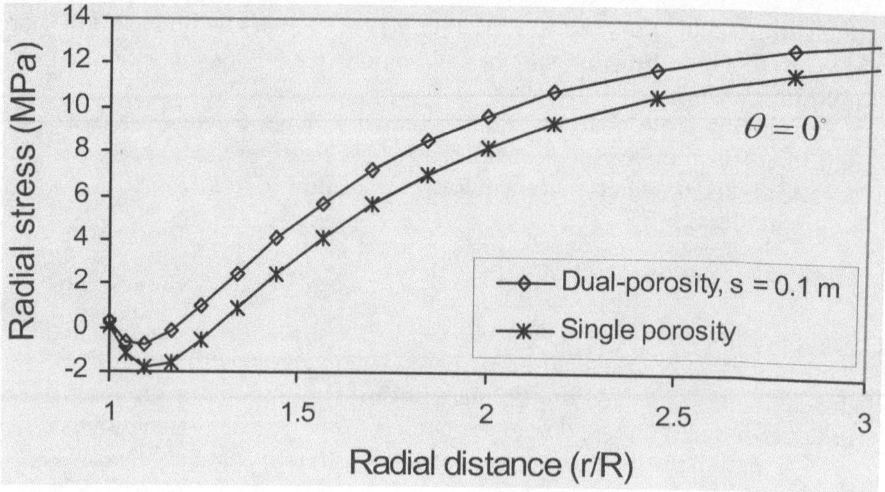

Fig. 6.11. Comparison of effective radial stress for single- and double-porosity models in the maximum stress direction at $t = 100$ s.

6.5.2 Time-dependent effects

In Figs. 6.12 through 6.14, the pore pressure distributions around the wellbore are presented at three different time domains. In the near-field, pore pressure concentrations occur at smaller times (for $t = 10$ s in Fig. 12 and $t = 100$ s in Fig. 13) around the minimum stress direction ($\theta = 90°$), and the pore pressures decrease as time increases, as shown in Figs. 6.12 - 6.14. The pore pressure in the near field is larger as θ increases, which is due to Skempton's effect (Detournay and Cheng 1993), because at 0° a larger far-field compressive normal stress prevails (Fig. 6.4). Note that non-monotonic pressure distributions and pressure peaks are found at a small distance inside the wall at small times, which is attributed to the poroelastic effect. At larger times this effect disappears. At large distances, the pore pressure approaches asymptotically the far-field pore pressure value of 10 MPa, and as time increases this poroelastic effect becomes negligible.

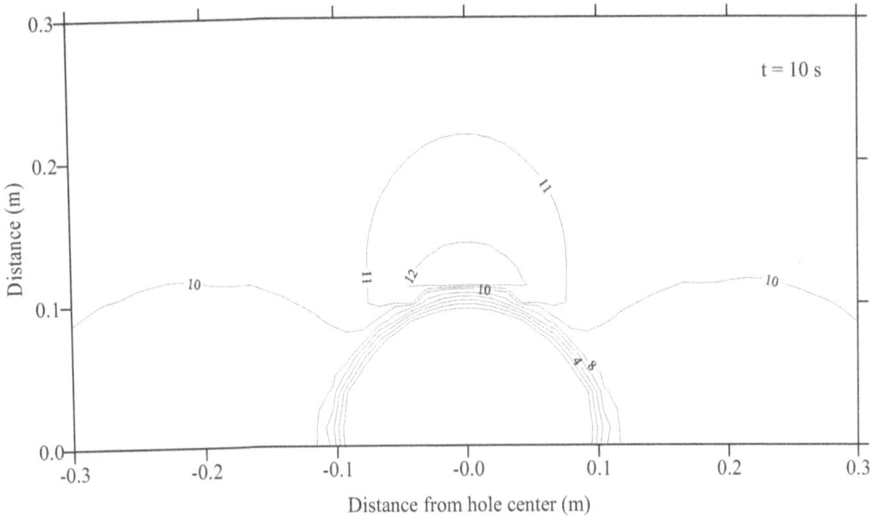

Fig. 6.12. Pore pressure distribution around the wellbore at t = 10 s.

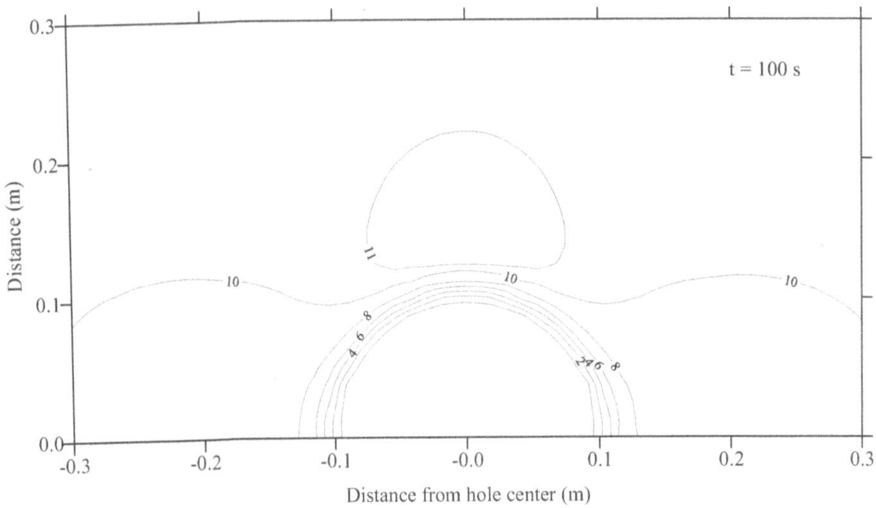

Fig. 6.13. Pore pressure distribution around the wellbore at t = 100 s.

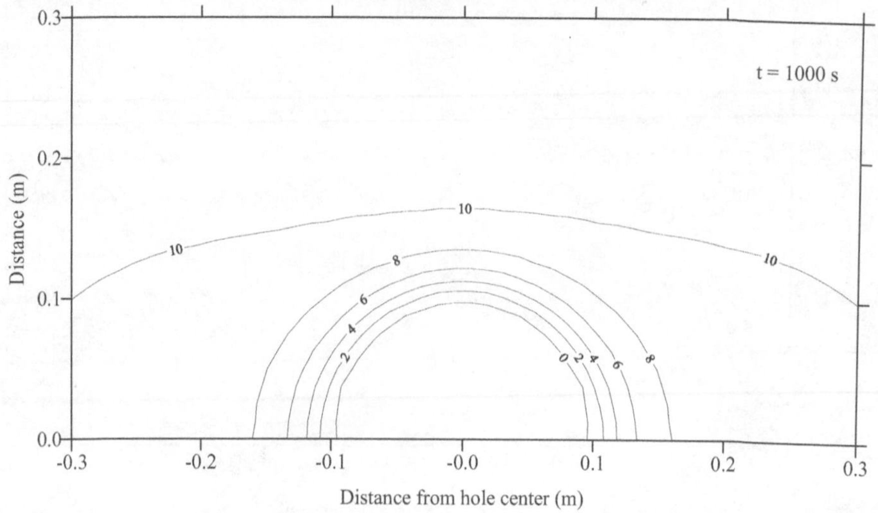

Fig. 6.14. Pore pressure distribution around the wellbore at t = 1000 s.

The Terzaghi's effective radial stress distributions for four different times are plotted in Figs. 6.15 and 6.16. The results clearly show a tensile region near the wellbore at small times. The tensile radial stress magnitude in the near field is much larger in the minimum stress direction (θ = 90°) than that in the maximum stress direction (θ = 0°). This is due to higher pore pressure in the minimum stress direction (refer to Figs. 6.12 and 6.13). It is evident that the radial stress has a very strong time effect; i.e., the tensile stress reduces significantly as time elapses. Tensile stress disappears as time is large enough.

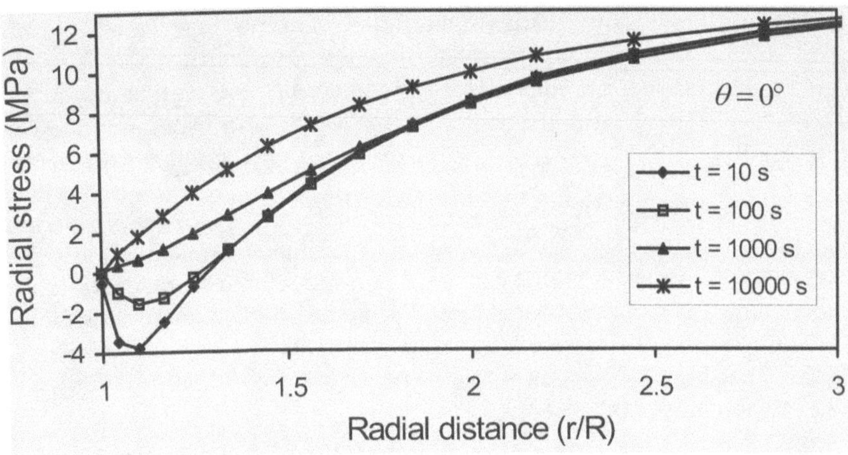

Fig. 6.15. Effective radial stress in the maximum stress direction for different times.

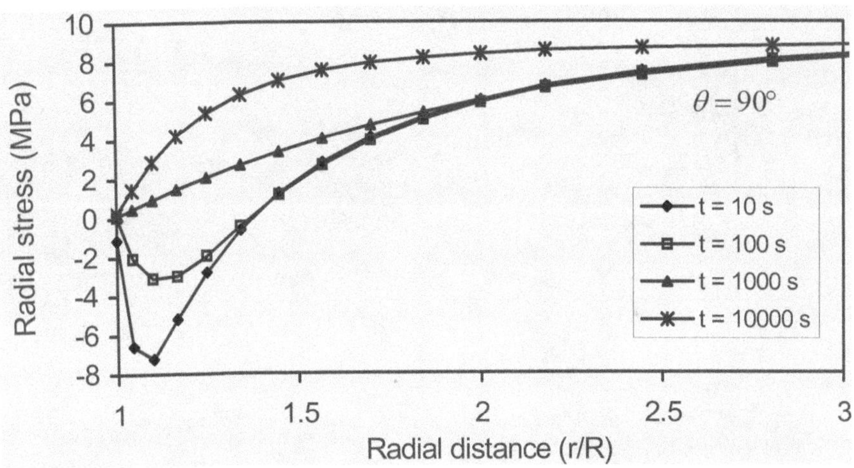

Fig. 6.16. Effective radial stress in the minimum stress direction for different times.

6.5.3 Mud weight effects

Rock failure, such as spalling and breakout, is likely to take place when a borehole is drilled with air or without sufficient mud pressure to support it before a casing is placed. However, too high mud pressures may induce borehole instability by tensile fracturing leading to unacceptable mud losses. In the following analysis, the borehole true vertical depth is assumed to be 1000 m, and four values of mud density are examined, i.e., $\rho_w = 0.006$, $\rho_w = 0.01$, $\rho_w = 0.02$ and $\rho_w = 0.025$ MN/m^3 corresponding to mud pressures of $p_w = 6$, $p_w = 10$, $p_w = 20$, and $p_w = 25$ MPa. Other parameters remain the same, as in previous analyses.

Due to the mud pressure, there are no tensile radial stresses inside the borehole except for the case of a very small mud pressure value (e.g. $p_w = 6$ MPa), as shown in Fig. 6.17; instead, a non-monotonic stress distribution appears for higher mud pressures. Furthermore, the increasing mud pressure causes an increase of the compressive radial stress; this is due to the fact that the mud weight acts on the wellbore wall as an additional radial stress component.

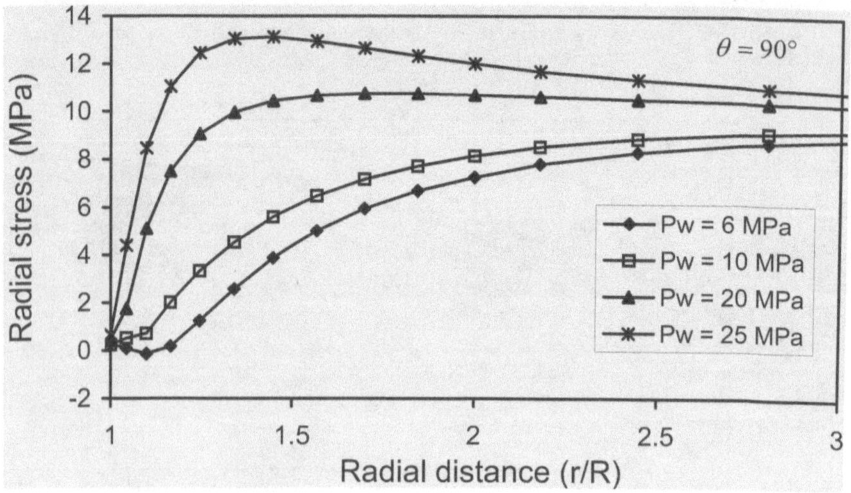

Fig. 6.17. Effective radial stress along the minimum stress direction at $t = 100$ s for different mud pressures.

Figure 6.18 shows the response of the effective compressive tangential stresses along the maximum stress direction ($\theta = 0°$) at t = 100 s. It is obvious that the compressive tangential stress decreases as the mud pressure increases, which reduces the high stress concentration around the wellbore. It is noted that the effective tangential stress becomes tensile along the local maximum stress direction for high mud pressures (e.g. $p_w = 25$ MPa). This illustrates that the borehole will fail in tension or fracturing when the tensile tangential stress exceeds the formation tensile strength.

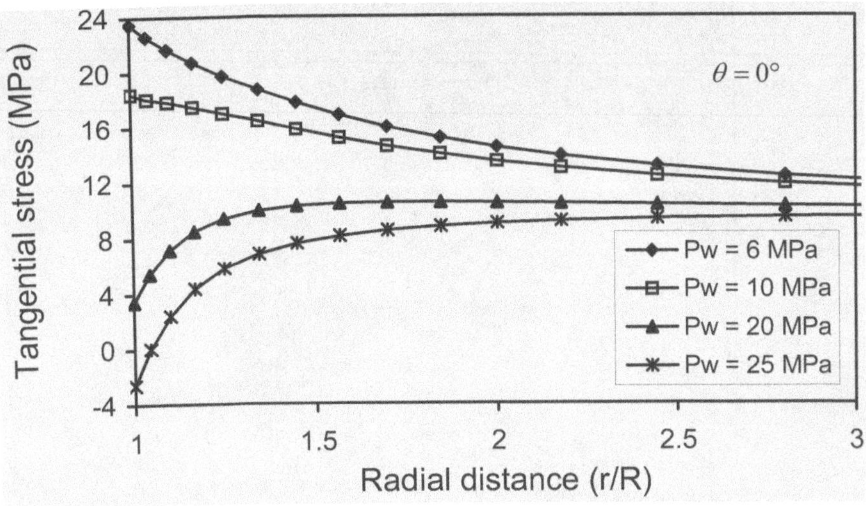

Fig. 6.18. Effective tangential stress along the maximum stress direction at $t = 100$ s for different mud pressures.

6.5.4 Fracture stiffness effects

Figure 6.19 presents the pore pressure responses for different fracture stiffnesses along the minimum stress direction ($\theta = 90°$). It shows that the pore pressure increases as the fracture stiffness becomes increasingly smaller. When the stiffness is large enough, representing nearly no deformation in the fractures, the pore pressure no longer varies with fracture stiffness.

The comparisons of the effective radial stresses indicate that the compressive stress increases, while the tensile stress reduces as the stiffness decreases, as shown in Fig. 6.20. The radial tensile stress becomes insignificant when the fracture stiffness is very small.

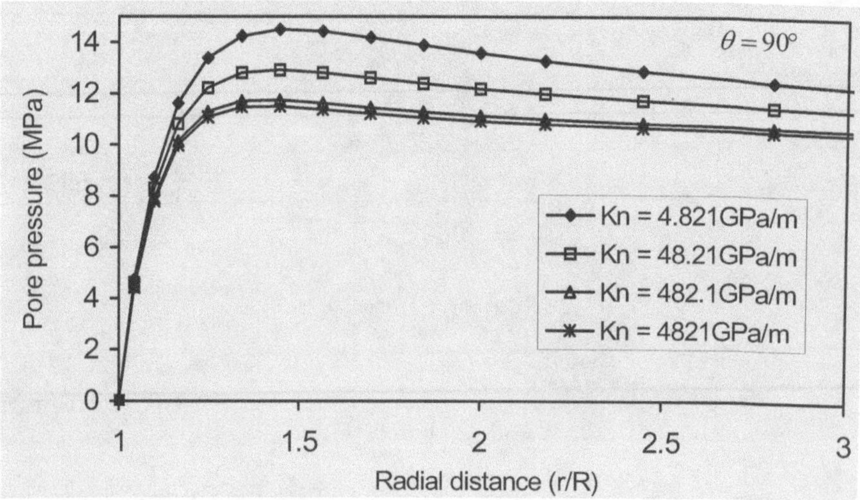

Fig. 6.19. Pore pressure around the wellbore in the minimum stress direction at $t = 100$ s for different fracture stiffnesses.

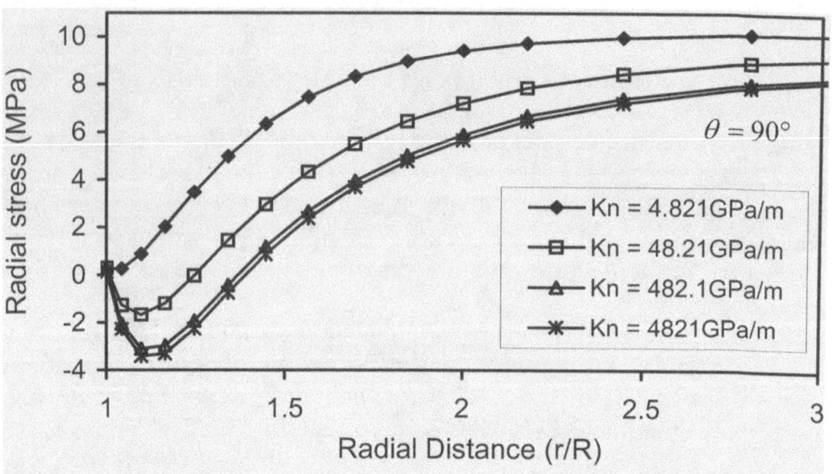

Fig. 6.20. Effective radial stress around the wellbore in the minimum stress direction at $t = 100$ s for different fracture stiffnesses.

6.5.5 Borehole inclination effects

To assess the effect of borehole inclination, the pore pressure responses at $t = 100$ s, $\theta = 90°$ for four different inclination angles, 0°, 40°, 70° and 90° are examined in Fig. 6.21. Note that a hole inclination angle of $\varphi_z = 0°$ represents a vertical hole. It shows that the inclination decreases the pore pressure magnitude in this example. This results from the extra pore pressure generated by Skempton's effect. At $\varphi_z = 0°$ the relevant local far-field compressive stresses at $\theta = 0°$ are $\sigma_x = 29$ MPa and $\sigma_y = 20$ MPa, which are changed to $\sigma_x = 25$ MP and $\sigma_y = 20$ MPa at $\varphi_z = 90°$ due to the borehole inclination (refer to Fig. 6.22).

Figure 6.23 explores the influence of hole inclination on radial stresses at $t = 100$ s, $\theta = 0°$. The inclination causes a reduction in the compressive radial stress and an increase in the tensile radial stress. This trend is more pronounced at the azimuthal angle $\theta = 0°$, because the local far-field total stress varies from $\sigma_x = 29$ MPa (effective stress of 19 MPa) at a hole inclination $\varphi_z = 0°$ to $\sigma_x = 25$ MPa (effective stress 15 MPa) compared to a hole inclination $\varphi_z = 90°$ (Zhang and Roegiers 2005); however, at an angle of $\theta = 90°$, the local far-field total stress does not change ($\sigma_y = 20$ MPa) from a hole inclination $\varphi_z = 0°$ to $\varphi_z = 90°$ (refer to Figs. 6.4 and 6.22).

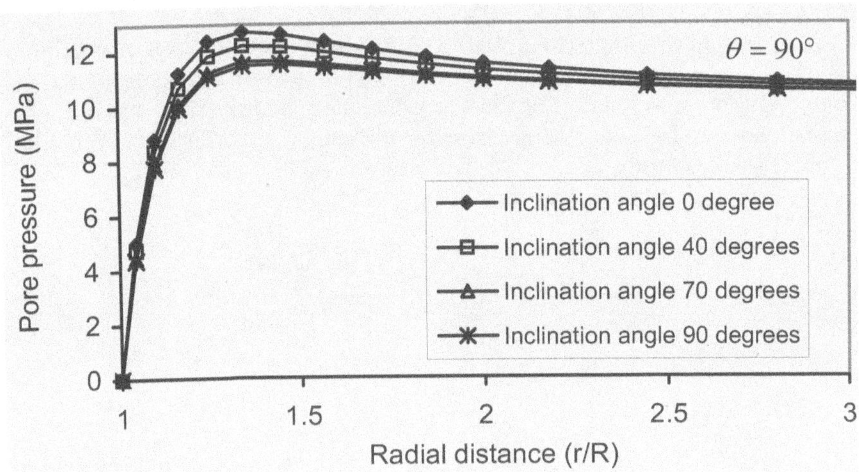

Fig. 6.21. Pore pressure distribution at $t = 100$ s, $\theta = 90°$ for different borehole plunges, φ_z.

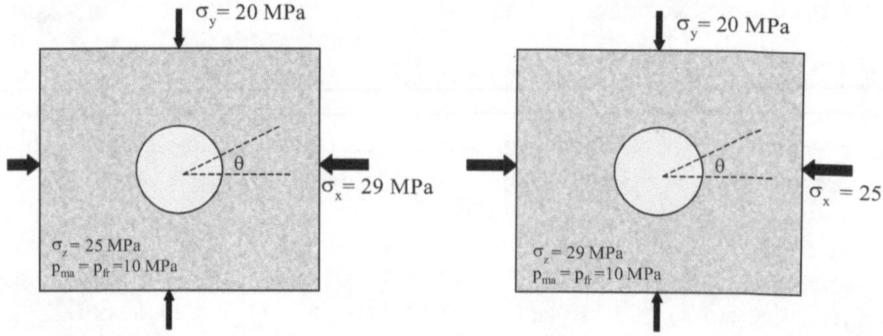

(a) Hole inclination $\varphi_z = 0°$ (b) Hole inclination $\varphi_z = 90°$

Fig. 6.22. State of stress in the local coordinate systems after 0° and 90° inclinations.

Figures 6.24 and 6.25 demonstrate the effect of borehole inclination on the effective tangential and axial stresses at the azimuthal angle of $\theta = 90°$. It is clear that at larger inclination angles the tangential stress decreases and the axial stress increases because the inclination from $\varphi_z = 0°$ to $\varphi_z = 90°$ causes the local far-field total stress to vary from $\sigma_x = 29$ MPa, $\sigma_z = 25$ MPa to $\sigma_x = 25$ MPa, $\sigma_z = 29$ MPa (see Fig. 6.22).

From the above analyses the effects of borehole inclination on wellbore pressures and stresses depend strongly upon the far-field stress magnitudes. By the same token, for the given boundary and far-field stress conditions, the borehole inclination can be optimized in order to avoid high stress concentrations.

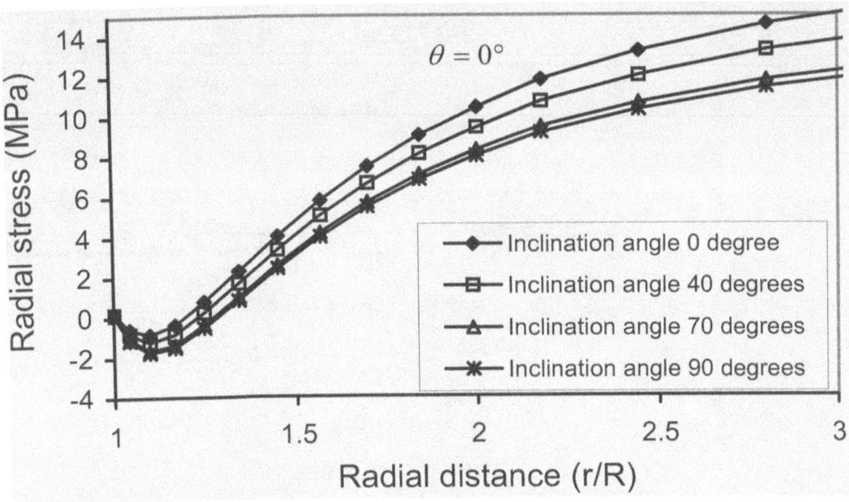

Fig. 6.23. Effective radial stress distributions, at t = 100 s, θ = 0°, for different borehole plunges, φ_z.

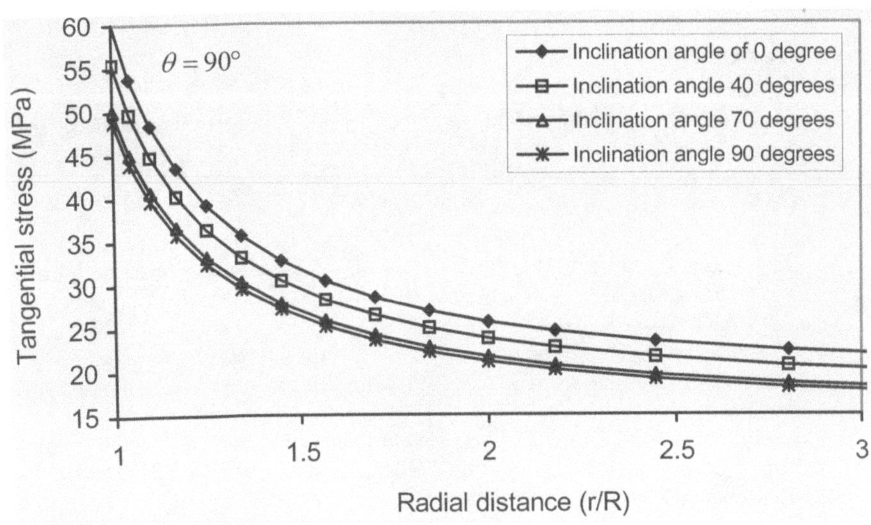

Fig. 6.24. Effective tangential stress distributions, at t = 100 s, θ = 90°, for different plunges, φ_z.

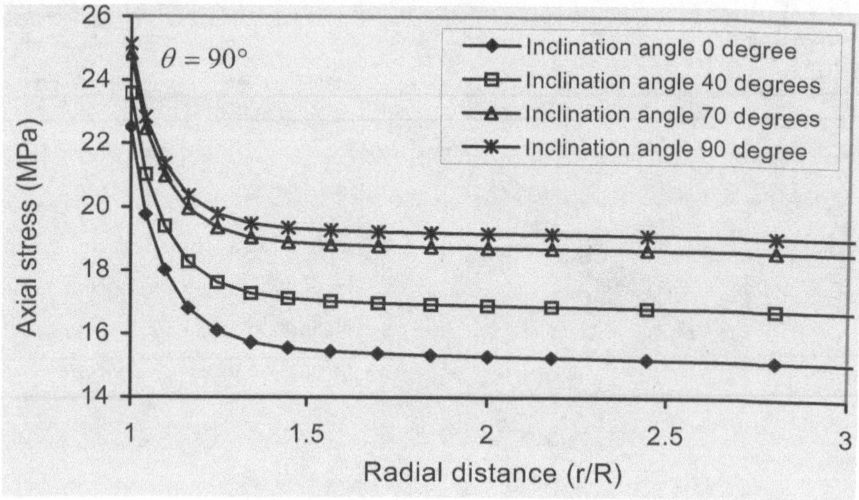

Fig. 6.25. Effective axial stress distributions, at t = 100 s, θ = 90°, for different borehole plunges, φ_z.

References

Aadnøy BS (1987) Modeling of the stability of highly inclined boreholes in anisotropic rock formations. Paper SPE 16526 presented at SPE Offshore European Conf Aberdeen

Aifantis EC (1977) Introducing a multi-porous medium. Developments in Mechanics 9:201-211

Aoki T, Tan CP, Bamford WE (1993) Effects of deformation and strength anisotropy on borehole failures in saturated shales. Int J Rock Mech Min Sci & Geomech Abstr 30:1031-1034

Bai M, Elsworth D, Roegiers JC (1993) Multi-porosity/multi-permeability approach to the simulation of naturally fractured reservoirs. Water Resour Res 29:1621-1633

Bai M, Abousleiman Y, Cui L, Zhang J (1999a) Dual-porosity porous elastic modeling of generalized plane strain. Int J Rock Mech Min Sci 36:1087-1094

Bai M, Meng F, Elsworth D, Abousleiman Y, Roegiers JC (1999b) Numerical modeling of coupled flow and deformation in fractured rock specimens. Int J Numer Anal Meth Geomech 23:141-160

Barenblatt GI, Zheltov IP, Kochina IN (1960) Basic concepts in the theory of seepage of homogeneous liquids in fissured rocks. Prikl Mat Mekh 24:852-864

Biot MA (1941) General theory of three-dimensional consolidation. J Appl Phys 12:155-164
Bradley WB (1979) Failure of inclined boreholes. Trans ASME 101:232-239
Charlez PA (1999) Rock Mechanics, Petroleum Application vol 2. Editions Technip Paris
Cui L, Cheng AHD, Abousleiman Y (1997a) Poroelastic solution of an inclined borehole. ASME J Applied Mechanics 64:32-38
Cui L, Kaliakin VN, Abousleiman Y, Cheng AHD (1997b) Finite element formulation and application of poroelastic generalized plane strain problems. Int J Rock Mech Min Sci 34:953-962
Detournay E, Cheng AHD (1988) Poroelastic response of a borehole in a non-hydrostatic stress field. Int J Rock Mech Min Sci Geomech Abstr 25:171-182
Detournay E, Cheng AHD (1993) Fundamental of poroelasticity. Ch.5 in Comprehensive Rock Engineering Vol 2. Fairhurst C (ed), Pergamon Press
Elsworth D, Bai M (1992) Flow-deformation response of dual-porosity media. J Geotech Eng 118:107-124
Fairhurst C (1968) Methods of determining in situ rock stress at great depth. TRI-68 Missouri River Div. Corps of Engineering
Goodman RE (1966) On the distribution of stresses around circular tunnels in nonhomogeneous rocks. In: Proc 1st Int Congress ISRM, Lisbon, pp249-255
Hubbert MK, Willis DG (1957) Mechanics of hydraulic fracturing. JPT Trans ASME 210:153-66
Kazemi H (1969) Pressure transient analysis of naturally fractured reservoirs with uniform fracture distribution. Soc Pet Eng J 9:451-462
Kovari K (1977) The elasto-plastic analysis in the design practice of underground openings. In: Gudehus G (ed), Finite Elemnets in Geomechaincs, John Wiley Sons, pp377-412
Maury V, Zurdo C (1996) Drilling-induced lateral shifts along pre-existing fractures: a common cause of drilling problems. SPE Drilling Completion, March
Peng S, Meng Z (1999) Numerical modeling on coal roof and floor stability in longwall mining. J China Univ Min Tech 28(1):41-45 (in Chinese)
Pruess K, Tsang YW (1990) On two-phase relative permeability and capillary pressure of rough-walled rock fractures. Water Resour Res 26:1915-1926
Warren JE, Root PJ (1963) The behavior of naturally fractured reservoirs. Soc Pet Eng J Trans ASME 228:245-255
Wilson RK, Aifantis EC (1982) On the theory of consolidation with double porosity. Int J Eng Sci 20:1009-1035
Willson SM, Last NC, Zoback MD, Moos D (1999) Drilling in South America: a wellbore stability approach for complex geologic conditions. Paper SPE 53940 presented at Latin America and Caribbean Petroleum Engineering Conference held in Caracas, Venezuela
Woodland DC (1990) Borehole instability in the western Canadian overthrust belt. SPE Drilling Engineers 5:23-33
Terzaghi K (1943) Theory soil mechanics. John Wiley Sons, New York
Zhang J (2002) Dual-porosity approach to wellbore stability in naturally fractured reservoirs. PhD Dissertation, Univ of Oklahoma

Zhang J, Bai M, Roegiers JC (2003) Dual-porosity poroelastic analyses of wellbore stability. Int J Rock Mech Min Sci 40:473-483

Zhang J, Roegiers JC (2002) Borehole stability in naturally deformable fractured reservoirs – a fully coupled approach. Paper SPE 77355 presented at SPE Annual Technical Conference Exhibition.

Zhang J, Roegiers JC (2005) Double porosity finite element method for borehole modeling. Rock Mech Rock Engng 38:217-242

Zienkiewicz OC (1977) The finite element method. McGraw-Hill, 3rd ed, New York

Appendix 6.1 Combined elasticity matrix and compliance matrices

The combined elasticity matrix D_{mfijkl} is given as (Bai et al. 1999b):

$$D_{mfijkl} = \frac{1}{|D_{mfijkl}|}\begin{bmatrix} d_{1111} & d_{1122} & d_{1133} & 0 & 0 & 0 \\ d_{2211} & d_{2222} & d_{2233} & 0 & 0 & 0 \\ d_{3311} & d_{3322} & d_{3333} & 0 & 0 & 0 \\ 0 & 0 & 0 & d_{4444} & 0 & 0 \\ 0 & 0 & 0 & 0 & d_{55555} & 0 \\ 0 & 0 & 0 & 0 & 0 & d_{6666} \end{bmatrix} \tag{6.1.1}$$

The compliance matrices are given as follows:

$$C_{maijkl} = \frac{1}{E}\begin{bmatrix} 1 & -\nu & -\nu & 0 & 0 & 0 \\ -\nu & 1 & -\nu & 0 & 0 & 0 \\ -\nu & -\nu & -\nu & 0 & 0 & 0 \\ 0 & 0 & 0 & 2(1+\nu) & 0 & 0 \\ 0 & 0 & 0 & 0 & 2(1+\nu) & 0 \\ 0 & 0 & 0 & 0 & 0 & 2(1+\nu) \end{bmatrix} \tag{6.1.2}$$

$$C_{frijkl} = \frac{1}{E}\begin{bmatrix} \frac{1}{K_n s} & 0 & 0 & 0 & 0 & 0 \\ 0 & \frac{1}{K_n s} & 0 & 0 & 0 & 0 \\ 0 & 0 & \frac{1}{K_n s} & 0 & 0 & 0 \\ 0 & 0 & 0 & \frac{1}{K_{sh} s} & 0 & 0 \\ 0 & 0 & 0 & 0 & \frac{1}{K_{sh} s} & 0 \\ 0 & 0 & 0 & 0 & 0 & \frac{1}{K_{sh} s} \end{bmatrix} \tag{6.1.3}$$

and,

$$|D_{mfijkl}| = \left[\frac{1}{K_{sh}s} + \frac{2(1+\nu)}{E}\right]^3 \left[\left(\frac{1}{E} + \frac{1}{K_n s}\right)^3 - \frac{3\nu^2}{E^2}\left(\frac{1}{E} + \frac{1}{K_n s}\right) - \frac{2\nu^2}{E^2}\right]$$

$$d_{1111} = d_{2222} = d_{3333} = \left[\frac{1}{K_{sh}s} + \frac{2(1+\nu)}{E}\right]^3 \left[\left(\frac{1}{E} + \frac{1}{K_n s}\right)^2 - \frac{\nu^2}{E^2}\right]$$

$$d_{1122} = d_{2211} = d_{1133} = d_{3311} = d_{2233} = d_{3322} = \left[\frac{1}{K_{sh}s} + \frac{2(1+\nu)}{E}\right]^3 \left[\frac{\nu}{E}\left(\frac{1}{E} + \frac{1}{K_n s}\right)^2 - \frac{\nu^2}{E^2}\right]$$

$$d_{4444} = d_{55555} = d_{6666} = \left[\frac{1}{K_{sh}s} + \frac{2(1+\nu)}{E}\right]^2 \left[\left(\frac{1}{E} + \frac{1}{K_n s}\right)^2 - \frac{3\nu^2}{E^2}\left(\frac{1}{E} + \frac{1}{K_n s}\right) - \frac{2\nu^3}{E^3}\right]$$

where E is the Young's modulus, and K_n , K_{sh} are the fracture normal and shear stiffnesses, respectively.

Appendix 6.2 Terms in the finite element matrices

$$\mathbf{K} = \int_\Omega \mathbf{B}^{\mathrm{T}} \mathbf{D}_{\mathbf{mf}} \mathbf{B} d\Omega \tag{6.2.1}$$

$$\mathbf{R}_1 = \alpha_{ma} \int_{\Omega} \mathbf{B}^{\mathrm{T}} \mathbf{D}_{\mathbf{mf}} \mathbf{C}_{\mathbf{ma}} \mathbf{m} \mathbf{N} d\Omega \tag{6.2.2}$$

$$\mathbf{R}_2 = \alpha_{fr} \int_{\Omega} \mathbf{B}^{\mathrm{T}} \mathbf{D}_{\mathbf{mf}} \mathbf{C}_{\mathbf{fr}} \mathbf{m} \mathbf{N} d\Omega \tag{6.2.3}$$

$$\mathbf{L}_1 = \frac{1}{\mu} \int_{\Omega} \nabla \mathbf{N}^{\mathrm{T}} \mathbf{k}_{\mathbf{ma}} \nabla \mathbf{N} \mathrm{d}\Omega \tag{6.2.4}$$

$$\mathbf{L}_2 = \frac{1}{\mu} \int_{\Omega} \nabla \mathbf{N}^{\mathrm{T}} \mathbf{k}_{\mathbf{fr}} \nabla \mathbf{N} \mathrm{d}\Omega \tag{6.2.5}$$

$$\mathbf{N}_1 = \beta_{ma} \int_{\Omega} \mathbf{N}^{\mathrm{T}} \mathbf{N} d\Omega \tag{6.2.6}$$

$$\mathbf{N}_2 = \beta_{fr} \int_{\Omega} \mathbf{N}^{\mathrm{T}} \mathbf{N} d\Omega \tag{6.2.7}$$

$$\mathbf{Q} = -\omega \int_{\Omega} \mathbf{N}^{\mathrm{T}} \mathbf{N} d\Omega \tag{6.2.8}$$

$$\mathbf{M}_1 = \alpha_{ma} \int_{\Omega} \mathbf{N}^{\mathrm{T}} \mathbf{D}_{\mathbf{mf}} \mathbf{C}_{\mathbf{ma}} \mathbf{m} \mathbf{B} d\Omega \tag{6.2.9}$$

$$\mathbf{M}_2 = \alpha_{fr} \int_{\Omega} \mathbf{N}^{\mathrm{T}} \mathbf{D}_{\mathbf{mf}} \mathbf{C}_{\mathbf{fr}} \mathbf{m} \mathbf{B} d\Omega \tag{6.2.10}$$

$$\mathbf{F} = \int_{S} \mathbf{N} \mathbf{f} dS \tag{6.2.11}$$

$$\mathbf{Q}_{\mathbf{ma}} = \int_{\Gamma} \mathbf{N}^{\mathrm{T}} \mathbf{N} \mathbf{q}_{\mathbf{ma}} d\Gamma \tag{6.2.12}$$

$$\mathbf{Q}_{\mathbf{fr}} = \int_{\Gamma} \mathbf{N}^{\mathrm{T}} \mathbf{N} \mathbf{q}_{\mathbf{fr}} d\Gamma \tag{6.2.13}$$

7 Wellbore/borehole stability

7.1 Wellbore stability while drilling

The drilling for oil and gas exploration and production under increasingly difficult geological conditions has revealed a need for better understanding of borehole stability issues. It is estimated that wellbore instability results in substantial economic losses of about US$ 8 billion per year worldwide.

Many innovative technologies have been applied in the oil and gas industry, such as underbalanced drilling, high pressure jet drilling, re-entry horizontal wells, and multilaterals from a single well. These have definitely increased the demand for wellbore stability studies. Recently, technological advances have been pushing the boreholes to reach beyond 34,100 ft (10,400 m) below the sea level in deepwater of the Gulf of Mexico. Highly inclined, extended-reach wellbores may have to remain open for prolonged time periods, not only during the drilling stage but also over the life of a reservoir. New challenges also emerged since the increasing use of horizontal wells, drilling in naturally fractured media, in very deep formations, and in difficult geological conditions, where wellbore stability is of major concern (Willson and Willis 1986). For example, a 8,715 m deep well was drilled in crystalline rock in Germany and some types of wellbore instabilities (breakouts, washout, undergauged sections) were observed (Hoffers et al. 1994). Some wellbore instabilities associated with complex geologic conditions, where the stress regime was controlled by active faults, are reported in the Cusiana field (Colombia), the Pedernales field (Venezuela), the Alberta Basin (Canada), the Tarim Basin (China), certain areas of the Norwegian Sea, and offshore Indonesia (Willson et al. 1999, Plumb et al. 1998, Wiprut and Zoback 1998, Ramos et al. 1998).

When a borehole is drilled in a naturally fractured formation, excessively high mud density allows the drilling fluid to penetrate into fractures, mobilizing the rock blocks, and intensifying ovalization (Charlez 1999). When this occurs, the fractured blocks are no longer subject to the mud overbalance pressure, and the destabilized blocks can cave into the wellbore as a result of swabbing the formation when tripping (Willson et al. 1999). When a borehole crosses a fault, drilling mud may invade the discontinuity plane. Apart from mud losses, penetration of the fluid reduces the normal stress and induces a displacement along the crack planes which shears the well, as shown in Fig. 7.1. The consequences can quickly become dramatic and could lead to partial or even total loss of a well. Two case histories in Aquitaine, France were described that resulted in the loss of the wells and the need for the drilling of two new wells, costing in the range of US$30 million (Maury and Zurdo 1996).

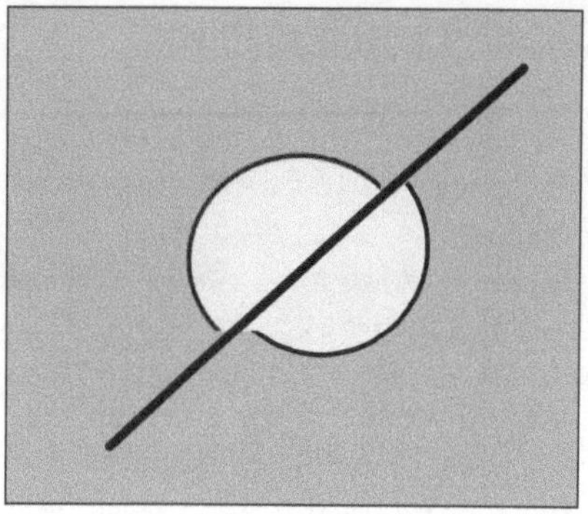

Fig. 7.1. Schematic representation of borehole shearing when crossing a discontinuity.

Wellbore instability can result in lost circulation where tensile stresses have occurred due to high drilling mud pressure (Fig. 7.2a); breakouts and hole closure in case of compressive and shear failures (Fig. 7.2b). During drilling stage an open hole is supported by drilling mud pressure to keep wellbore from collapse. If the mud weight is lower than the shear failure stress or collapse stress, the shear failure and compressive failure occur in the wellbore in the minimum far-field stress (S_h) direction, causing hole collapse or breakout. If the mud weight exceeds the rock tensile strength, the tensile fracture is induced in the maximum far-field stress (S_H) direction. Consequently, this may cause drilling fluid losses or lost circulation. Figure 7.3 shows a typical wellbore instability due to breakout and drilling induced tensile fracture. For a circular opening with large diameter the hole/tunnel breakout has a similar behavior as small boreholes. Figure 7.4 presents hole breakout in a circular tunnel with a radius of R = 1.75 m in the Underground Research Laboratory of Canada (Martino and Chandler 2004). The fully developed notch (breakout) in the roof was caused by stress redistribution due to excavation. The notch is stable, owing to its naturally formed shape, which develops a confining pressure at the notch tip. The notch will remain stable unless disturbed by changing conditions, such as increased temperature, small stress changes caused by nearby excavations.

In severe cases the borehole instability can lead to loss of the open hole section. The borehole stability problem can be considered by separating the potential rock failure mechanisms into the following four categories (Roegiers1990):

- failures related to pre-existing or drilling-induced formation damage;
- failures caused by the induced stress, pore pressure concentrations, and temperature redistributions;
- failures attributed to deliberate or unintentional additional stresses; and,
- failures related to shock-wave loading.

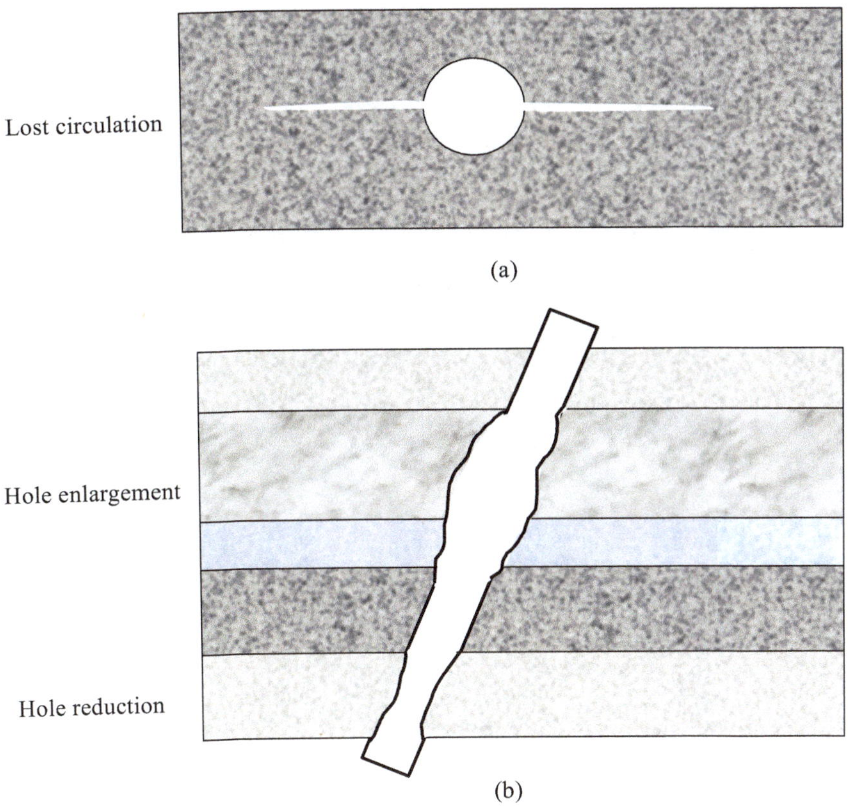

Fig. 7.2. Schematic plot of stress-induced wellbore instability. **a.** tensile failure due to high mud weight; **b.** shear failure and deformation.

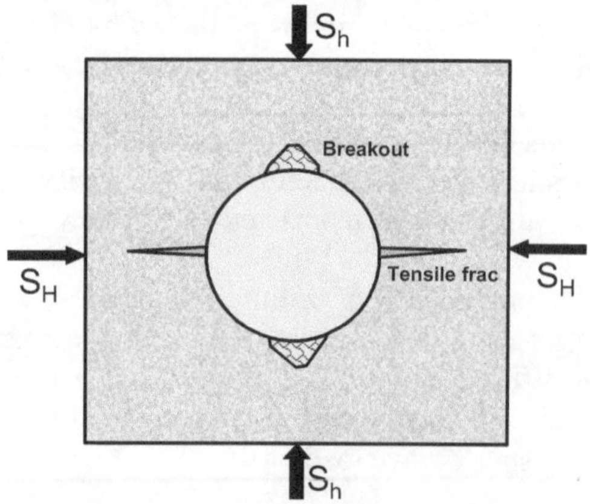

Fig. 7.3. Schematic representation of stress-induced wellbore instability for a cross section in a vertical hole - breakout due to shear and compressive failure, and drilling induced tensile fracture due to high mud pressure.

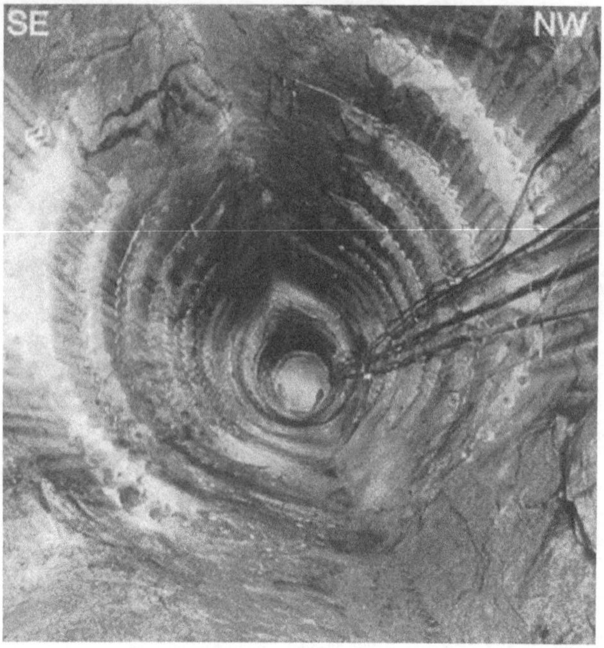

Fig. 7.4. Breakout in a circular tunnel with a radius of R = 1.75 m in the Underground Research Laboratory, Canada. The maximum breakout occurred in the minimum stress direction and the maximum hole radius is 1.3R. The in-situ stresses are σ_3 = 14 MPa and σ_1 = 55MPa (Martino and Chandler 2004 with permission of Elsevier).

Borehole instabilities are the main cause of drilling difficulties, resulting in an expensive loss of time, sometimes in a loss of part of or even whole boreholes. Wellbore instabilities make logging very difficult to perform and to interpret (Maury and Sauzay 1987). A bad condition of the borehole wall alters artificially the annulus zone corresponding to the depth of investigation of most of the logging tools. The shape of the borehole can be strongly modified giving an elongated hole in one direction, diameter reduction in the other direction and also almost circular cavings in places. In the Cusiana field in Colombia, even though some measures to prevent borehole instability were taken, extensive breakouts in fissile and naturally fractured shales – of up to 44" in $12^1/_4$" hole – occurred (Willson et al. 1999). Approximately 10% of the well costs in the Cusiana field were spent coping with bad holes, mainly because of abnormally high tectonic stresses induced by an active thrust-faulting environment (Addis et al. 1993, Last and McLean 1995). In addition to the cost associated with borehole instability while drilling, borehole stability also has a substantial impact on reservoir productions (Bradley 1979).

There are several stages in the life of a well, i.e., drilling, completion, stimulation, flow tests, production, and depletion. Borehole instabilities can be encountered in all these stages (Ramos et al. 1996). In the drilling stage, the main concerns are to determine the mud composition and density (mud weight) which will maintain the integrity of the well, without the loss of drilling fluids. During the completion and stimulation stage, the reservoir must be connected to the well via perforations. This operation could fail if the rock adjacent to the cemented casing is non-brittle. Prior to full production, downhole tests include open-hole logging, fluid sampling, build-up, drawdown, injection, and deliverability tests. It is possible to induce wellbore failure and casing collapse during these processes (Peng et al. 2007). As hydrocarbons are depleted, the drained region compacts which could induce solids production, casing damage, surface subsidence and wellbore failure. All these stages in the life of a well, integrated borehole stability analyses are important to ensure the reservoir economical production and minimize the costly problems induced by the wellbore instabilities.

In order to predict wellbore stability, stress components and distributions near the wellbore due to drilling perturbation need to be analyzed. Applying rock failure criteria wellbore stress and rock strength can, then, be compared to determine rock failures, such as shear failure, tensile failure etc. Finally, a mud weight range can be determined to avoid wellbore shear, compressive, and tensile failures. This mud weight range is the safe mud weight window for drilling.

7.2 Wellbore stability – elastic solution

7.2.1 Vertical borehole in an isotropic stress field

The plane strain solution (i.e. strain in z-direction $\varepsilon_z = 0$) can be applied to a wellbore loaded on its external boundary by an isotropic stress, σ, and on its internal boundary by a drilling mud pressure, p_m (Fig. 7.5). This is known as Lame's problem.

In cylindrical coordinates, the stress components at a distance r from the wellbore are given by:

$$\begin{cases} \sigma_r = \sigma\left(1 - \dfrac{R^2}{r^2}\right) + \dfrac{p_m R^2}{r^2} \\ \sigma_\theta = \sigma\left(1 + \dfrac{R^2}{r^2}\right) - \dfrac{p_m R^2}{r^2} \\ \sigma_z = \sigma_v \\ \tau_{r\theta} = \tau_{rz} = \tau_{\theta z} = 0 \end{cases} \tag{7.1}$$

where R is the wellbore radius; r is the distance from the wellbore center; σ_r, σ_θ and σ_z are the radial, tangential and vertical stress components, respectively; $\tau_{r\theta}$, τ_{rz}, and $\tau_{\theta z}$ are the shear stress components; σ and σ_v are the horizontal and vertical stresses in the far-field stress field, respectively.

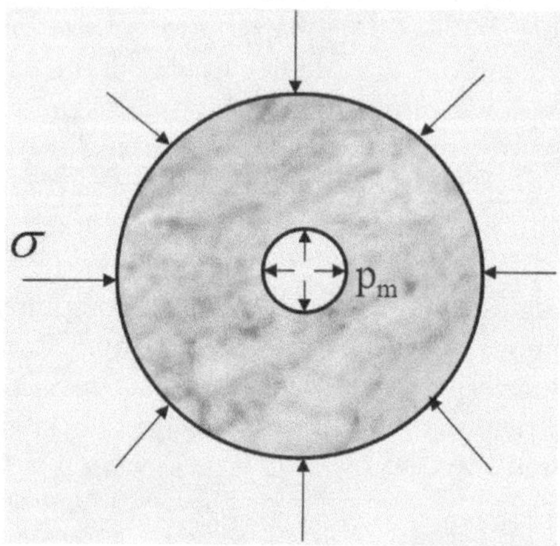

Fig. 7.5. A borehole in an isotropic stress field.

At the wellbore wall, $r = R$, one obtains:

$$\begin{cases} \sigma_r = p_m \\ \sigma_\theta = 2\sigma - p_m \\ \sigma_z = \sigma_v \end{cases} \tag{7.2}$$

It can be seen that stress concentrations around the wellbore occur due to drilling. The radial displacement or well convergence, can be computed using the following equation:

$$u_r = \frac{\sigma - p_m}{2G}\frac{R^2}{r} \tag{7.3}$$

where G is the shear modulus.

7.2.2 Vertical borehole in an anisotropic stress field

Kirsch's problem refers to a borehole subjected to a bi-axial stress state σ_H, σ_h ($\sigma_H > \sigma_h$) (Fig. 7.6). The solution for the various stress components is expressed as follows:

$$\begin{cases} \sigma_r = \frac{(\sigma_H + \sigma_h)}{2}\left(1 - \frac{R^2}{r^2}\right) + \frac{(\sigma_H - \sigma_h)}{2}\left(1 - \frac{4R^2}{r^2} + \frac{3R^4}{r^4}\right)\cos 2\theta + p_m \frac{R^2}{r^2} \\ \sigma_\theta = \frac{(\sigma_H + \sigma_h)}{2}\left(1 + \frac{R^2}{r^2}\right) - \frac{(\sigma_H - \sigma_h)}{2}\left(1 + \frac{3R^4}{r^4}\right)\cos 2\theta - p_m \frac{R^2}{r^2} \\ \sigma_z = \sigma_v - \nu \frac{4R^2}{r^2}\frac{(\sigma_H - \sigma_h)}{2}\cos 2\theta \\ \tau_{r\theta} = -\frac{(\sigma_H - \sigma_h)}{2}\left(1 + \frac{2R^2}{r^2} - \frac{3R^4}{r^4}\right)\sin 2\theta \\ \tau_{z\theta} = \tau_{rz} = 0 \end{cases} \tag{7.4}$$

where σ_H and σ_h are the maximum and minimum far-field horizontal stresses; θ is defined in Fig. 7.6.

In particular, at the borehole wall, the tangential and radial stress components can be expressed as follows:

$$\sigma_\theta = (\sigma_H + \sigma_h) - 2(\sigma_H - \sigma_h)\cos 2\theta - p_m \tag{7.5}$$
$$\sigma_r = p_m$$

The maximum and minimum tangential stress components at the borehole wall can be obtained as follows:

$$\begin{cases} \sigma_\theta = 3\sigma_h - \sigma_H - p_m & \text{at} \quad \theta = 0^\circ \\ \sigma_\theta = 3\sigma_H - \sigma_h - p_m & \text{at} \quad \theta = 90^\circ \end{cases} \tag{7.6}$$

It can be concluded from Fig. 7.6 and Eq. 7.6 that the maximum tangential stress occurs in the minimum far-field horizontal stress direction ($\theta = 90^\circ$), and the minimum tangential stress is in the maximum far-field horizontal stress direction ($\theta = 0^\circ$). That is the reason why the shear failure first occurs at the borehole wall of the minimum horizontal stress direction.

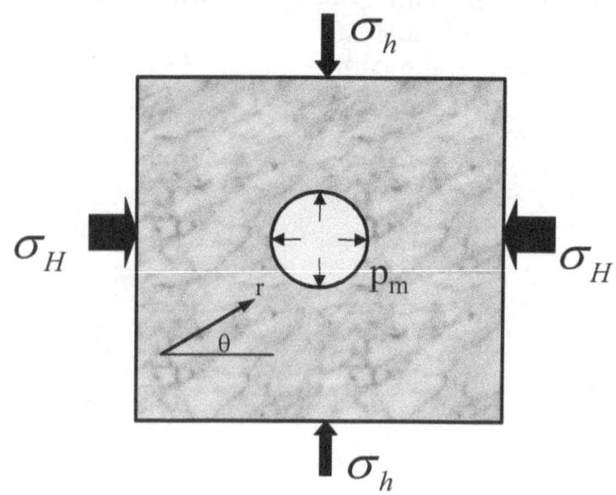

Fig. 7.6. A vertical borehole in an anisotropic stress field.

7.2.3 Inclined borehole in an anisotropic stress field

For an inclined borehole, the stress components can be calculated by referring to Eq. 7.4 by considering far-field stress change due to the hole deviation. This introduces shear stress components of far-field stress to the hole. In a cross section perpendicular to the axial direction of the hole, the stress components in the far-field need to be calculated. The wellbore stresses then can be calculated. Bradley (1979) gave the stress distribution around a borehole located in an arbitrary far-field stress field:

$$\begin{cases} \sigma_r = \dfrac{(\sigma_x+\sigma_y)}{2}\left(1-\dfrac{R^2}{r^2}\right)+\dfrac{(\sigma_x-\sigma_y)}{2}\left(1-\dfrac{4R^2}{r^2}+\dfrac{3R^4}{r^4}\right)\cos 2\theta + \\ \tau_{xy}\left(1-\dfrac{4R^2}{r^2}+\dfrac{3R^4}{r^4}\right)\sin 2\theta + p_m\dfrac{R^2}{r^2} \\ \sigma_\theta = \dfrac{(\sigma_x+\sigma_y)}{2}\left(1+\dfrac{R^2}{r^2}\right)-\dfrac{(\sigma_x-\sigma_y)}{2}\left(1+\dfrac{3R^4}{r^4}\right)\cos 2\theta - \\ \tau_{xy}\left(1+\dfrac{3R^2}{r^4}\right)\sin 2\theta - p_m\dfrac{R^2}{r^2} \\ \sigma_z = \sigma_{zz} - 2\nu(\sigma_x-\sigma_y)\dfrac{R^2}{r^2}\cos 2\theta - 4\nu\tau_{xy}\dfrac{R^2}{r^2}\sin 2\theta \\ \tau_{r\theta} = \left(\dfrac{\sigma_x-\sigma_y}{2}\sin 2\theta + \tau_{xy}\cos 2\theta\right)\left(1+\dfrac{2R^2}{r^2}-\dfrac{3R^4}{r^4}\right) \\ \tau_{rz} = (\tau_{yz}\sin\theta + \tau_{xz}\cos\theta)\left(1-\dfrac{R^2}{r^2}\right) \\ \tau_{\theta z} = (-\tau_{xz}\sin\theta + \tau_{yz}\cos\theta)\left(1+\dfrac{R^2}{r^2}\right) \end{cases} \tag{7.7}$$

where σ and τ with subscript of r and θ are the normal and shear stresses in a cylindrical coordinate system with the z-direction parallel to the drilling direction; while σ and τ with subscript of x, y and z are the normal and shear stresses in a Cartesian coordinate system which has the same z-axis as the cylindrical system, with the z-direction parallel to the drilling direction; θ is the azimuthal angle measured from the x-axis.

The conversion of this Cartesian coordinate system's stresses from in-situ principal stresses can be performed through coordinate transformation (refer to the previous chapter in Sec. 6.3.4).

The stresses at the wellbore wall can be written as:

$$\begin{cases} \sigma_r = p_m \\ \sigma_\theta = \sigma_x + \sigma_y - p_m - 2(\sigma_x - \sigma_y)\cos 2\theta - 4\tau_{xy}\sin 2\theta \\ \sigma_z = \sigma_{zz} - \nu[2(\sigma_x - \sigma_y)\cos 2\theta + 4\tau_{xy}\sin 2\theta] \\ \tau_{\theta z} = 2(-\tau_{xz}\sin\theta + \tau_{yz}\cos\theta) \\ \tau_{r\theta} = 0 \\ \tau_{rz} = 0 \end{cases} \tag{7.8}$$

7.2.4 Vertical borehole considering pore pressure effects

The effective stress components around a vertical borehole with consideration of pore pressure and temperature effects are expressed as:

$$\begin{cases} \sigma_r' = \dfrac{(\sigma_H + \sigma_h - 2p_p)}{2}\left(1 - \dfrac{R^2}{r^2}\right) \\ + \dfrac{(\sigma_H - \sigma_h)}{2}\left(1 - \dfrac{4R^2}{r^2} + \dfrac{3R^4}{r^4}\right)\cos 2\theta + (p_m - p_p)\dfrac{R^2}{r^2} \\ \sigma_\theta' = \dfrac{(\sigma_H + \sigma_h - 2p_p)}{2}\left(1 + \dfrac{R^2}{r^2}\right) \\ - \dfrac{(\sigma_H - \sigma_h)}{2}\left(1 + \dfrac{3R^4}{r^4}\right)\cos 2\theta - (p_m - p_p)\dfrac{R^2}{r^2} - \sigma_T \\ \sigma_z' = \sigma_v - p_p - \nu\dfrac{2R^2}{r^2}(\sigma_H - \sigma_h)\cos 2\theta - \nu\sigma_T \\ \tau_{r\theta} = -\dfrac{(\sigma_H - \sigma_h)}{2}\left(1 + \dfrac{2R^2}{r^2} - \dfrac{3R^4}{r^4}\right)\sin 2\theta \\ \tau_{z\theta} = \tau_{rz} = 0 \end{cases} \tag{7.9}$$

where p_p is the pore pressure; σ_T is the thermal stress induced by the difference (ΔT) between the mud temperature and the formation temperature. The tangential thermal stress can be calculated by the following equation:

$$\sigma_T = \frac{\alpha_T E \Delta T}{1-\nu} \tag{7.10}$$

where α_T is the thermal constant; E is the Young's modulus; and ν is the Poisson's ratio.

The effective stress at the wellbore wall can be written as:

$$\begin{cases} \sigma'_r = p_m - p_p \\ \sigma'_\theta = \sigma_H + \sigma_h - p_p - p_m - \sigma_T - 2(\sigma_H - \sigma_h)\cos 2\theta \\ \sigma'_z = \sigma_v - p_p - \nu[2(\sigma_H - \sigma_h)\cos 2\theta] - \nu\sigma_T \end{cases} \tag{7.11}$$

If the temperature effect is negligible, the effective tangential and radial stress components can be expressed as:

$$\begin{cases} \sigma'_r = p_m - p_p \\ \sigma'_{\theta\max} = 3\sigma_H - \sigma_h - p_p - p_m, \quad \theta = 90° \\ \sigma'_{\theta\min} = 3\sigma_h - \sigma_H - p_p - p_m, \quad \theta = 0° \end{cases} \tag{7.12}$$

where θ is defined in Fig. 7.6, and $\theta = 90°$ is the minimum far-field direction.

Assuming that the effective maximum tangential and radial stresses are the principal maximum and minimum stresses, the Mohr-Coulomb failure criterion can be written as follows, when the hole is stable:

$$\sigma'_{\theta\max} \leq UCS + q\sigma'_r \tag{7.13}$$

Inserting Eq. 7.12 into Eq. 7.13, the minimum mud pressure, p_m, can be calculated:

$$p_m \geq \frac{3\sigma_H - \sigma_h - UCS + (q-1)p_p}{q+1} \tag{7.14}$$

Introducing $UCS = (2c\cos\phi)/(1-\sin\phi)$, $q = (1+\sin\phi)/(1-\sin\phi)$, the minimum mud pressure, p_m, can be rewritten as the following form:

$$p_m \geq \frac{1}{2}(3\sigma_H - \sigma_h)(1-\sin\phi) - c\cos\phi + p_p \sin\phi \tag{7.15}$$

where ϕ is the angle of internal friction; c is the cohesion.

The minimum mud pressure is also called collapse pressure or shear failure pressure, which is the minimum mud pressure to prevent wellbore from shear failures. The minimum mud pressure gradient is named the minimum mud weight, or shear failure gradient (SFG).

It can be seen that the minimum mud weight depends highly on the pore pressure. Therefore, precise prediction of the pore pressure is of vital importance for determining drilling mud weight.

Figure 7.7 plots the minimum mud pressure and different failure areas for a vertical borehole at depth of 2000 m. The minimum effective tangential (hoop) stress and effective radial stress are also plotted in the same figure to determine the rock splintering/spalling (when radial stress is less than tensile strength) and drilling-induced tensile failure (tangential stress exceeds tensile strength).

Figure 7.8 shows a relationship of the mud pressure and wellbore failures. It can be seen from Figs. 7.7 and 7.8 that when the mud pressure is less than the pore pressure, the wellbore has splintering failure or spalling. When the mud pressure is less than the shear failure gradient, the wellbore has shear failure or breakout. If the mud pressure is too high, the drilling induced hydraulic fractures are generated, which may cause drilling mud losses.

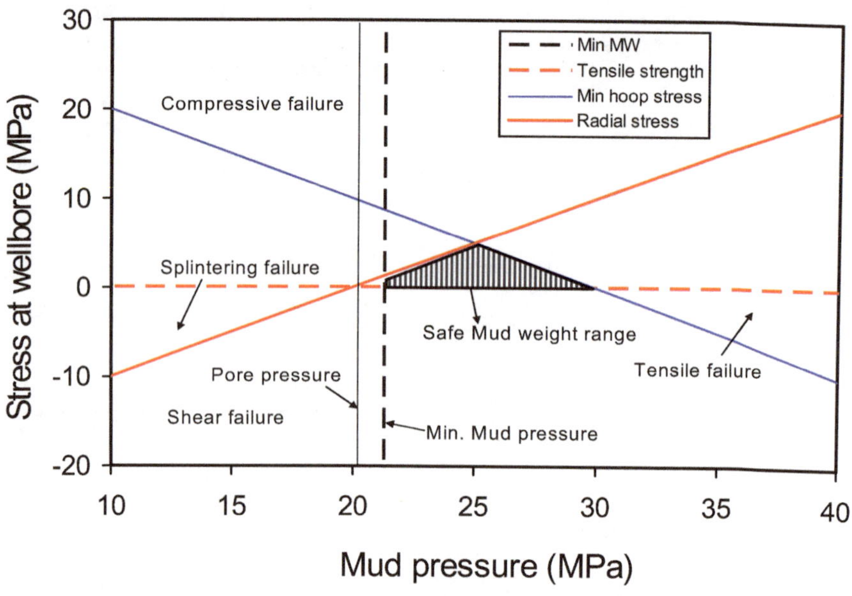

Fig. 7.7. Mud pressure window for a wellbore drilled at depth of 2000 m with far-field stress of $\sigma_v = 46$ MPa, $\sigma_H = 40$ MPa, $\sigma_h = 30$ MPa and pore pressure of $p_p =$ 20 MPa. The rock strength parameters are UCS = 36 MPa, $\phi = 30°$, and tensile strength of zero. The calculated safe mud pressure ranges from 21.3 to 30 MPa.

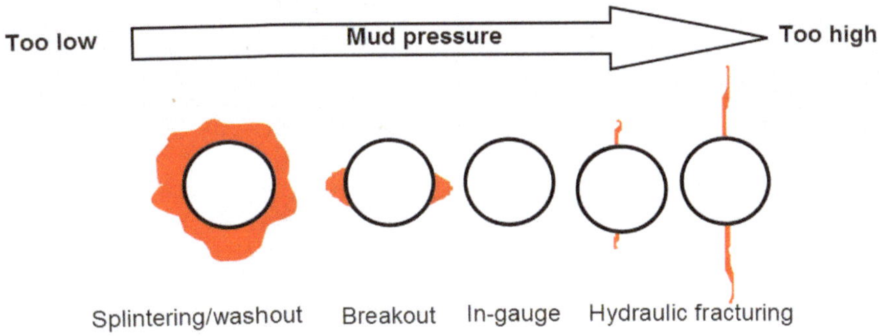

Fig. 7.8. Schematic relationship of mud pressure (mud weight) and wellbore failure behaviors. To keep wellbore stability, the mud weight should be in appropriate range (modified from Madge et al. 2006).

7.3 Wellbore stability – elastoplastic solution

An ideal case is considered, i.e. a plane strain circular infinite domain loaded on its internal boundary by a mud pressure p_m and at infinity by an isotropic far-field stress σ (Fig. 7.9) in an ideal elastoplastic material. In Fig. 7.9 a failed/plastic zone with radius of R_p is formed due to low mud pressure. Assuming that the tangential stress (σ_θ) and radial stress (σ_r) satisfy $\sigma_\theta > \sigma_r$, an ideal plastic rock formation with a Mohr-Coulomb yield criterion in effective stress form in this failed zone can be expressed as follows:

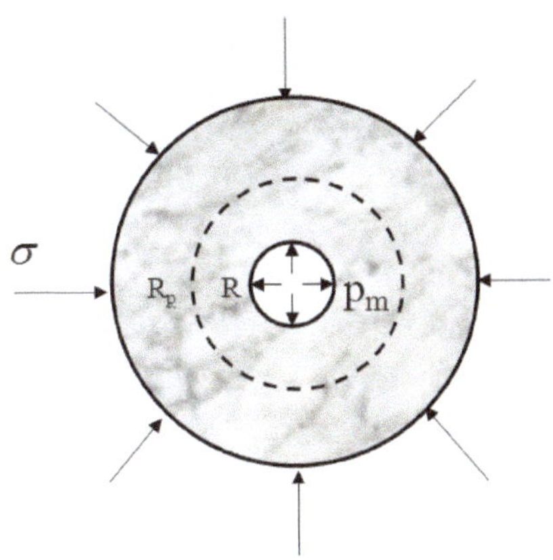

Fig. 7.9. A borehole in an ideal elastoplastic material.

$$\sigma_\theta - p_p = \sigma_c' + q_f(\sigma_r - p_p) \tag{7.16}$$

where σ_c' is the uniaxial compressive strength in the failed rock; p_p is the pore pressure; $q_f = (1+\sin\phi_f)/(1-\sin\phi_f)$ and ϕ_f is the angle of internal friction of the failed rock.

The equation of equilibrium is given by:

$$\frac{d\sigma_r}{dr} + \frac{\sigma_r - \sigma_\theta}{r} = 0 \tag{7.17}$$

Substituting Eq. 7.16 into Eq. 7.17 and solving it, one obtains:

$$\sigma_r = \frac{\sigma_c' + (1-q_f)p_p}{1-q_f} + C_1 r^{q_f - 1} \tag{7.18}$$

$$\sigma_\theta = \frac{\sigma_c' + (1-q_f)p_p}{1-q_f} + C_1 q_f r^{q_f - 1} \tag{7.19}$$

Using the boundary condition (for $r = R$, $\sigma_r = p_m$), constant C_1 can be expressed as:

$$C_1 = p_m - \frac{\sigma_c' + (1-q_f)p_p}{1-q_f} R^{1-q_f} \tag{7.20}$$

Therefore, in the failed zone the stresses can be obtained (Roegiers 2002):

$$\begin{cases} \sigma_r = \dfrac{\sigma_c' + (1-q_f)p_p}{1-q_f} + \left(p_m - \dfrac{\sigma_c' + (1-q_f)p_p}{1-q_f} \right) \left(\dfrac{r}{R} \right)^{q_f - 1} \\ \sigma_\theta = \dfrac{\sigma_c' + (1-q_f)p_p}{1-q_f} + q_f \left(p_m - \dfrac{\sigma_c' + (1-q_f)p_p}{1-q_f} \right) \left(\dfrac{r}{R} \right)^{q_f - 1} \end{cases} \tag{7.21}$$

In the elastic zone, the stress can be obtained from elasticity theory, i.e.:

$$\begin{cases} \sigma_{re} = \sigma + \dfrac{C_2}{r^2} \\ \sigma_{\theta e} = \sigma - \dfrac{C_2}{r^2} \end{cases} \tag{7.22}$$

When $r = R_p$ (R_p is the radius of the failed zone), the radial stress can be obtained both from Eqs. 7.21 and 7.22. Using the continuity of the radial stress for $r = R_p$, the constant C_2 can be obtained.

$$\sigma_r\big|_{r=R_p} = \frac{\sigma_c' + (1-q_f)p_p}{1-q_f} + \left(p_m - \frac{\sigma_c' + (1-q_f)p_p}{1-q_f}\right)\left(\frac{r}{R}\right)^{q_f-1} = \sigma + \frac{C_2}{r^2} \tag{7.23}$$

and,

$$C_2 = R_p^2\left[\frac{\sigma_c' + (1-q_f)p_p}{1-q_f} - \sigma + \left(p_m - \frac{\sigma_c' + (1-q_f)p_p}{1-q_f}\right)\left(\frac{R_p}{R}\right)^{q_f-1}\right] \tag{7.24}$$

Therefore, the stresses in the elastic zone are as follows:

$$\begin{cases} \sigma_{re} = \sigma + \left[\dfrac{\sigma_c' + (1-q_f)p_p}{1-q_f} - \sigma + \left(p_m - \dfrac{\sigma_c' + (1-q_f)p_p}{1-q_f}\right)\left(\dfrac{R_p}{R}\right)^{q_f-1}\right]\left(\dfrac{R_p}{r}\right)^2 \\ \sigma_{\theta e} = \sigma - \left[\dfrac{\sigma_c' + (1-q_f)p_p}{1-q_f} - \sigma + \left(p_m - \dfrac{\sigma_c' + (1-q_f)p_p}{1-q_f}\right)\left(\dfrac{R_p}{R}\right)^{q_f-1}\right]\left(\dfrac{R_p}{r}\right)^2 \end{cases} \tag{7.25}$$

In the elastic region, when $r = R_p$, the stresses also satisfy Mohr-Coulomb's failure criterion expressed in effective stress form; i.e.:

$$\sigma_{\theta e} - p_p = \sigma_c + q(\sigma_{re} - p_p) \tag{7.26}$$

where σ_c is the uniaxial compressive strength in the elastic rock zone, $q = (1+\sin\phi)/(1-\sin\phi)$ and ϕ is the angle of internal friction of the rock.

Substituting Eq. 7.25 into Eq. 7.26, the radius of failed zone can be obtained:

$$R_p = R\left\{\frac{(2\sigma-\sigma_c)(1-q_f) - [\sigma_c' + (1-q_f)p_p](q+1) + (q-1)(1-q_f)p_p)}{\{p_m(1-q_f) - [\sigma_c' + (1-q_f)p_p]\}(q+1)}\right\}^{\frac{1}{q_f-1}} \tag{7.27}$$

Equation 7.27 shows that the entire rock block will remain purely elastic, if $R_p \le R$. Therefore, the mud pressure (p_m) to keep wellbore from plastic deformation can be solved from Eq. 7.27:

$$p_m \ge \frac{2\sigma - \sigma_c + (q-1)p_p}{(q+1)} \tag{7.28}$$

7.4 Wellbore stability – fracture gradient determinations

Fracture gradient is one of important parameters for wellbore stability modeling, mudweight design, and real-time application. Fracture gradient is the upper bound of the mudweight, while shear failure gradient is the lower bound of the mudweight. If the mudweight is lower than the rock fracture gradient, then the wellbore will have tensile failure, which may cause losses of drilling mud, or even lost circulation. There are several approaches to calculate fracture gradient. The following two methods are commonly used in the drilling industry; i.e. the minimum stress method and tensile failure method.

7.4.1 The minimum stress method for fracture gradient

The minimum stress method is based on that most formations are naturally fractured media. Therefore, in naturally fractured formations with closed fractures the drilling mud loss occurs, once the mudweight is higher than the minimum stress. The reason is that at this case the minimum stress is the closure pressure to make the fracture closed.

Hubbert and Willis (1957) presented the following equation to estimate fracture gradient:

$$FG = \frac{\nu}{1-\nu}(OBG - PP) + PP \tag{7.29}$$

where FG is fracture gradient; OBG is the overburden stress gradient; PP is the pore pressure gradient; ν is the Poisson's ratio.

Daines (1982) developed this method by introducing a superimposed tectonic stress correction:

$$FG = \frac{\nu}{1-\nu}(OBG - PP) + PP + S_t \tag{7.30}$$

where S_t is the superimposed tectonic stress gradient. This value may be evaluated from the first leak off test on drilling. It is regarded as being constant for the rest of the well.

Matthews and Kelly (1967) introduced a variable of effective stress coefficient into fracture gradient prediction:

$$FG = K_0(OBG - PP) + PP \tag{7.31}$$

where K_0 is the effective stress coefficient, $K_0 = \sigma_h'/\sigma_v'$; σ_h' is the minimum effective stress; σ_v' is the maximum effective stress.

In this method the values of K_0 were established on the basis of fracture threshold values derived empirically in the field. The effective stress coefficient is variable and dependent on depth (Mouchet and Mitchell, 1989). The application of this method in a new area requires local variations in K_0 to be determined in relation to depth. The K_0 can be obtained from leak-off tests (refer to Chap. 3 at Sect. 3.3).

Compared to Eqs. 7.29 and 7.31, one obtains that Matthews and Kelly method and Hubbert and Willis method are very similar. They are identical, when the Poisson's ratio and the K_0 have the following relationship:

$$K_0 = \frac{\nu}{1-\nu} \tag{7.32}$$

It should be noted that Eqs. 7.29 - 7.32 are based on that the overburden stress is the maximum principal stress, which is the normal faulting stress regime. In other stress regimes these equations need to be modified.

Figure 7.10 presents a case to predict fracture gradient and shear failure gradient (Madge et al. 2006) using Drillworks software, in which both lithology and formation depth effects are considered for fracture gradient prediction. It can be seen that the limestone has very low fracture gradient because of its low Poisson's ratio.

7.4.2 The tensile failure for fracture gradient prediction

For intact rocks, only when the wellbore appears tensile failures, drilling mud loss can occur. In this case, fracture gradient can be calculated from Eq. 7.12, when the minimum tangential stress is equal to the tensile strength by noticing that the tensile stress is negative. For a vertical well fracture pressure can be expressed as follows:

$$F_p = 3\sigma_h - \sigma_H - p_p - \sigma_T + T_0 \tag{7.33}$$

where F_p is the fracture pressure; p_p is the pore pressure; σ_T is the thermal stress induced by the difference between the mud temperature and the formation temperature, and T_0 is the tensile strength of the rock. In some cases the temperature effect can be neglected ($\sigma_T = 0$).

This fracture pressure is also called the fracture breakdown pressure. For the naturally fractured formation (with pre-existing fractures) this method will overestimate the fracture gradient.

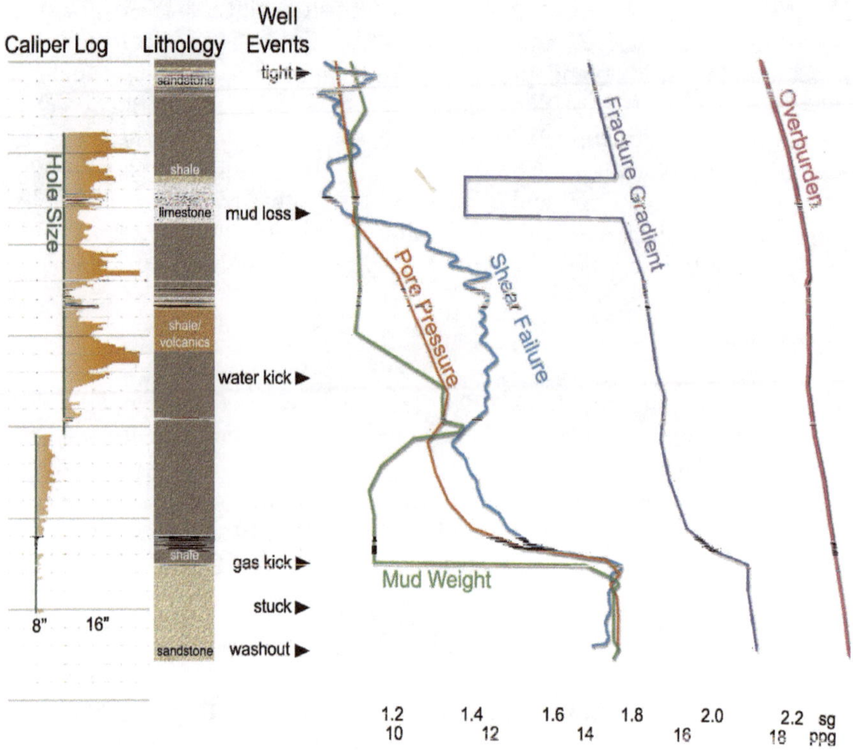

Fig. 7.10. Predicted fracture gradient (FG), pore pressure gradient (PP) and shear failure gradient (SFG). The drilling events, such as drilling mud losses, tight hole, stuck etc. coincide to the predicted fracture gradient and shear failure gradient. From left to right, the curves are mudweight (MW), PP, SFG, FG and OBG, respectively (from Madge et al. 2006). To keep wellbore stability, the mud weight should be in between SFG and FG in each hole sections.

7.5 Wellbore stability – the finite element poroelastic solution

Naturally fractured formations are often treated ideally as double porosity media, which can be solved by using the finite element method (Zhang and Roegiers 2005), as described in Chap. 6. The first step in the solution of the coupled fluid flow and solid deformation problem is to discretize the problem domain. This can be done by replacing the problem domain with a collection of nodes and elements, referred to as the finite element mesh. The second step is to derive an integral formulation for the governing equations. This integral formulation leads to a system of algebraic equations that can be solved for values of the field variable at each node in the mesh (Zhang et al. 2003). This section introduces boundary conditions, boundary forces, and several failure criteria for the finite element modeling of wellbore stability. A number of numerical examples are also given.

7.5.1 Boundary stresses and forces determinations

In the finite element analysis, the common approach is to subtract the constant far-field stresses and original pore pressures from the field quantities. The boundary conditions are thus modified; i.e., on the outer boundary surfaces of the finite domain, all the tractions and pore pressures vanish. At the borehole wall equivalent stresses and pressures are given. After solving this problem, the final solution can be obtained by adding back the constant in-situ stresses and original pressures (Zhang and Roegiers 2002).

The equivalent stresses and pressures at any node at the borehole wall can be calculated by using the following equations (Zhang et al. 2003):

$$\begin{cases} \sigma_x = lS_x + mS_{yx} + nS_{zx} \\ \sigma_y = mS_y + nS_{zy} + lS_{xy} \\ \tau_{xy} = nS_z + lS_{xz} + mS_{yz} \\ p_{ma} = -p_{ma}^0 \\ p_{fr} = -p_{fr}^0 \end{cases} \tag{7.34}$$

where l, m, and n are the direction cosines between the normal to the inclined plane and the x-, y-, and z-axes, respectively.

The boundary nodal forces along the wellbore can be determined by using the following equations:

$$\begin{cases} F_x = \sigma_x A \\ F_y = \sigma_y A \\ F_z = \tau_{xy} A \end{cases} \tag{7.35}$$

where A is the elemental area shown in Fig. 7.11.

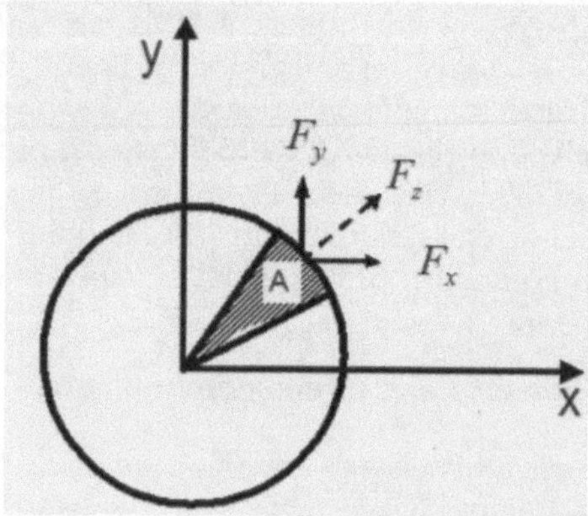

Fig. 7.11. Boundary nodal forces at the wellbore wall.

7.5.2 Stress transformation

When analyzing stress and pore pressure distributions in and around wellbores, the polar coordinate system is generally adopted. For the generalized plane strain formulation, the stresses in polar coordinates are related to the Cartesian coordinate stresses according to the following rules:

$$\begin{cases} \sigma_r = \sigma_x \cos^2\theta + \sigma_y \sin^2\theta + 2\tau_{xy} \sin\theta\cos\theta \\ \sigma_\theta = \sigma_x \sin^2\theta + \sigma_y \cos^2\theta - 2\tau_{xy} \sin\theta\cos\theta \\ \sigma_z = \sigma_z \\ \tau_{r\theta} = (\sigma_y - \sigma_x)\sin\theta\cos\theta + \tau_{xy}(\cos^2\theta - \sin^2\theta) \\ \tau_{rz} = \tau_{xz}\cos\theta + \tau_{yz}\sin\theta \\ \tau_{\theta z} = \tau_{yz}\cos\theta - \tau_{xz}\sin\theta \end{cases} \tag{7.36}$$

where r and θ are the radius and the angle with reference to the center of the wellbore in the polar coordinate system, respectively.

The strains relate to the Cartesian coordinate strains as:

$$\begin{cases} \varepsilon_r = \varepsilon_x \cos^2\theta + \varepsilon_y \sin^2\theta + 2\gamma_{xy} \sin\theta\cos\theta \\ \varepsilon_\theta = \varepsilon_x \sin^2\theta + \varepsilon_y \cos^2\theta - 2\gamma_{xy} \sin\theta\cos\theta \\ \varepsilon_z = \varepsilon_z \\ \gamma_{r\theta} = (\varepsilon_y - \varepsilon_x)\sin\theta\cos\theta + \gamma_{xy}(\cos^2\theta - \sin^2\theta) \\ \gamma_{rz} = \gamma_{xz}\cos\theta + \gamma_{yz}\sin\theta \\ \gamma_{\theta z} = \gamma_{yz}\cos\theta - \gamma_{xz}\sin\theta \end{cases} \tag{7.37}$$

7.5.3 Mud weight Considerations at the Wellbore

Permeable boundary

A suitable mud weight of the drilling fluid during drilling plays a very important role in helping to protect the borehole wall from breakouts and other failures. The mud weight pressure, p_m, acts both as a radial stress and as fluid pressure at the borehole wall when the wellbore formation is permeable. Considering mud pressure effects, the stresses at the borehole wall can be expressed as (Zhang 2002):

$$\begin{cases} \sigma_x = l(S_x - S_{mx}) + m(S_{yx} - S_{myx}) + nS_{zx} \\ \sigma_y = m(S_y - S_{my}) + nS_{zy} + l(S_{xy} - S_{mxy}) \\ \tau_{xy} = nS_z + lS_{xz} + mS_{yz} \\ p_{ma} = -p^0_{ma} + p_m \\ p_{fr} = -p^0_{fr} + p_m \end{cases} \tag{7.38}$$

where S_{mx}, S_{my}, and S_{myx} are stresses induced by the mud pressure (p_m) as the radial stresses, which can be obtained by:

$$\begin{cases} S_{mx} = l^2 p_m \\ S_{my} = m^2 p_m \\ S_{mxy} = lmp_m \end{cases} \tag{7.39}$$

Impermeable boundary

Mud cake can be formed during drilling which may make the borehole wall impermeable. In this case, the flow rate at the borehole wall is zero and the mud weight pressure, p_w, acts only as a radial stress at the borehole wall under such conditions (Zhang and Standifird 2007). Therefore, the stresses at the borehole wall can be written as:

$$\begin{cases} \sigma_x = l(S_x - S_{mx}) + m(S_{yx} - S_{myx}) + nS_{zx} \\ \sigma_y = m(S_y - S_{my}) + nS_{zy} + l(S_{xy} - S_{mxy}) \\ \tau_{xy} = nS_z + lS_{xz} + mS_{yz} \\ p_{ma} = -p^0_{ma} \\ p_{fr} = -p^0_{fr} \end{cases} \tag{7.40}$$

7.5.4 Failure Criteria

Borehole walls may fail when the surrounding stress exceeds the tensile, the compressive, or the shear strengths of the rock formation, whichever is reached first. The effective stress concentration near the wellbore due to the far-field stresses and the specific borehole orientation need to be considered for failure initiation. It is commonly accepted that wellbore failure is controlled by Terzaghi's effective stress; i.e.

$$\sigma'_{ij} = \sigma_{ij} - \delta_{ij} p_{ma} \tag{7.41}$$

where σ'_{ij} is the effective stress tensor, p_{ma} is the matrix pore pressure and δ is the Kronecker delta.

Various failure criteria, under tensile, Mohr-Coulomb and Drucker-Prager, are considered and incorporated into the dual-porosity finite element model, as individually described below.

Tensile Failure

There are two possible modes of tensile failure. One is due to radial fracturing at the borehole wall as the mud pressure in the wellbore is increased. The other is the outburst failure (or spalling) due to an effective radial tensile stress caused by a very rapid depressurization of the borehole.

Spalling failure

As discussed above, spalling (or splintering) will occur when the effective radial tensile stress equals or is greater than the formation tensile strength; that is:

$$\sigma_r' = \sigma_r - p_{ma} = T_0 \tag{7.42}$$

where σ_r is the total radial stress; σ'_r is the effective radial tensile stress; and T_0 is the tensile strength of the formation.

Usually, the formation tensile strength is very small, especially for fractured formations. Therefore, it is usually considered to be zero. It should be pointed out that, due to poroelastic effects, the tensile fracture does not necessarily initiate at the borehole wall. In addition, tensile failure conditions vary with time.

Fracturing failure

If mud pressure is too high, radial tensile failure will occur when the minimum principal effective stress or effective circumferential stress is equal to the formation tensile strength, as shown in Eq. 7.43. In this condition the drilling induced tensile fractures will occur due to high mud weight, which may induce hydraulic fracturing in the wellbore and cause drilling mud losses.

$$\sigma_3' = T_0 \tag{7.43}$$

where σ'_3 is the minimum effective principal stresses (note that σ'_3 is negative).

Shear Failure

Shear failure occurs when the shear strength of the formation is exceeded. The Mohr-Coulomb failure criterion can be used to determine such a failure mechanism; i.e.:

$$\tau = c + \sigma' \tan\phi \tag{7.44}$$

where ϕ is the angle of internal friction and c is the cohesion.

In the principal space, $(\sigma'_1, \sigma'_2, \sigma'_3)$, the Mohr-Coulomb failure criterion is given by:

$$\sigma_1' = \sigma_c + q\sigma_3' \tag{7.45}$$

where σ'_1, σ'_3 are the maximum and minimum effective principal stresses, respectively; and $q = (1+\sin\phi)/(1-\sin\phi)$.

Collapse Failure

The effect of the intermediate principal stress is not considered in the failure criteria mentioned above. Laboratory data have shown that the intermediate principal stress plays an important role in the failure of rocks and can be conveniently represented by the Drucker-Prager yield condition; that is:

$$\sqrt{J_2} = \alpha I_1' + \kappa \tag{7.46}$$

where α and κ are material constants; and, I'_1 and J_2 are the stress invariants, which can be expressed as:

$$I_1' = \sigma_x' + \sigma_y' + \sigma_z' \tag{7.47}$$

$$J_2 = \frac{1}{6}[(\sigma_x - \sigma_y)^2 + (\sigma_y - \sigma_z)^2 + (\sigma_z - \sigma_x)^2] + \tau_{xy}^2 + \tau_{yz}^2 + \tau_{zx}^2 \tag{7.48}$$

7.6 Finite element application - Case studies

7.6.1 Wellbore in a strike slip faulting stress regime

For this case study the in-situ stresses and initial pore pressures are obtained from Woodland (1990): $S_{x'} = \sigma_H = 29$ MPa, $S_{y'} = \sigma_h = 20$ MPa, $S_{z'} = \sigma_v = 25$ MPa, and $p_0 = 10$ MPa. Two wellbore inclinations are considered: $\varphi_z = 0$ and 90°. The wellbore radius is $R = 0.1$ m. Biot's effective stress coefficients of rock matrix and fractures are, $\alpha_{ma} = 0.771$ and $\alpha_{fr} = 0.91$. The corresponding parameters (Table 7.1) for the dual-porosity poroelastic model are similar as the case described in Chap. 6, Sect. 6.4.

Table 7.1. Input parameters for inclined borehole analysis

Parameter	Unit	Magnitude
Elastic modulus (E)	GN/m^2	20.6
Poisson's ratio (ν)	-	0.19
Fracture stiffness (K_n , K_{sh})	$MN/m^2/m$	4.82×10^5
Fluid bulk modulus (K_f)	MN/m^2	419.17
Grain bulk modulus (K_s)	GN/m^2	48.21
Matrix porosity (n_{ma})	-	0.02
Fracture porosity (n_{fr})	-	0.002
Matrix mobility (k_{ma}/μ)	$m^4/MN\ s$	10^{-10}
Fracture mobility (k_{fr}/μ)	$m^4/MN\ s$	10^{-9}
Fracture spacing (s)	m	0.1
Compressive strength (σ_c)	MN/m^2	41
Internal friction angle (φ)	degree	30
Material strength parameter (α)	-	0.14
Material strength parameter (κ)	MN/m^2	12
Tensile strength	MN/m^2	1.5

Wellbore failures in elastic, single- and dual-porosity media

The comparisons between elastic, single porosity, and dual-porosity solutions are given for the inclined borehole without mud weight support. The total tangential and radial stress distributions at $t = 100$ s and $\theta = 90°$ for the elastic and single-porosity models as well as the one for the dual-porosity model are compared in Fig. 7.12. Note that the elastic solution is obtained from Eq. 7.7, and it has no pore pressure variations and time effects. It can be seen that at and near wellbore wall the tangential stress in dual-porosity solution has the least magnitude compared to the elastic and single porosity solutions. For the radial stress, the dual-porosity solution has the maximum value. The reason for this phenomenon is that the total deformation increases due to the introduction of a fracture elastic modulus in the dual-porosity governing equations, which leads to an increase in the total radial stress. Therefore, at and near wellbore wall (Fig. 7.12) the stress difference between the tangential and radial stresses is the minimum for the dual porosity solution. The different results for the three solutions are primarily caused by the pore pressure changes induced by the borehole perturbation.

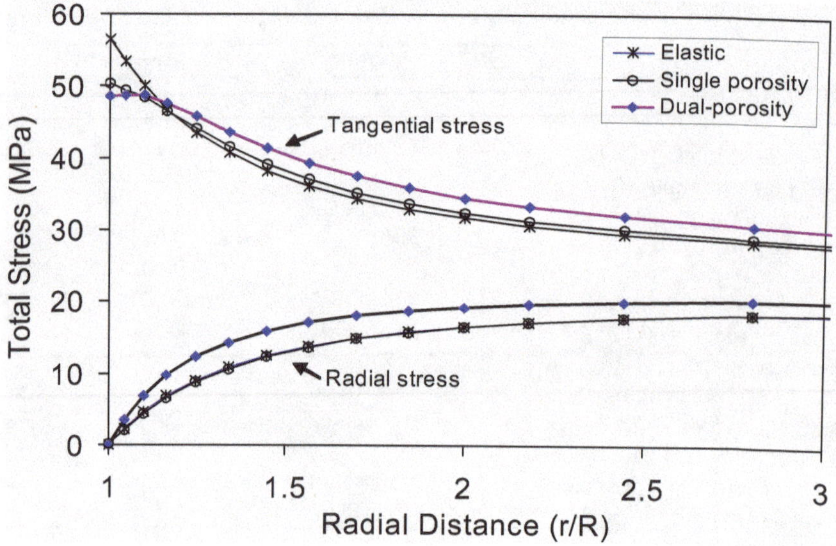

Fig. 7.12. Total tangential and radial stresses around an inclined wellbore for elastic, single porosity, and dual-porosity solutions ($\theta = 90°$). Refer to Fig. 7.6 for the definition of θ.

Figure 7.13 represents the spalling/splintering area around the borehole for a zero mud weight with an inclined wellbore of 70° and for $t = 100$ s. It can be observed that the spalling failure does not start at the borehole wall, but a short distance away inside the formation where the borehole is subjected to larger radial tensile stresses, mainly around the minimum stress direction. Figure 7.14 gives the counterpart of Fig. 7.13 for the single-porosity solution with an assumption that the two media have the same strength parameters. Comparing Fig. 7.13 to Fig. 7.14, it can be observed that the single porosity solution has a much larger spalling area. This is due to the fact that the single porosity solution has much larger radial tensile stresses (Zhang et al. 2003).

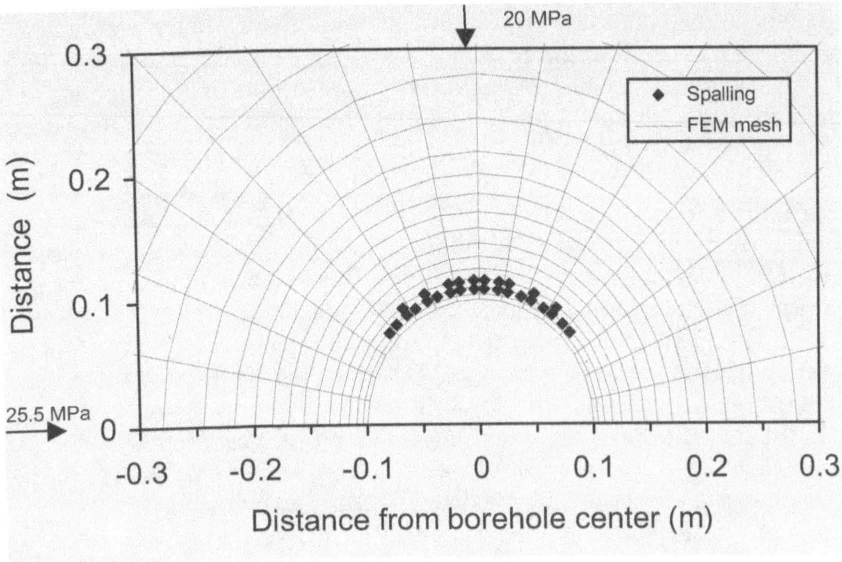

Fig. 7.13. Spalling/splintering area in the dual-porosity medium at $t = 100$ s for a borehole inclination of 70°.

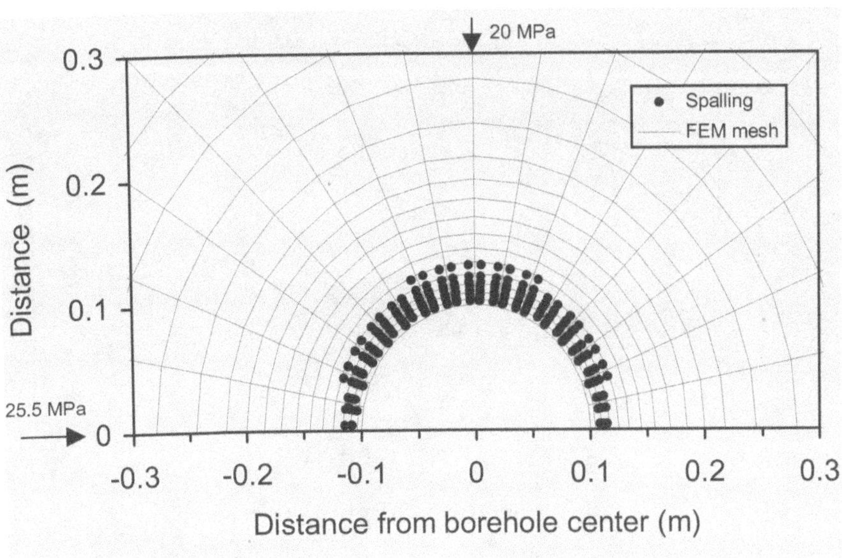

Fig. 7.14. Spalling/splintering area in single-porosity medium at $t = 100$ s for a borehole inclination of 70°.

Figure 7.15 represents the Mohr-Coulomb (shear) failure around the borehole for an open wellbore inclined at an angle of 70° and at t = 100 s in a dual-porosity medium. The shear failure occurs mainly around the minimum stress direction; while no failure occurs in the maximum stress direction (borehole sidewalls). This is similar to what has been observed in the field, as shown in Fig. 7.4 (Martin et al. 1994). Figure 7.16 gives the shear failure in a single porosity medium. It is clear that the single-porosity solution has a much larger shear failure area, since the single porosity solution has a much larger radial tensile stress. This causes a much smaller minor principal stress (note that tensile stress is negative).

With sufficient mud pressure support, wellbore spalling and compressive failure can usually be controlled. However, tensile failure or hydraulic fracturing may be induced in the wellbore when the mud pressure is too high. Figure 7.17 shows the fracturing area for a hole inclination of 70° at t = 100 s and for a mud pressure p_m = 25 MPa. It is seen that the fracturing takes place mainly around the maximum stress direction.

Figure 7.18 presents the fracturing area for the single-porosity solution. One observes that the single-porosity solution has a smaller fracturing area, since the tensile tangential stress that induces wellbore fracturing has a smaller magnitude in the single porosity solution.

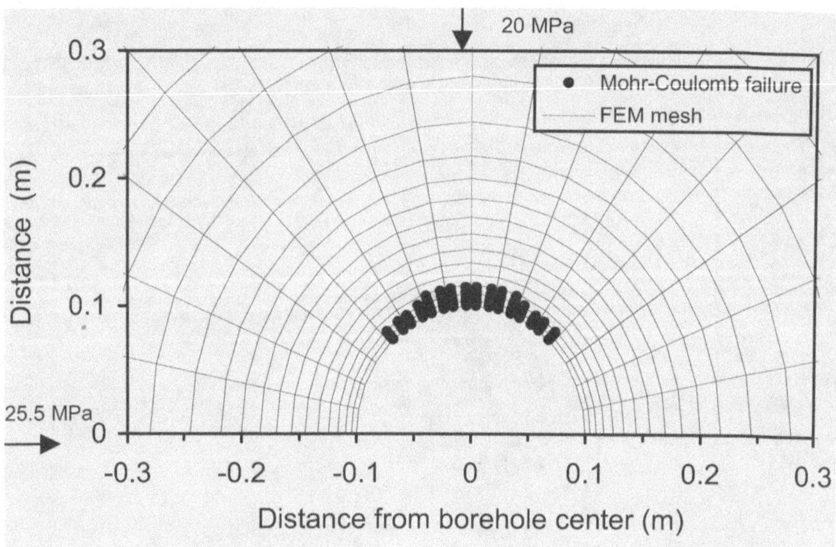

Fig. 7.15. Shear failure area in the dual-porosity medium at t = 100 s for a borehole inclination of 70°.

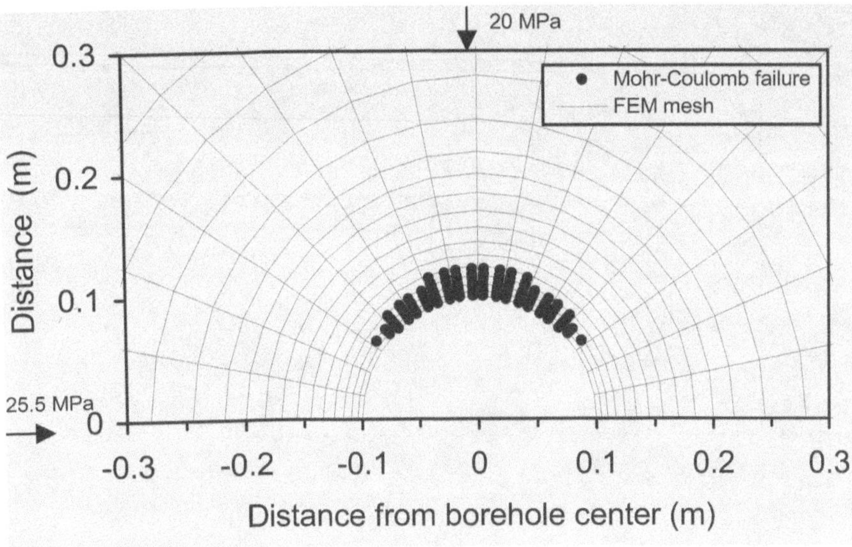

Fig. 7.16. Shear failure area in the single-porosity medium at $t = 100$ s for a borehole inclination of 70°.

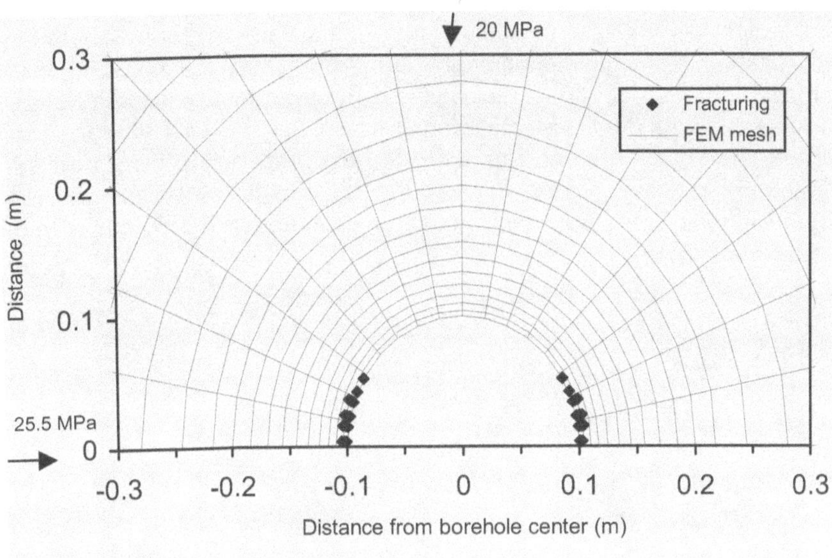

Fig. 7.17. Fracturing area in the dual-porosity medium at $t = 100$ s, $p_m = 25$ MPa for a borehole inclination of 70°.

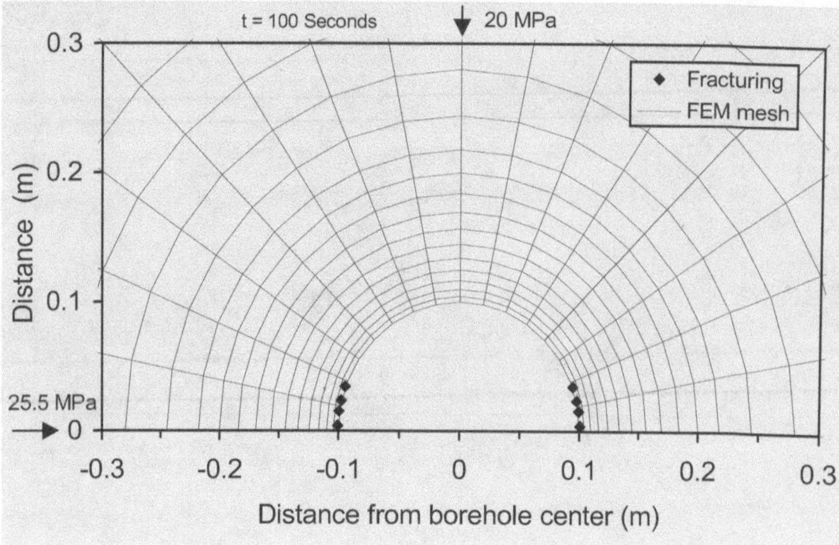

Fig. 7.18. Fracturing area in the single-porosity medium at $t = 100$ s, $p_m = 25$ MPa for a borehole inclination of 70°.

Figure 7.19 represents the envelopes that define the safe range of the mud weight for collapse (Druker-Prager failure) and hydraulic fracturing as a function of the inclination angles at t =100 s for both single and dual-porosity solutions. It shows that the safe range of mud weight increases considerably with the hole inclination. Hence, in a strike slip faulting stress regime, the higher the borehole deviation the more stable the boreholes. Comparing to the single porosity solution, the dual-porosity solution has a larger stable area, a quite smaller collapse area, and a slightly larger fracturing area.

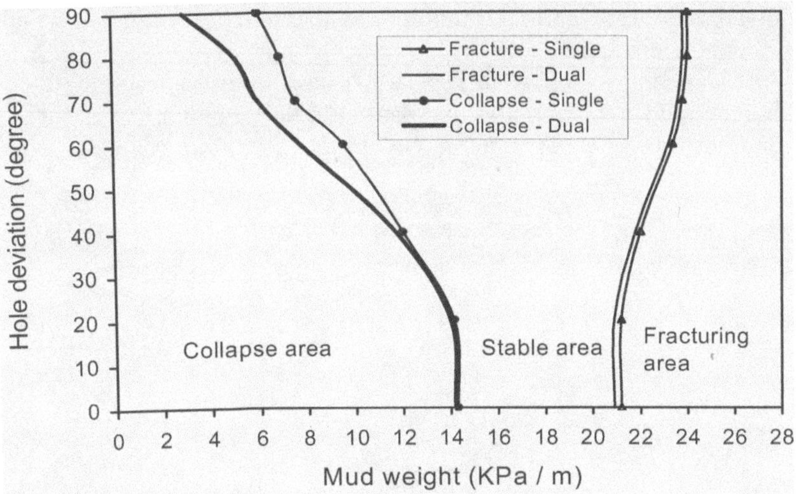

Fig. 7.19. Mud weight range varying with the hole inclinations for collapse and fracturing at $t = 100$ s in a strike slip faulting stress regime for both single-porosity and dual-porosity solutions. In the figure, the vertical hole is 0° and the horizontal hole is 90°.

Horizontal boreholes in dual-porosity media

Many oil companies provide horizontal drilling strategy guidelines that call for drilling in the direction of the minimum horizontal stress. Their approach for best trajectory selections and drilling guidelines is to minimize the maximum value of stress concentration along the borehole wall. However, such an approach may not be universally advisable. The following analyses examine wellbore failures in dual-porosity media for the following two options: parallel and perpendicular to the maximum horizontal stress directions.

The Mohr-Coulomb (shear) failure areas are given in Figs. 7.20 and 7.21 for drilling parallel to the maximum and to the minimum horizontal stresses without any internal support. Comparing to these two figures, it is obvious that there is a much larger failure area for the hole drilled parallel to the minimum stress direction. Comparing to the local stress configurations around the wellbores, it is observed that the stress difference for the borehole in the maximum stress direction is $\Delta\sigma = 5$ MPa (Fig. 7.20), while it is only $\Delta\sigma = 4$ MPa in the minimum stress direction (Fig. 7.21) but, the latter has a much larger failure area. Therefore, the bigger stress difference does not induce larger borehole failure which is not consistent with the conventional understanding.

Therefore, it can be concluded that in a strike slip stress regime a horizontal borehole drilled parallel to the maximum horizontal stress direction has a smaller shear failure area, as shown in Fig. 7.22 (Zhang et al. 2006).

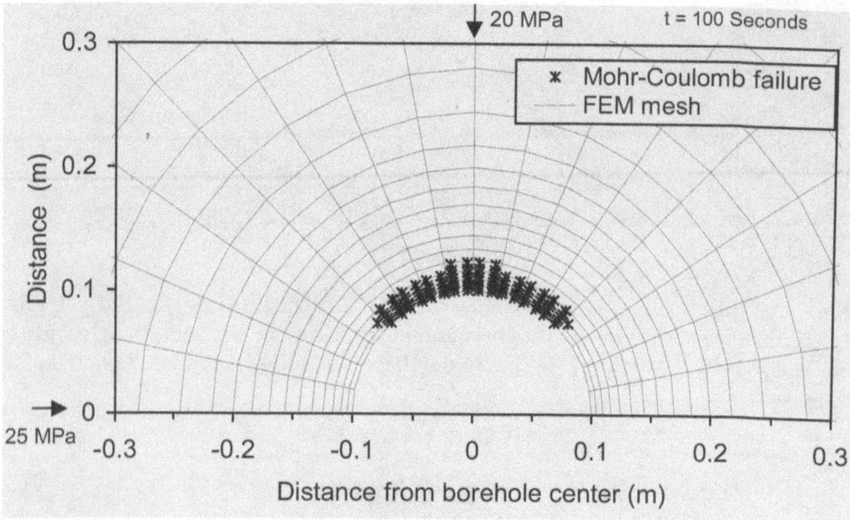

Fig. 7.20. Shear failure area in the strike slip faulting stress regime at $t = 100$ s for a horizontal borehole drilled parallel to the maximum stress direction.

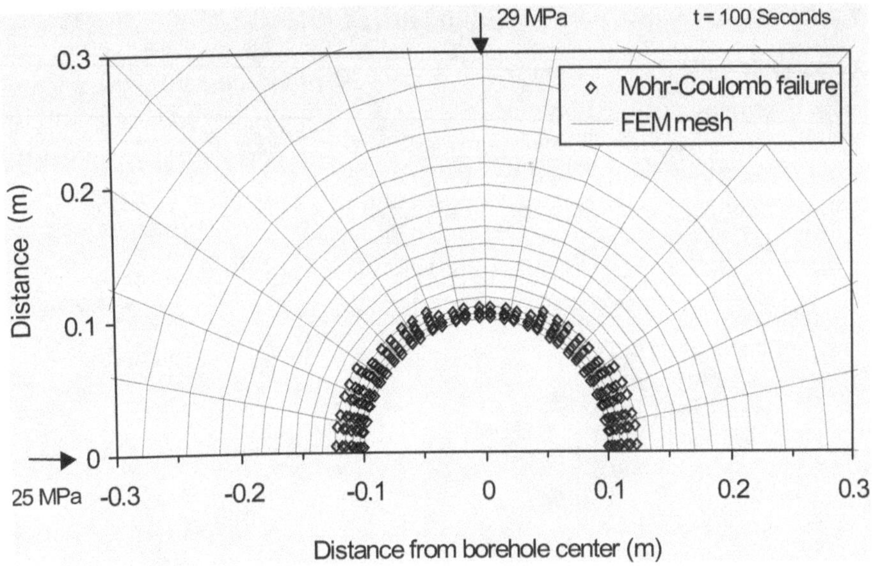

Fig. 7.21. Shear failure area in the strike slip faulting stress regime at t = 100 s for a horizontal borehole drilled parallel to the minimum stress direction.

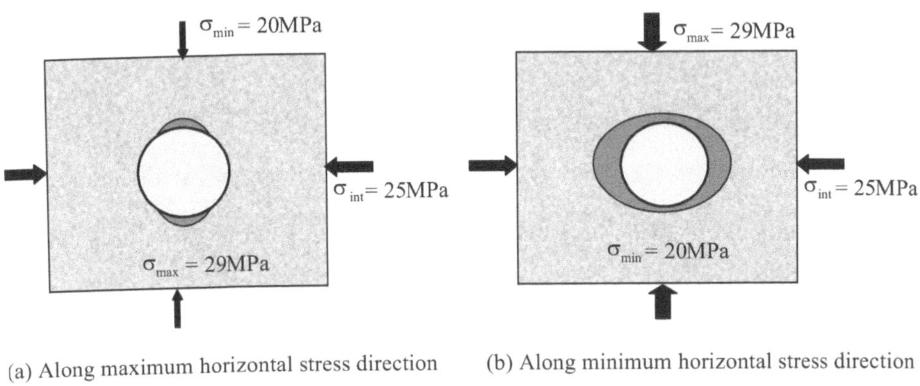

Fig. 7.22. Schematic collapsed zone in a strike slip faulting stress regime for horizontal wells drilled in different horizontal stress directions.

7.6.2 Wellbore in a normal faulting stress regime

The following specific geometry and material properties are used for this case. The radius of the borehole is again $R = 0.1$ m. The rock is a Gulf of Mexico shale with the following properties: $G = 7.6 \times 10^2$ MPa, $\nu_u = 0.461$, $k = 1\times 10^{-7}$ Darcy with $\mu = 0.001$ Pa·s (Cui et al. 1997). Table 7.2 lists the poromechanical parameters used in the analysis.

Table 7.2. Parameters for analyses of the borehole in normal stress regime

Parameter	Unit	Magnitude
Elastic modulus (E)	GN/m^2	18.53
Poisson's ratio (ν)	-	0.22
Fracture stiffness (K_n , K_{sh})	MN/m^2/m	4.82×10^5
Fluid bulk modulus (K_f)	MN/m^2	173.45
Grain bulk modulus (K_s)	GN/m^2	323.23
Matrix porosity (n_{ma})	-	0.02
Fracture porosity (n_{fr})	-	0.002
Matrix mobility (k_{ma}/μ)	m^4/MN s	10^{-10}
Fracture mobility (k_{fr}/μ)	m^4/MN s	10^{-9}
Fracture spacing (s)	m	0.1
Uniaxial compressive strength (σ_c)	MN/m^2	41
Internal friction angle (φ)	degree	30
Material strength parameter (α)	-	0.14
Material strength parameter (κ)	MN/m^2	12

The far-field stress and pore pressure gradients are: $S_{x'} = \sigma_H = 18$ kPa/m, $S_{y'} = \sigma_h = 14$ kPa/m, $S_{z'} = \sigma_v = 22.6$ kPa/m, $p_w = 10.4$ kPa/m (Bradley 1979). All the analyses are carried out at a true vertical depth of 1000 m, and concentrated on the borehole failure and mud weight selection.

In order to demonstrate the characteristics of borehole failures, a zero mud pressure in the wellbore is examined first. Boreholes with three different inclination angles, i.e., $\varphi_z = 0$, 50, and 90° are analyzed.

Figures 7.23 and 7.24 show collapse (Drucker-Prager failure) areas around the borehole wall at $t = 100$ s for hole inclinations of 50° and 90°. Note that no failure occurs when the borehole inclination is 0°. The collapsed area increases with the borehole inclination, since the stresses and stress differences increase in the case of a normal stress regime.

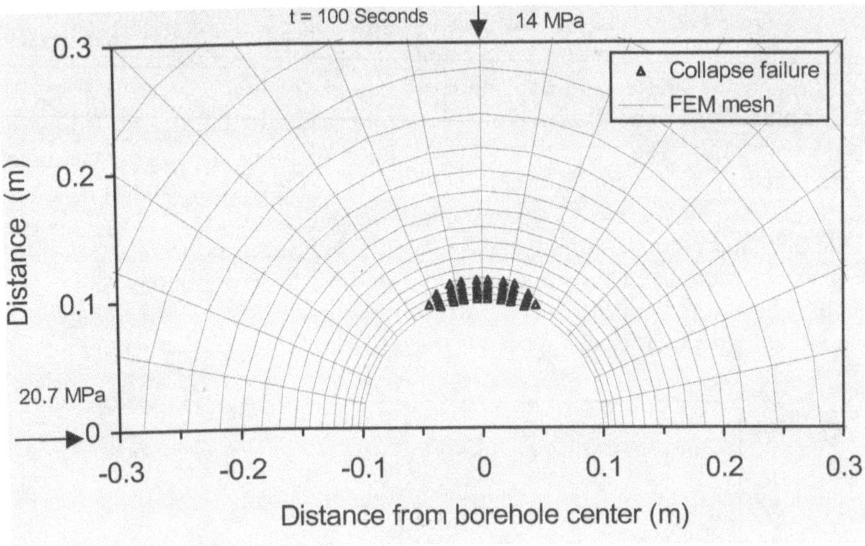

Fig. 7.23. Collapse area in the normal faulting stress regime at $t = 100$ s for a hole inclination of 50°.

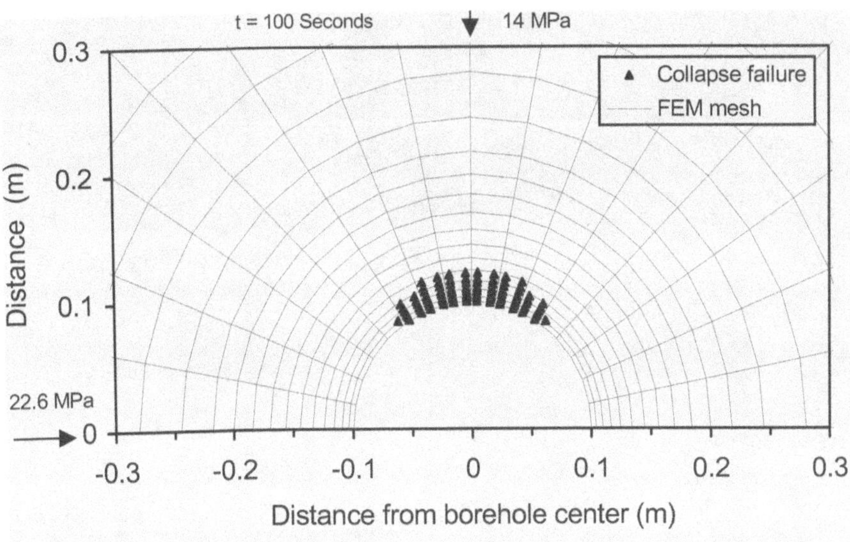

Fig. 7.24. Collapse area in the normal faulting stress regime at $t = 100$ s for a hole inclination of 90°.

A range of mud weights for a stable borehole is evaluated at different inclination angles. Figure 7.25 presents the envelopes that define such a safe range for single- and dual-porosity media (Zhang et al. 2003). From this figure some general observations can be made for boreholes drilled in a normal faulting stress regime. As the angle of inclination increases, the safe range of mud weight for collapse (Druker-Prager failure) and hydraulic fracturing decreases. At lower inclination angles, i.e., $\varphi < 35°$, the borehole does not collapse for any mud weight. Therefore, in the normal faulting stress regime the lower deviational and vertical boreholes are more stable. In comparison, the dual-porosity solution has a larger stable area due to its quite smaller collapse area and slightly larger fracturing area.

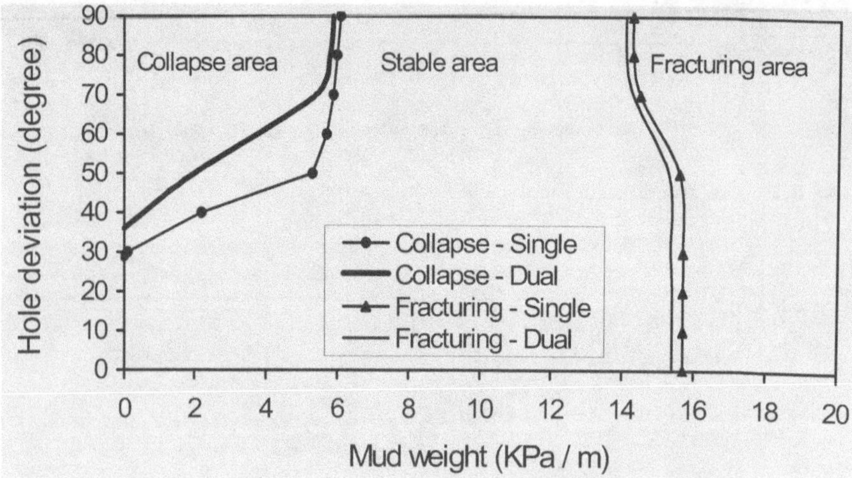

Fig. 7.25. Mud weight range varying with the hole inclinations for collapse and fracturing at $t = 100$ s in the normal faulting stress regime for both single-porosity and dual-porosity solutions.

References

Addis T, Boulter RRL, Plumb D (1993) The question for borehole stability in the Cusiana field, Colombia. Schlumbeger Oilfield Review, April/July, pp33-43

Bradley WB (1979) Failure of inclined boreholes. Trans of ASME 101:232-239

Charlez PA (1999) Rock Mechanics, Petroleum Application, Vol 2. Editions Technip Paris

Cui L, Kaliakin VN, Abousleiman Y, Cheng AHD (1997) Finite element formulation and application of poroelastic generalized plane strain problems. Int J Rock Mech Min Sci 34:953-962

Daines SR (1982) Prediction of fracture pressures for wildcat wells. JPT 34:863-872

Fu J, Peng S, et al. (2002) Statistical analysis and mechanism of casing damages in Gudao and Gudong oilfileds. J Jianghan Inst of Petroleum 4(2):84-85

Hubbert MK, Willis DG (1957) Mechanics of hydraulic fracturing. Pet Trans AIME 210:153-163

Hoffers B, Engeser B, Rischmuller H (1994) Wellbore stability of a superdeep borehole in crystalline rock, the KTB-Hauptbohrung, Germany. Eurock'94, Balkema, Rotterdom, pp371-378

Last NC, McLean MR (1995) Assessing the impact of trajectory on wells drilled in an overthrust region. Paper SPE 30465 presented at Annual Technical Conference, Dallas, pp161-171

Madge C, Zhang J, Standifird WB (2006) Predict predrill pressure and stress. E&P, May, pp81-83

Martin CD, Martino JB, Dzik EJ (1994) Comparison of borehole breakouts from laboratory and field test. In: SPE/ISRM Eurock'94, Belkema, Rotterdam, pp183-90

Martino JB, Chandler NA (2004). Excavation-induced damage srudies at the Underground Research Laboratory. Int J Rock Mech Min Sci 41:1413-1426

Maury V, Zurdo C (1996) Drilling-induced lateral shifts along pre-existing fractures: a common cause of drilling problems. SPE Drilling Completion, March

Matthews WR, Kelly J (1967) How to predict formation pressure and fracture gradient. Oil Gas J 65:92-106

Peng S, Fu J, Zhang J (2007) Borehole casing failure analysis in unconsolidated formations in Shengli oilfield. J Petroleum Sci Eng (in press)

Plumb R, Papanastasiou P, Last NC (1998) Constraining the state of stress in tectonically active settings. Paper SPE/ISRM 47240 presented at the SPE/ISRM Eurock'98 Conference, Trondheim, Norway

Ramos GG, Mouton DE, Wilton BS, Leksmono (1998) Integrating rock mechanics with drilling strategies in a Tectonic belt, offshore Bali, Indonesia. Paper SPE/ISRM 47286 presented at the SPE/ISRM Eurock'98 Conference, Trondheim, Norway

Roegiers JC (1990) Stability and failure of circular openings. In: Rock at Great Depth, Maury & Fourmaintraux (ed), Balkema, Rotterdam, pp1115-1121

Roegiers JC (2002) Well modeling: an overview. Oil Gas Sci Tech 57:569-577

Wilson RC, Willis DN (1986) Successful high angle drilling in the Stratfjord Field. Paper SPE 15465 presented at the 61st Annual Technical Conference Exhibition, New Orleans

Willson SM, Last NC, Zoback MD, Moos D (1999) Drilling in South America: a wellbore stability approach for complex geologic conditions. Paper SPE 53940 presented at Latin America and Caribbean Petroleum Engineering Conference held in Caracas, Venezuela

Wiprut DJ, Zoback MD (1998) High horizontal stress in the Visund field, Norwegian North Sea: consequences for borehole stability and sand production. Paper SPE/ISRM 47244 presented at the SPE/ISRM Eurock'98 Conference, Trondheim, Norway

Woodland DC 1990 Borehole instability in the western Canadian overthrust belt. SPE Drilling Engineers 5:23-33

Zhang J (2002) Dual-porosity approach to wellbore stability in naturally fractured reservoirs. PhD Dissertation, Univ of Oklahoma

Zhang J, Bai M, Roegiers JC (2003) Dual-porosity poroelastic analyses of wellbore stability. Int J Rock Mech Min Sci 40:473-483

Zhang J, Roegiers JC (2002) Borehole stability in naturally deformable fractured reservoirs – a fully coupled approach. Paper SPE 77355 presented at SPE Annual Technical Conference Exhibition

Zhang J, Roegiers JC (2005) Double porosity finite element method for borehole modeling. Rock Mech Rock Engng 38:217-242

Zhang J, Bai M, Roegiers JC (2006) On drilling directions for optimizing horizontal well stability using dual-porosity poroelastic approach. J Petrol Sci Eng 53:61-76

Zhang J, Standifird WB (2007) Impermeable drilling mud to strengthen wellbore stability. 1st Canada-US Rock Mechanics Symposium, Vancouver, Canada

8 Stress-dependent permeability

8.1 Rock core tests on stress and permeability

8.1.1 Stress, strain and permeability tests

Permeability of a rock mass is a physical property of extreme importance in petroleum engineering, civil and environmental engineering, and various areas of geology. Permeability in a fractured porous medium is mainly controlled by the geometry and interconnectionedness of the pores and fractures as well as stress state. Many research efforts have led to the permeability expressions for porous and fractured media. However, most approaches ignored the influences of the pore pressure and stress changes. Various attempts have found that stress-deformation behavior of fractures and pores present in rock masses is a key factor governing rock permeability and fluid flow through the rock masses.

To examine permeability in rock samples with respect to a complete stress-strain path, stress and permeability tests can be conducted under triaxial loading conditions (Peng 1998). A lab experiment is given to illustrate how to determine stress-permeability relation. The rock samples were cored from coal seam floor strata in Yangzhuang coal mine, Huaibei Coalfield, Anhui Province, China (Wang et al. 2001). The specimens included shale, sandy shale, fine-grained sandstone, and medium-grained sandstone. The lab tests were conducted in MTS 815 servo-controlled triaxial test machine (Peng and Meng 2002). The confining pressure was 4 MPa, and fluid pressure was 3.8 MPa at the top of the specimen, and 2.3 MPa at the bottom of the specimen. Fluid flow through the rock samples was in the axial direction (Peng and Wang 2001). The loading direction was also in the axial direction. The specimen sizes and experimental results are listed in Table 8.1.

Figure 8.1 presents the complete stress-strain and permeability-strain curves in shale. It shows that permeability has a slight increase as the axial stress increases within the elastic deformation range until rock yielding. The yielding leads to a jump in the permeability magnitude due to sudden creation of micro-fractures. After reaching the maximum stress, macroscopic fracture planes develop, causing a significant increase in permeability. After the maximum stress, the permeability continues to increase before reaching its peak value. This is likely due to the fact that the fracture planes extend towards the ends of the specimen and new cracks continue to appear, as reported by acoustic emissions (Jaeger and Cook 1979) and numerical analyses (Diederichs 2003). Upon reaching a maximum value, the permeability drops again since the failed rock undergoes a second phase of compaction. The subsequent permeability changes are the result of the post-failure deformations, which continue until the axial stress reaches its minimum magnitude.

Table 8.1. Experimental results for triaxial compressive stress-strain-permeability test in sedimentary rock specimens

Lithology	Diameter (mm)	Height (mm)	Peak load (MPa)	Peak Permeability (10^{-7} Darcy)	Initial Permeability (10^{-7} Darcy)
Shale	54.3	74.5	69.8	1.16	0.044
Sandy shale	53.8	72.0	28.2	3.6	0.19
Fine-grained sandstone	54.3	64.7	93.2	3.14	6.81
Medium-grained sandstone	53.2	62.3	95.7	8.86	3.66

In the sandy shale specimen (Fig. 8.2) the permeability, strain, and stress have the similar relationship as described in the shale specimen. The difference is that the sandy shale has a larger initial permeability and a larger permeability induced by triaxial loading (Peng et al. 2000).

Figure 8.3 exhibits the complete stress-strain and permeability-strain curves in medium-grained sandstone. It can be seen that, during the initial elastic deformation, rock permeability reduces due to compaction of existing cracks. However, permeability starts to increase as the rock begins to dilate. When the stress reaches the peak load, permeability has a significant increase. After the peak load, the permeability continues to increase before reaching its peak value. After reaching a maximum value, the permeability drops again since the failed rock undergoes a second phase of compaction. The fine-grained sandstone as shown in Fig. 8.4 has a similar stress-strain-permeability behavior.

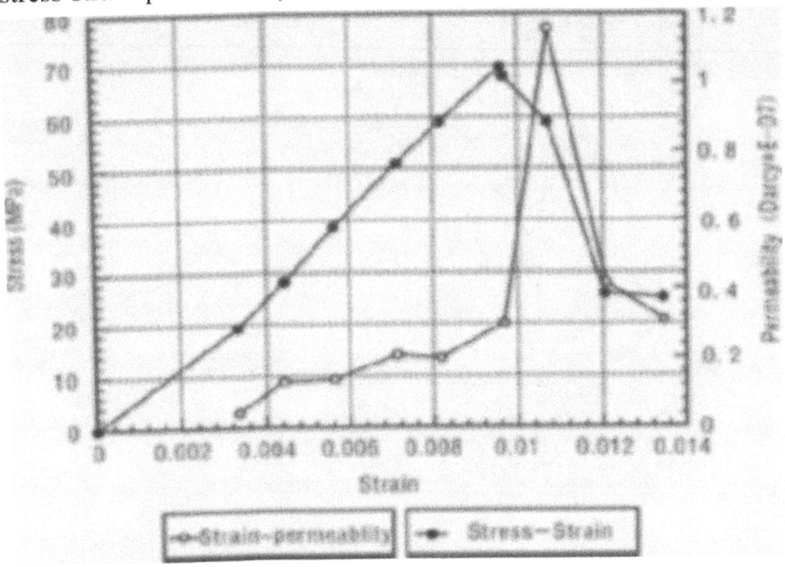

Fig. 8.1. Stress, permeability, and strain relationships for shale in the triaxial compressive test with a confining pressure of 4MPa.

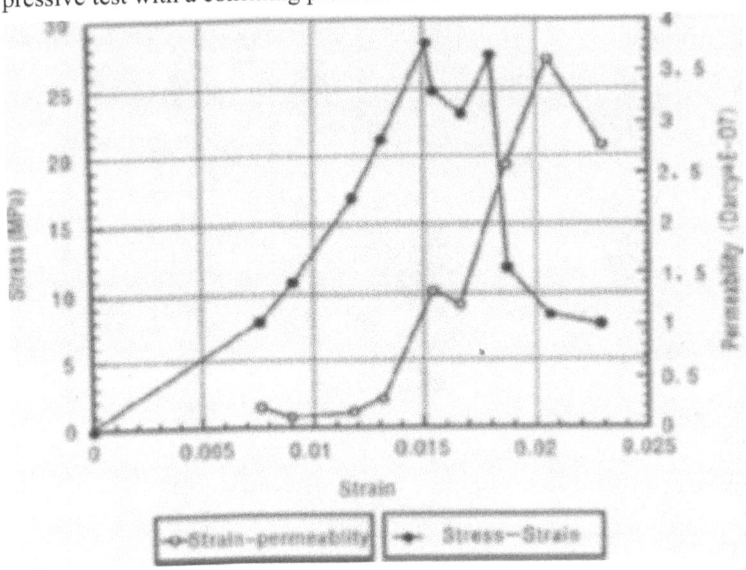

Fig. 8.2. Stress, permeability, and strain relationships for sandy shale in the triaxial compressive test with a confining pressure of 4MPa.

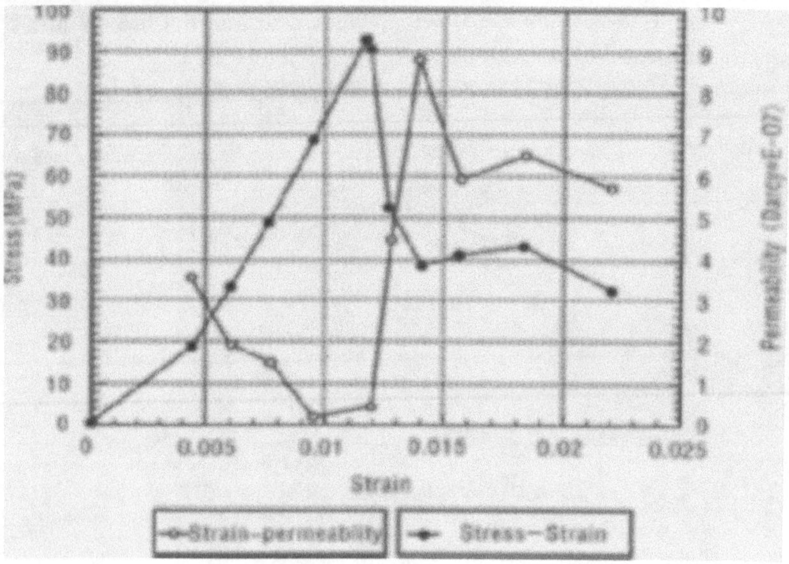

Fig. 8.3. Stress, permeability, and strain relationships for medium-grained sandstone in the triaxial compressive test with a confining pressure of 4MPa.

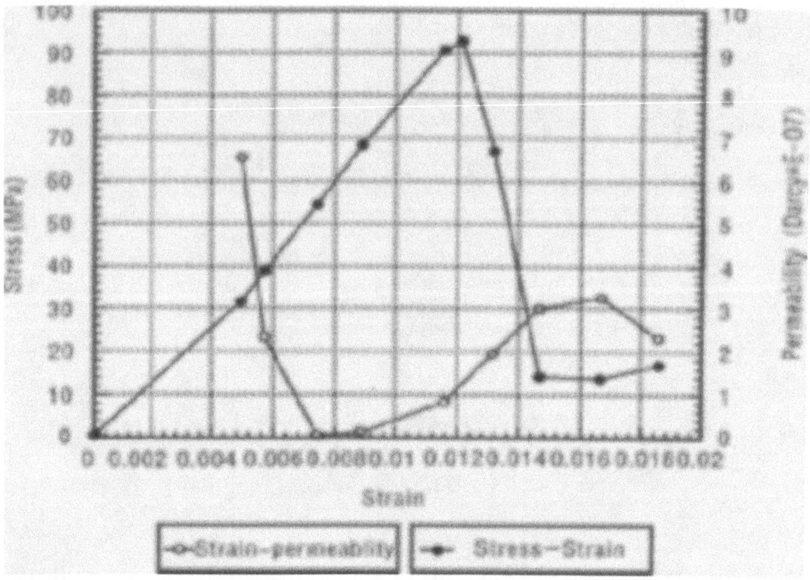

Fig. 8.4. Stress, permeability, and strain relationships for fine-grained sandstone in the triaxial compressive test with a confining pressure of 4MPa.

Compared to sandstones and shales, the sandstone has a larger initial permeability, and also has a much larger stress-induced peak permeability (refer to Table 8.1). In other words, the sandstone has a more pronounced stress-dependent permeability, which is important for engineering design.

8.1.2 Permeability and strain relationship

The stress-strain relationships and rock failure mechanisms have been investigated by many researchers (e.g. Li et al. 1994, Zhu and Wong 1997, Zhang et al. 1999, Zhang et al. 2000, Heiland 2003). The stress-strain curves can be divided into several regions (Martin and Chandler 1994). For the purpose of studying stress-permeability relationships, two major regions can be distinguished, as shown in Fig. 8.5 (Zhang et al. 2007). Region I commences from the origin until the crack-initiation stress level (σ_{ci}) is reached. This first region corresponds to the closure of existing microcracks pervading the rock sample. The permeability decreases gradually and the sample volume contracts as the load increases. Region II corresponds to the initiation of new cracks, which induces a dramatic increase in permeability and volumetric dilation. It is obvious from Fig. 8.5 that changes of permeability and volumetric strain are coherent in trend. This phenomenon illustrates that the volumetric strain and permeability are related by some kind of functional relationships, although it is difficult to determine a correlation between stress and permeability after the peak strength (σ_c) is reached.

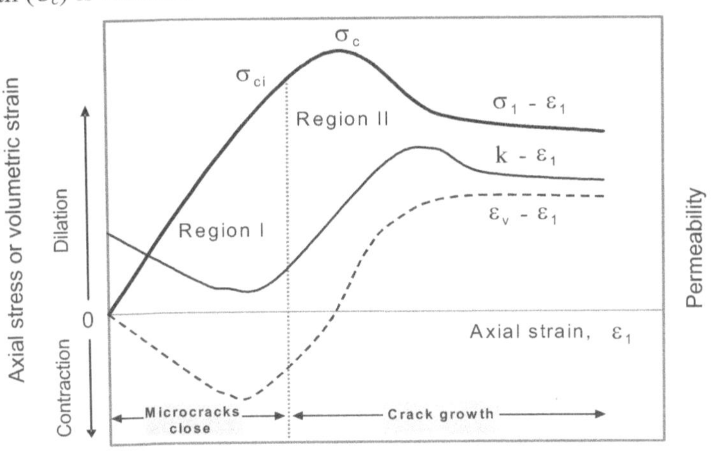

Fig. 8.5. Schematic relationships among axial stress, volumetric strain, and permeability.

8.2 Concrete block tests on stress and fluid flow

Concrete blocks containing arbitrary fractures were tested to determine the relationships of stress and fluid flow (Zhang et al. 2007). Figure 8.6 shows one of those concrete blocks under biaxial loading. Water at constant pressure infiltrated from one side of the fracture plane, and discharged from the other side, where flow rates were measured. An initial flowrate was measured before the loading was applied. After loading, the flow magnitude was monitored continuously. The width changes of the fractures were recorded before and after each loading step. The biaxial loading ratios varied from 1 to 5 (Zhang et al. 2007).

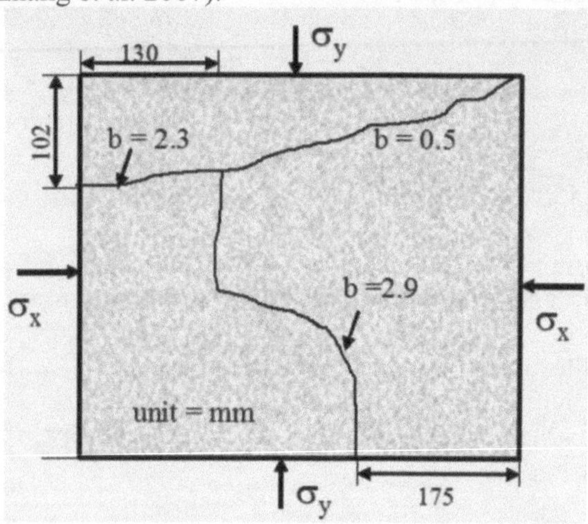

Fig. 8.6. Schematic representation of testing blocks showing the fracture networks and loading directions. The water flow is measured in the *z*-direction, which is perpendicular to the block surface.

For a ratio of $\sigma_x/\sigma_y = 3$, with a maximum initial fracture aperture of 2.9 mm, the increase in applied loads results in a decrease of flowrate and fracture aperture, as indicated in Figs. 8.7 and 8.8. For this particular case the flowrate-stress and aperture-stress are well described by:

$$Q_z = 3.9[1 - 0.14(\sigma_x + \sigma_y)]^3 \tag{8.1}$$

$$b = 3.38[1 - 0.096(\sigma_x + \sigma_y)] \tag{8.2}$$

Equations 8.1 and 8.2 imply a correlation between flowrate and aperture, which can be characterized as (Zhang et al. 2007):

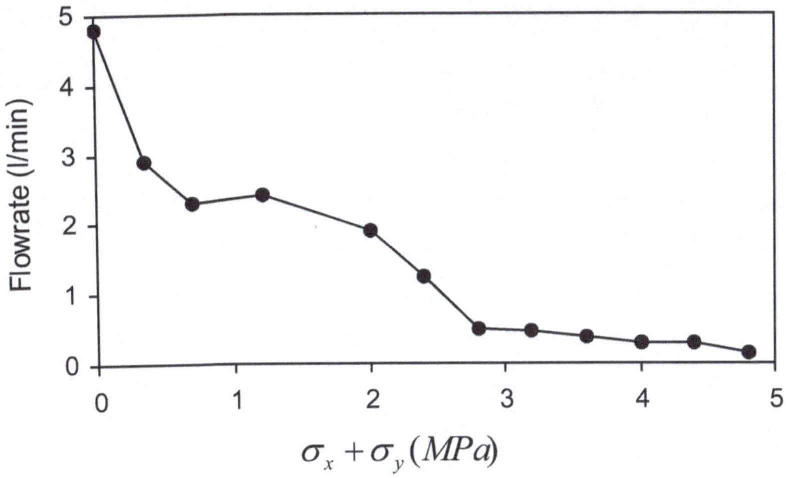

Fig. 8.7. Flowrate versus biaxial loading ($\sigma_x / \sigma_y = 3$) in the concrete block.

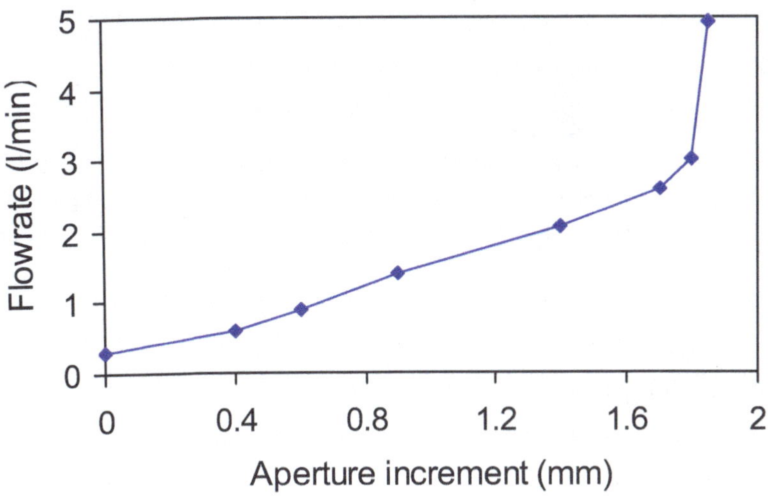

Fig. 8.8. Flowrate versus changes of fracture aperture for biaxial loading ($\sigma_x / \sigma_y =$ 3) in the concrete block.

$$Q_z = 0.27[1 + 0.704\Delta b]^3 \tag{8.3}$$

where Δb is the aperture change ($\Delta b = b - b_0$); b_0 and b are the aperture before and after load application.

The experimental data of flowrate versus fracture aperture is shown graphically in Fig. 8.8. This relationship verifies that the cubic law is still valid for a medium with randomly distributed multiple fractures and with confining stresses.

The flowrate and fracture aperture relationship from Eq. 8.3 can be written as the following general form:

$$Q = C_1\left(1 + \frac{\Delta b}{C_2}\right)^3 \tag{8.4}$$

where C_1 and C_2 are constants.

8.3 Stress-dependent permeability in fractured media

Researchers have developed the parallel model to describe fluid flow in a single fracture. In the model a single fracture is conceptualized as a channel between two parallel plates, and the vertical separation between the plates, or aperture, is constant throughout the length of the fracture. The flow rate in this model is governed by the "cubic law", i.e. the volumetric flow rate of a fluid through a cross section of the fracture is proportional to the applied pressure gradient and the cube of the vertical separation (Waite et al. 1999):

$$Q = -w\frac{b^3}{12\mu}\frac{dp}{dx} \tag{8.5}$$

where Q is the volumetric flow rate, w is the width of the fracture perpendicular to flow, b is the fracture aperture, μ is the dynamic viscosity of the fluid.

Using Dacy's Law, the fracture permeability k_f can be defined as:

$$k_f = \frac{b^2}{12} \tag{8.6}$$

The parallel plate model for a single fracture can be extended to multiple fracture systems by considering regular families of parallel fractures

(Bear et al. 1993). For a set of parallel fractures of equal aperture, oriented parallel to the flow direction, the permeability through this set is given by:

$$k_f = \frac{b^3}{12s} \tag{8.7}$$

where b is the fracture aperture; and s is the mean fracture spacing.

Since natural rock fractures are neither smooth nor parallel, investigators have questioned the accuracy of applying the cubic law to natural fractures. Investigations show that the cubic law is valid when corrected by considering the fracture tortuosity, correction factor or using effective aperture (Witherspoon et al. 1980). Waite et al. (1999) examined experimentally the parallel plate geometry and confirmed that the rate of fluid movement through a parallel plate fracture can be predicted by the cubic law with the traditional estimations of aperture and pressure gradient. However, for the sinusoidal geometries (Fig. 8.9), the cubic law is valid only when the fracture aperture is measured normal to the flow path and averaged according to a harmonic averaging scheme. The normal aperture can be calculated at any point in the cycle of the sine wave using the vertical separation b and the local slope of the average flow path (dy/dx) as follows:

$$b_n = b\cos\left[\tan^{-1}\left(\frac{dy}{dx}\right)\right] \tag{8.8}$$

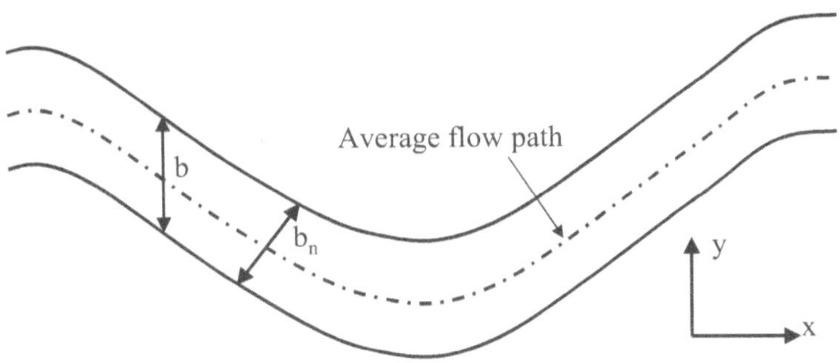

Fig. 8.9. Schematic sinusoidal geometry of a natural fracture.

Along the fracture, an average normal aperture $<b_n>$ can be determined by averaging over the length of the fracture and used in the cubic law, i.e.:

$$< b_n >= \left[L \Big/ \sum_{i=1}^{k} \left(\frac{L_i}{b_{ni}^3} \right) \right]^{\frac{1}{3}} \tag{8.9}$$

where L is the total length of the fracture; L_i is the interval length; and, b_{ni} is the normal aperture of the ith interval.

Iwai (1976) showed that the validity of the cubic laws is dependent on the stress applied onto the fracture. After measuring water flowrates through granite fractures under an applied normal stress, Raven and Gale (1985) found an increasing deviation from the cubic law associated with stress and larger samples. Many experimental and theoretical research of stress-dependent permeability has been presented (such as; Louis 1974, Bawden et al. 1980, Witherspoon et al. 1980, Barton et al. 1985, Bai et al. 1994, Zimmerman and Bodvarsson 1996, Zhang et al. 1999, Zhang and Wang 2000, Zhang 2002).

Applying the cubic law, the permeability change in the z-direction due to an aperture change for only one set of fractures can be expressed as follows (Zhang et al. 2007):

$$k_z = k_0 \left(1 + \frac{\Delta u_x}{b_{x0}} \right)^3 \tag{8.10}$$

where k_z is the permeability change due to the aperture increment Δu_x; b_{xo} is the average normal aperture of the original fracture in the x-direction as defined by Eq. 8.9, and k_0 is the original permeability in the z-direction.

The permeability change due to the aperture changes for two sets of fractures (Fig. 8.10) can be expressed as:

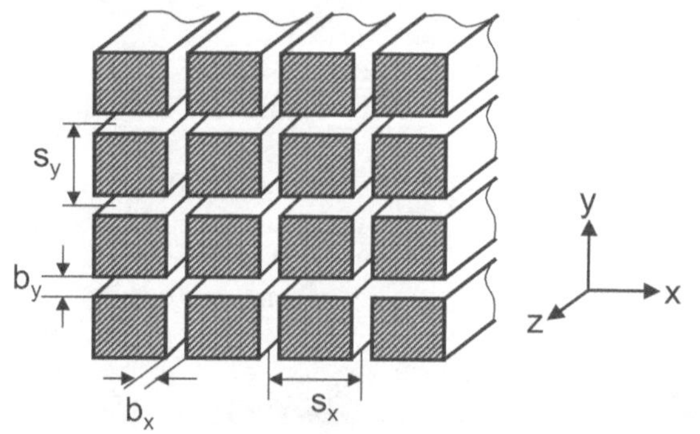

Fig. 8.10. Multiple fracture system for two sets of parallel fractures in the z-direction.

$$k_z = k_0\left[\left(1+\frac{\Delta u_x}{b_{x0}}+\frac{\Delta u_y}{b_{y0}}\right)^3\right] \tag{8.11}$$

where k_z is the permeability change due to the aperture increment Δu_x and Δu_y.

In order to determine stress-dependent permeability in fractured media, both the fracture aperture variation and the rock matrix deformation need to be considered. To derive permeability and stress relation, an idealized three-dimensional regularly spaced fracture-matrix system can be assumed, as illustrated in Fig. 8.11.

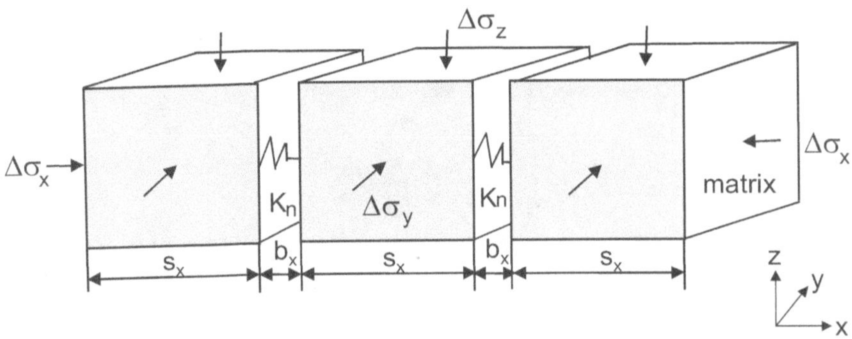

Fig. 8.11. Coupled fracture-matrix system with 3-D stresses.

The change of total displacement along x-direction is, therefore, the sum of the fracture displacement and the matrix displacement changes; i.e.:

$$\Delta u_{tx} = \Delta u_{fx} + \Delta u_{rx} \tag{8.12}$$

where Δu_{tx}, Δu_{fx}, and Δu_{rx} are the displacements of total fracture-matrix system, fracture and rock matrix, respectively.

Therefore, the displacement across the fracture can be obtained by:

$$\Delta u_{fx} = \Delta u_{tx} - \Delta u_{rx} \tag{8.13}$$

The above equation can be expressed as the strain form:

$$\Delta u_{fx} = (s_x + b_x)\Delta\varepsilon_{tx} - s_x\Delta\varepsilon_{rx} \tag{8.14}$$

where s_x and b_x are the fracture spacing and aperture along the x-direction, respectively; $\Delta\varepsilon_{tx}$, $\Delta\varepsilon_{rx}$ are the total and matrix strains, respectively.

The total strain along the x-direction can be obtained according to Hooke's law:

$$\Delta\varepsilon_{tx} = \frac{1}{E_{mx}}\left[\Delta\sigma_x - \nu(\Delta\sigma_y + \Delta\sigma_z)\right] \tag{8.15}$$

where E_{mx} is the Young's modulus of the rockmass in the x-direction.

According to Hooke's law, the matrix strain along the x-direction may be described by:

$$\Delta\varepsilon_{rx} = \frac{1}{E_r}\left[\Delta\sigma_x - \nu(\Delta\sigma_y + \Delta\sigma_z)\right] \tag{8.16}$$

where E_r is the Young's modulus of the rock matrix; $\Delta\sigma_x$, $\Delta\sigma_y$, and $\Delta\sigma_z$ are the stress changes in the x-, y- and z-directions, respectively.

Substituting Eqs. 8.15 and 8.16 into Eq. 8.14, the change of the fracture aperture due to the stress variation can be obtained; i.e.:

$$\Delta u_{fx} = \left(\frac{(s_x + b_x)}{E_{mx}} - \frac{s_x}{E_r}\right)\left[\Delta\sigma_x - \nu(\Delta\sigma_y + \Delta\sigma_z)\right] \tag{8.17}$$

The Young's modulus of the rockmass (E_{mx}) related to the properties of the intact rock matrix and the fractures can be expressed as follows:

$$\frac{1}{E_{mx}} = \frac{1}{E_r} + \frac{1}{K_{nx}s_x} \tag{8.18}$$

where K_{nx} is the fracture normal stiffness in the x-direction.

Substituting Eq. 8.18 into Eq. 8.17, the fracture displacement in x-direction can be obtained:

$$\Delta u_{fx} = \left(\frac{1}{K_{nx}} + \frac{b_x}{K_{nx}s_x} + \frac{b_x}{E_r}\right)\left[\Delta\sigma_x - \nu(\Delta\sigma_y + \Delta\sigma_z)\right] \tag{8.19}$$

The fracture permeability in the direction perpendicular to the x-direction can be calculated directly from the parallel plate analog defined in Eq. 8.10. With reference to Eqs. 8.19 and 8.10, and considering the compressive strain as negative, the change of fracture permeability in z-direction can be expressed as:

$$k_z = k_{0z}\left\{1 - \left(\frac{1}{K_{nx}b_x} + \frac{1}{K_{nx}s_x} + \frac{1}{E_r}\right)\left[\Delta\sigma_x - \nu(\Delta\sigma_y + \Delta\sigma_z)\right]\right\}^3 \tag{8.20}$$

where k_{0z} and k_z are the permeabilities along the z-direction before and after stress change, respectively.

Hence, the generalized permeability-stress relation may be written according to Eq. 8.20:

$$k_k = k_{0k}\left\{1-\left(\frac{1}{K_{ni}b_i}+\frac{1}{K_{ni}s_i}+\frac{1}{E_r}\right)\left[\Delta\sigma_i - \nu(\Delta\sigma_j + \Delta\sigma_k)\right]\right\}^3 \tag{8.21}$$

where k_{0k} and k_k are the permeabilities along the k-direction before and after stress changes; $i = x, y, z$; $j = y, z, x$; $k = z, x, y$; and, $i \neq j \neq k$.

Only one set of fractures is examined in the above analysis. When three parallel fracture sets exist in the x-, y- and z-directions, as shown in Fig. 8.10, permeability changes along the k-direction by referring to Eq. 8.11 may be written as:

$$k_k = k_{0k}\left\{1-\left(\frac{1}{K_{ni}b_i}+\frac{1}{K_{ni}s_i}+\frac{1}{E_r}\right)\left[\Delta\sigma_i - \nu(\Delta\sigma_j + \Delta\sigma_k)\right] - \left(\frac{1}{K_{nj}b_j}+\frac{1}{K_{nj}s_j}+\frac{1}{E_r}\right)\left[\Delta\sigma_j - \nu(\Delta\sigma_i + \Delta\sigma_k)\right]\right\}^3 \tag{8.22}$$

8.4 Permeability-stress relation in a porous medium

For porous media, fluid flow moves mainly through pore spaces. The variation in grain sizes or pore spaces due to the effect of an applied stress results in a change of permeability. For a simple cubical grain packing structure, the permeability k can be expressed as (Bai and Elsworth 1994):

$$k = \frac{2R^2}{\pi^2} \tag{8.23}$$

where R is the grain radius.

Under a three-dimensional effective stress condition, the change in grain sizes within a cubical packing (Fig. 8.12) can be determined by analyzing the elastic contact of spheres. Applying the theory of Hertzian contact of spherical grains, the change of grain radius due to the effective compressive stresses changes of $\Delta\sigma_x$, $\Delta\sigma_y$, and $\Delta\sigma_z$ can be obtained by:

$$\begin{cases} \Delta R_x = -\dfrac{R_0}{2}\left[\dfrac{9(1-\nu^2)}{2}\left(\dfrac{\pi\Delta\sigma_x}{E}\right)^2\right]^{\frac{1}{3}} \\ \Delta R_y = -\dfrac{R_0}{2}\left[\dfrac{9(1-\nu^2)}{2}\left(\dfrac{\pi\Delta\sigma_y}{E}\right)^2\right]^{\frac{1}{3}} \\ \Delta R_z = -\dfrac{R_0}{2}\left[\dfrac{9(1-\nu^2)}{2}\left(\dfrac{\pi\Delta\sigma_z}{E}\right)^2\right]^{\frac{1}{3}} \end{cases} \tag{8.24}$$

The grain radius after stress changes can be represented as $R = R_0 + \Delta R_x + \Delta R_y + \Delta R_z$, i.e.:

$$R = R_0\left\{1 - \frac{1}{2}\left[\frac{9\pi^2(1-\nu^2)^2}{2E^2}\right]^{\frac{1}{3}}\left(\Delta\sigma_x^{2/3} + \Delta\sigma_y^{2/3} + \Delta\sigma_z^{2/3}\right)\right\}^2 \tag{8.25}$$

where R_0 is the original grain radius.

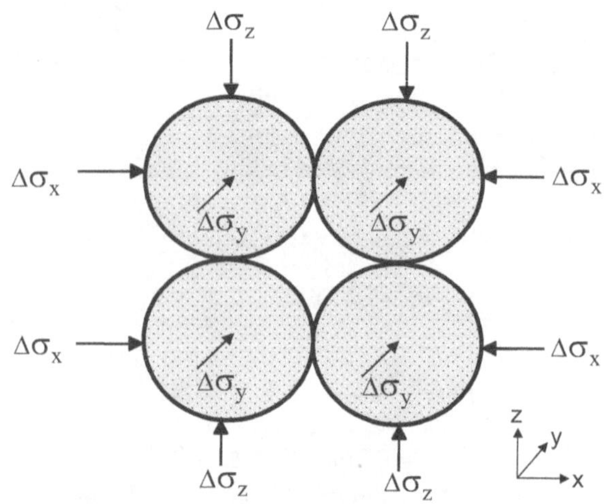

Fig. 8.12. Spherical contact of grains with 3-D stresses.

Substituting Eq. 8.25 into Eq. 8.23 and considering both effects of compressional and dilatational stresses, the change in permeability in a porous medium can be expressed as (Zhang and Wang 2006):

$$k = k_0 \left\{ 1 - \frac{1}{2} \left[\frac{9\pi^2 (1-\nu^2)^2}{2E^2} \right]^{\frac{1}{3}} \left(\pm\Delta\sigma_x^{2/3} \pm\Delta\sigma_y^{2/3} \pm\Delta\sigma_z^{2/3} \right) \right\}^2 \tag{8.26}$$

where a negative sign refers to compressional loading and a positive sign corresponds to dilatational loading; k_0 is the initial permeability.

8.5 Case application in mining panel

The solid deformation and fluid flow problems can be solved by using the finite element method (FEM) using governing equations for a dual-porosity poro-mechanical model found in Zhang (2002) and Zhang and Roegiers (2005). Introducing stress-permeability into the finite element model, permeability variations induced by stress changes can be obtained. A case example is given below to examine permeability changes due to mining.

When mining near aquifers, it is of critical importance to determine the changes of permeability due to mining (Zhang and Shen 2004). The coal mine considered here is located in the Yanzhou coalfield, Eastern China. The average mining depth is 305 m and the extraction thickness of the pertinent coal seam is 5 m. A water-bearing sand lies 75 m above the coal horizon, and the thickness of the sand layer is 30 m (Zhang et al. 2001, Zhang and Wang 2006). Figure 8.13 shows the finite element model and mesh of a section perpendicular to the mining direction. In this model, a half of the mining panel is considered due to geometric symmetry. The generalized plane strain model is adopted since the mining direction of the panel is much longer than the direction of the panel width. The model is laterally confined and impermeable. The bottom of the model is considered as a rigid and impermeable boundary. The strata gravity with average unit specific weight of γ = 23 kPa/m (or 2.3 sg) is considered as the far-field stress acting on the panel. The far-field stresses and pore pressure are respectively: σ_v = 7.0 MPa, $\sigma_H = \sigma_h$ =3.8 MPa, p_w = 3.0 MPa. The main parameters of these strata are listed in Table 8.2. In the table most rock parameters are based on laboratory experiments. However, the parameters for the sand aquifer and the mined area are simply estimated. The fracture spacing is assumed to be 1 m for all layers.

Table 8.2. Input rock parameters for the FEM calculations

Layer number in Fig. 8.13	Strata location	Layer thickness (m)	Elastic modulus (GPa)	Fracture stiffness (GPa)	Poisson's ratio
5	Sand aquifer	30	1	7	0.4
1	Roof	75	5	12	0.32
2	Immediate roof	25	2.4	10	0.38
3	Coal seam	5	1	10	0.38
4	Mined area	5	10^{-4}	10^{-3}	0.35
2	Immediate floor	20	2.4	10	0.38
1	Floor	50	5	12	0.32

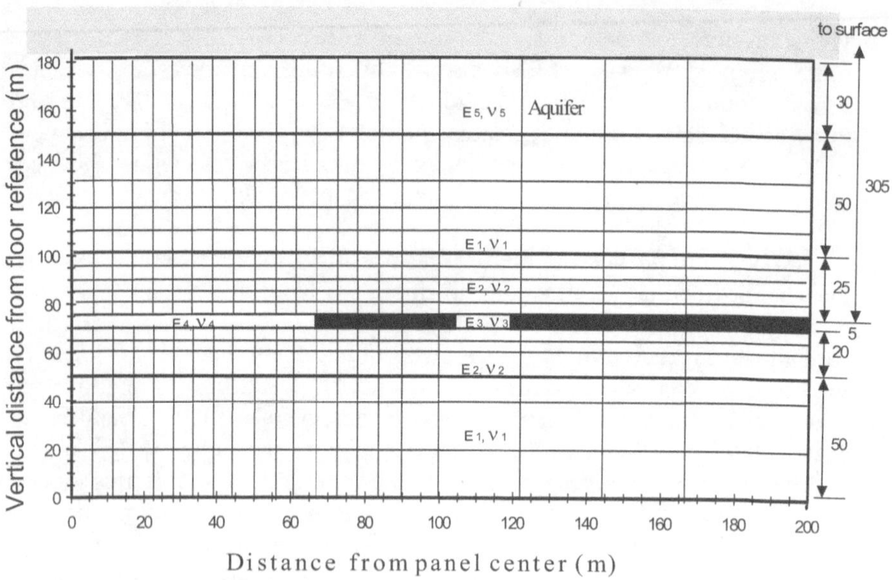

Fig. 8.13. Partial finite element mesh in a long-wall mining panel.

Figure 8.14 shows the FEM calculated contours of the permeability variations (permeability ratios of post- to pre-mining) in the vertical direction. In this figure, the mining width of the panel is 90 m, and the vertical axis is the central line of the panel. It can be seen that permeability increases in the strata around and beyond the mined area and decreases in some areas near the unmined coal seam. The maximum magnitude of increased permeability lies in the immediate roof and floor of the mined seam of the panel center. However, the maximum height of increased permeability zone appears in the strata over the unmined coal pillar.

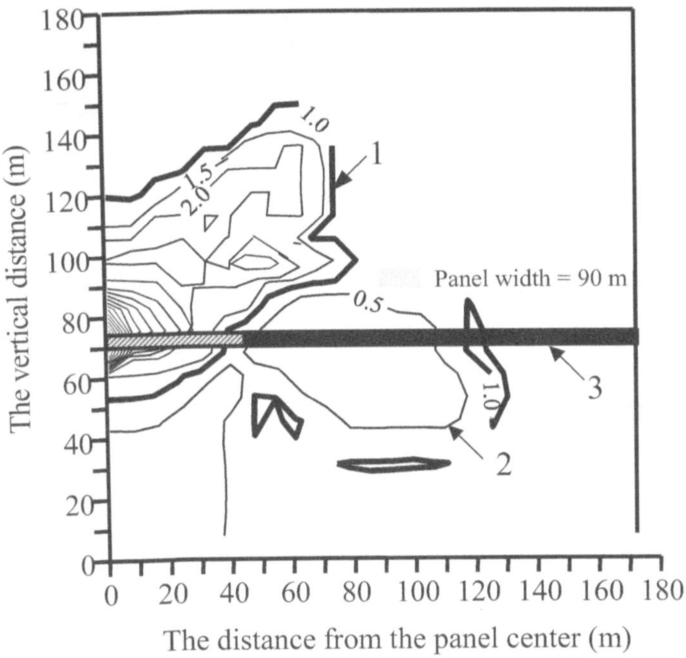

1 – Increased permeability zone; 2 – Decreased permeability zone;
3 – Coal seam.

Fig. 8.14. Contour of ratio of permeability (k_y/k_0) due to mining in the vertical direction.

Figure 8.15 presents the contours of permeability variations in the horizontal direction. The permeability variations are very different from the vertical one shown in Fig. 8.14, due to the different stress distributions between the vertical and horizontal directions. The height of the increased permeability zone in the horizontal direction is higher than that in the vertical direction. Furthermore, there are larger magnitudes of permeability changes in the horizontal direction (refer to Figs. 8.14 and 8.15).

It can also be seen that the maximum increased permeability zone occurs over the mining panel center for the horizontal permeability, while it occurs over the coal abutment of the mining panel in case of the vertical permeability. This implies that even if an aquifer is not exactly over the mining panel, water intrusion may still take place.

Figure 8.16 gives the field observed results for a long-wall mining face in similar geologic conditions as the numerical model. In order to measure permeability changes due to mining in the overburden strata, observation boreholes were drilled pre- and post-mining. The flowrates or drilling fluid circulation changes along the borehole during drilling were measured and well logs were run (Liu 1999). It can be seen from Fig. 8.16 that the observed increased permeability has a similar shape as the one predicted in Fig. 8.14.

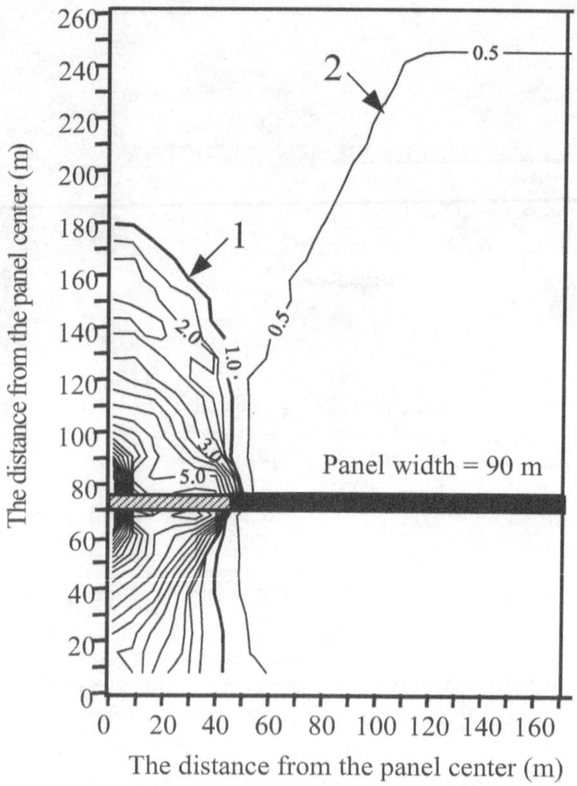

1 – Increased permeability zone; 2 – Decreased permeability zone.

Fig. 8.15. Contour of ratio of permeability (k_x/k_0) due to mining in the horizontal direction.

Parametric analyses using a finite element model are conducted to study the influences of thickness of extraction, mining width (*L* in Fig. 8.17), and depth of mining on permeability in the surrounding strata of the coal seam. The straight lines in Fig. 8.17 show that the permeability height increases with mining width and thickness of extraction. Figure 8.17 also compares the FEM results of increased permeability height in the overburden strata with field observed data and empirical formula (Liu et al. 1981). In the observed data, lift mining methods were used for mining thick coal seams found in China. It can be seen that the FEM results are coherent both in magnitude and in trend.

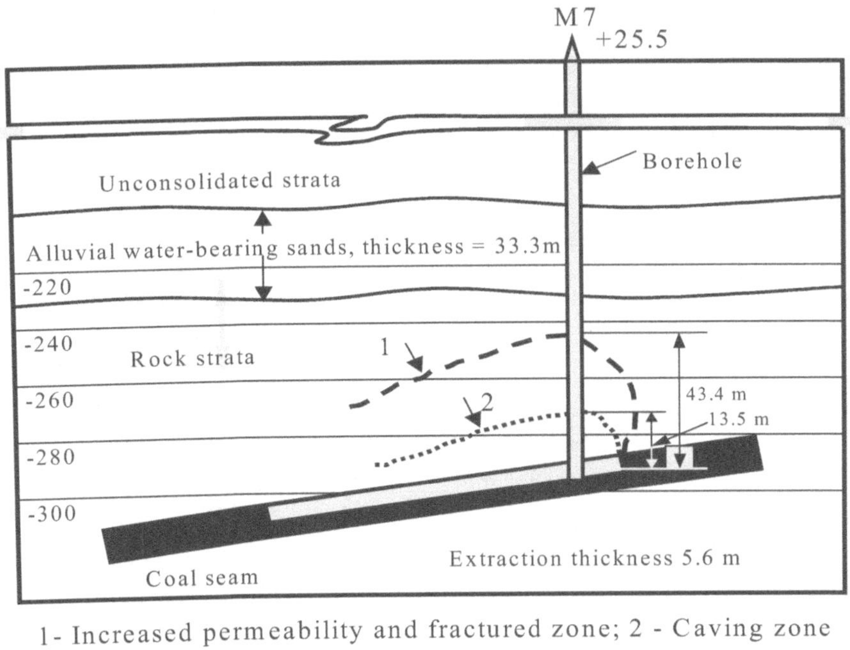

Fig. 8.16. Observed section of permeability change and strata failure by measuring borehole drilling fluid circulation and well logs in a long-wall extraction face. The range of the increased permeability defined in this observation was the area when the flowrate in the strata was greater than 0.1 l/s·m.

Figure 8.18 presents the increased permeability in the overburden strata with mining depth. Field observed height of strata with increased permeability, the FEM calculated result, and the one given by Alejano et al (1999) are compared in Fig. 8.18. The figure shows that as the depth of mining increases, the calculated height of the increased permeability zone decreases, which is coherent with the results given by Alejano et al. (1999). Compared to the observed results from Liu et al. (1981), again coherence is found in magnitude.

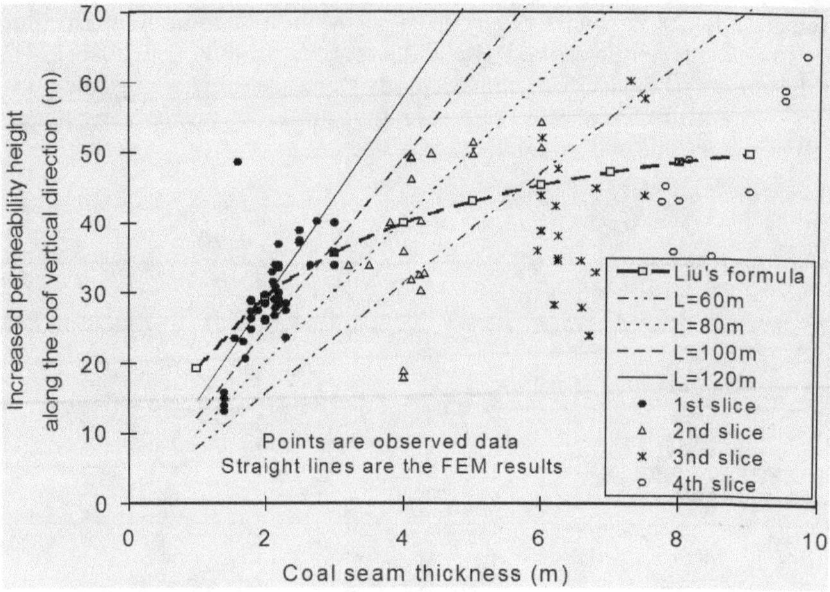

Fig. 8.17. FEM calculated increased vertical permeability height in the overburden strata versus mining thickness for different mining width (L = 60 - 120m), and the comparison with field observed data in China's coal mines for lift mining with different slices.

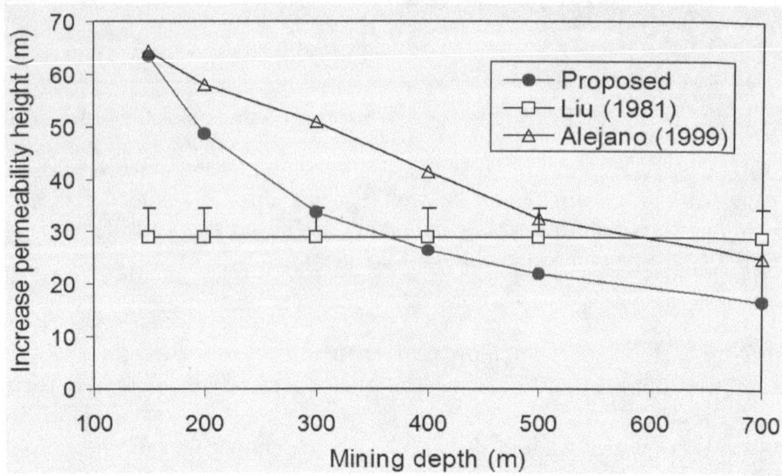

Fig. 8.18. FEM calculated increased vertical permeability height in the overburden strata versus mining depth, and the comparison with the works by Liu et al. (1981) and Alejano et al. (1999). In the figure the seam thickness of mining is 2 m. The error bar is added in each datum in the observed data in China's coal mines by Liu et al. (1981).

References

Alejano LR, Ramírez-Oyanguren P, Taboada J (1999) FDM predictive methodology for subsidence due to flat and inclined coal seam mining. Int J Rock Mech Min Sci 36:475-491

Bai M, Elsworth D (1994) Modeling of subsidence and stress-dependent hydraulic conductivity for intact and fractured porous media. Rock Mech Rock Engng 27:209-234

Barton N, Brandis S, Bakhtar K (1985) Strength, deformation and conductivity coupling of rock joints. Int J Rock Mech Min Sci Geomech Abstr 22:231-245

Bawden WF, Curran JH, Roegiers JC (1980) Influence of fracture deformation on secondary permeability – a numerical approach. Int J Rock Mech Min Sci Geomech Abstr 17:265-279

Bear J, Tsang C-F, de Marsily G (1993) Flow and contaminant transport in fractured rock. Academic, San Diego, California

Diederichs MS (2003) Rock fracture and collapse under low confinement conditions. Rock Mech Rock Engng 36(5):339-381

Heiland J (2003) Laboratory testing of coupled hydro-mechanical processes during rock deformation. Hydrogeology J 11:122-141

Iwai K (1976) Fundamental studies of fluid flow through a single fracture. PhD dissertation, Univ of Calif Berkeley

Jaeger JC, Cook NGW (1976) Fundamentals of Rock Mechanics. 2nd edition, Chapman Hall

Kranz RL, Frankel AD, Engelder T, Schulz CH (1979) The permeability of whole and jointed Barre granite. Int J Rock Mech Min Sci Geomech Abstr 16:225-234

Li SP, Li YS, Wu, ZY (1994) Permeability-strain equations corresponding to the complete stress-strain path of Yinzhuang sandstone. Int J Rock Mech Min Sci Geomech Abstr 31:383–391

Liu T (1981) Coal mine ground movement and strata failure. Coal Industry Publ House, Beijing, China (in Chinese)

Liu T (1999) Personal Communication

Louis C (1974) Rock hydraulics. In Müller L (ed) Rock Mechanics, Springer Verlag, Vienna pp 299-382

Martin JB, Chandler NA (1994) The progressive fracture of Lac du Bonnet granite. Int J Rock Mech Min Sci Geomech Abstr 31:643-659

Peng S (1998) Approach on mechanical properties of clastic rocks and rock microstructure. Proc. 5th Ann Conf of Young Scientists, China Coal Society. Coal Industry Press (in Chinese)

Peng S, Wang J (2001) Safe mining above confined aquifers. Coal Industry Press Beijing (in Chinese)

Peng S, Qu H, Luo L, Wang L, Duan Y (2000) Complete stress-strain and permeability experiments. J China Coal Society 25(2):113-116 (in Chinese)

Peng S, Meng Z (2002) Theory and practice of mining engineering geology. Geological Press (in Chinese)

Raven KG, Gale JE (1985) Water flow in a natural fracture as a function of stress and sample size. Int J Rock Mech Min Sci Geomech Abstr 22:251-261

Waite ME, Ge S, Spetzler H (1999) A new conceptual model for fluid flow in discrete fractures: an experimental and numerical study. J Geophys Res 104 (B6):13049-13059

Wang J, Peng S, Meng Z (2001) Permeability changes in triaxial complete stress-strain experiments. J Univ Beijing Sci Tech 23(6):489-491 (in Chinese)

Witherspoon PA, Wang JSY, Iwai K, Gale JE (1980) Validity of cubic law for fluid flow in a deformable rock fracture. Water Resour Res 16(6):1016-1024

Zimmerman RW, Bodvarsson GS (1996) Hydraulic conductivity of rock fractures. Transp Porous Media 23:1-30

Zhang J (2002) Dual-porosity Approach to Wellbore Stability in Naturally Fractured Reservoirs. PhD Dissertation, Univ of Oklahoma

Zhang J, Roegiers JC (2005) Double porosity finite element method for borehole modeling. Rock Mech Rock Engng 38:217-242

Zhang J, Bai M, Roegiers JC, Liu T (1999) Determining stress-dependent permeability in the laboratory. In Amadei B, Kranz RL, Scott GA, Smeallie PH (ed) Proc 37th US Rock Mech Symp Colorado. Rotterdam, Balkema pp 341-347

Zhang J, Bai M, Roegiers JC, Wang J, Liu T (2000) Experimental determination of stress-permeability relationship. Pacific Rocks, Girand et al. (eds), Balkema Rotterdam

Zhang J, Wang J (2000) Coupled stress and permeability behavior in fractured media and its application. In: Proc 4th N American Rock Mech Symp, Girard J, Liebman M, Breeds C, Doe T (ed) Seattle Washinton Rotterdam Balkema pp 811-816

Zhang J, Standifird WB, Roegiers JC, Zhang Y (2007) Stress-dependent fluid flow and permeability in fractured media: from lab experiments to engineering applications. Rock Mech Rock Engng 40(1):3-21

Zhang J, Roegiers JC, Bai M, Zhang Y (2001) Stress-dependent permeability variation and mine subsidence. In 38th US Rock Mechanics Sym Washington DC, Rotterdam Balkema

Zhang J, Shen B (2004) Coal mining under aquifers in China: a case study. Int J Rock Mech Min Sci 41:629-639

Zhang J, Wang J (2006) Coupled behavior of stress and permeability and its engineering applications. Chinese J Rock Mechanics Eng 25 (10):1982-1989

Zhu W, Wong TF (1997) The transition from brittle faulting to cataclastic flow: permeability evolution. J Geophys Res 102(B2):3027-3041

9 Strata failure and mining under surface and ground water

9.1 Threat of bodies of water on mining

Many coal mines are threatened by bodies of water during coal extractions. These bodies of water include rivers, lakes, reservoirs and groundwater. For example, there are about 125 rivers flowing through China's coal fields, and more than 200 coal mines are confronted with the problems of mining under rivers. For groundwater, there are three main possible water disasters affected safety operations of coal mines (Zhang et al. 1997, Peng and Meng 2002): (1) water inrushes from Ordovician limestone underlying the Permo-Carboniferous coal seams in Northern China; (2) water inrushes from low Permian limestone underlying coal seams and from Triassic limestone overlying coal seams of late Permian in Southern China; and (3) water inrushes from Cenozoic porous aquifers in the Yellow River and Huai River alluvial plain areas.

Table 9.1. Maximum water inflow of water inrushes in some China's coal mines

Coal field	Coal mine	Maximum flow rate (m^3/min)
Kailuan	Fangezhuang	2053
Feicheng	Taoyang Zhong No. 1	299
Feicheng	Yangzhuang	73.5
Fengfeng	No. 1	150
Jiaozhuo	Yanma	320
Jiaozhuo	Zhongmacun	105
Xuzhou	Zhangji	400
Hancheng	Magouqu	200
Chenhe	No. 2	335
Jingjing	No. 3	68
Zibo	Beidajing	566
Zhengzhou	Micun	75.2
Pingdingshan	No. 8	53.3

According to the incomplete official statistics, about 285 of 600 key coal mines in China are threatened by water inrushes during coal mining (Zhang et al. 1997). The total coal reserves threatened by bodies of water are estimated at 25 billion tons. During seam extraction from those coal mines, water inrushes from aquifers into the mining area have caused serious water disasters in China (Table 1). The consequences sometimes are dramatic and could lead to partial or even total loss of a mine (Peng 1997, 1998, 1999).

9.2 Overburden strata failure due to mining

9.2.1 Determinations of overburden strata failure

China has developed some specific methods of coal mining and experimental techniques under aquifers and surface water. Over the last 40 years, about 1000 longwall faces were extracted under surface and ground water, liberating millions of tons of coal reserves without disastrous consequences. Since coal extraction enhances hydraulic conductivity, it is desirable to determine accurately pre- and post-mining hydraulic conductivities in the overburden strata. To measure these conductivities, boreholes are drilled pre- and post-mining either on the surface or in underground observing roadways. The flow rate or circulation loss along the borehole during drilling is measured by pumping drilling mud into the borehole. Well logs are also applicable for the determination of mining induced fractures and permeability changes (Peng et al. 2002a).

In general, two failure zones that affect strata hydraulic conductivity are formed overlying the mined area: a caved zone and a water-conducting fractured zone (Liu et al. 1981, Zhang and Shen 2004). For mining under aquifers, the water-conducting fractured zone is more interesting, since it provides access for water inflow into the mine workings because of hydraulic conductivity enhancement in this zone. From in-situ testing of borehole flow rate, the water-conducting fractured zone can be divided into the following three subzones (Fig. 9.1):

(1) Slightly fractured subzone. Only little fractures are induced in the strata. Compared to the original strata, hydraulic conductivity in this zone increases slightly. The fluid circulation loss rates in the observing borehole are less than 0.1 l/s m;

(2) Moderately fractured subzone. The strata only have partial bed separations and fractures. Hydraulic conductivity in the strata increases moderately. The circulation loss rates are between 0.1 and 1.0 l/s m;
(3) Severely fractured subzone. Most of the strata have been fractured, and the fractures are interconnected. Hydraulic conductivity in the strata increases dramatically. The circulation loss rates are greater than 1.0 l/s m.

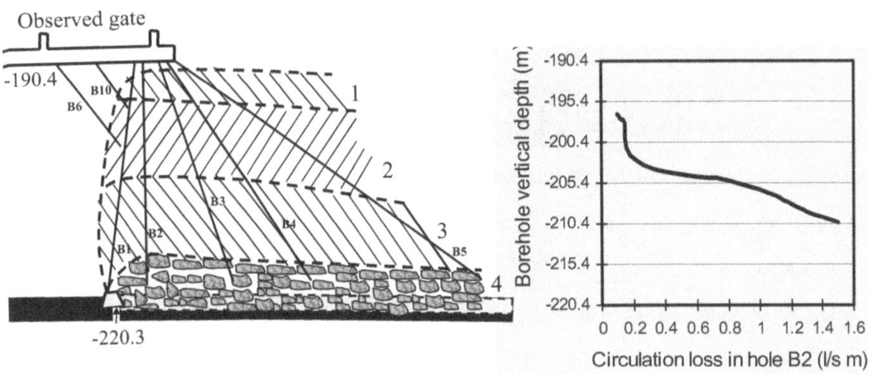

Fig. 9.1. Underground observation section of hydraulic conductivity changes and rock failure in overburden strata in Xingtai Colliery, Hebei Province, China. The numbers of 1, 2, 3, and 4 in the left figure represent the slightly, moderately, severely fractured subzones, and the caved zone, respectively. The figure in the right gives the observation results of water circulation loss rate in borehole B2.

Field observations by circulation loss measurements in boreholes while drilling have shown that the strata failure characteristics differ considerably for different inclinations of the extracted seams. For flat or slightly inclined coal seams (the dip angle, $\alpha < 30°$), the profile of the water-conducting fractured zone is broad in section with extended lobes over the headgate and tailgate, as shown in Fig. 9.2. For strong rocks, the failure zone has a different characteristic, as shown in Fig. 9.3, which is that the failure zones are much higher in the vertical direction and narrower in section.

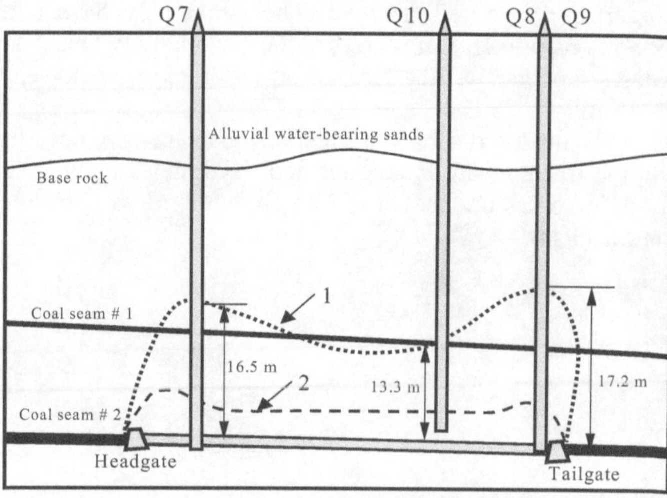

1. Water-conducting failure zone; 2. Caving zone

Fig. 9.2. Schematic observation section of hydraulic conductivity changes and rock failure in overburden strata for slightly inclined coal seam in Huangxian Colliery, Shandong Province, China.

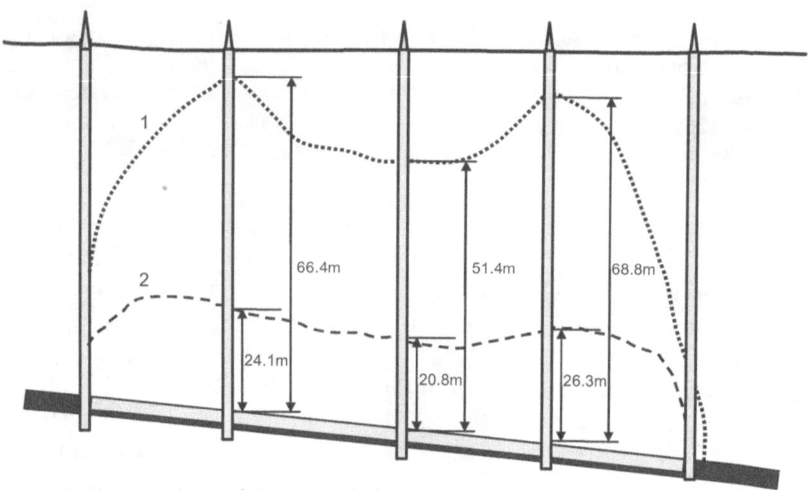

1. Water-conducting failure zone; 2. Caving zone.

Fig. 9.3. Schematic observation section of hydraulic conductivity changes and rock failure in strong overburden strata for slightly inclined coal seam in Datong Colliery, Shanxi Province, China.

For inclined coal seams (30°< α < 60°), failures zones propagate upwards in an asymmetric manner in the upseam direction as shown in Fig. 9.4. The extents of the failure zones reduce gradually from updip to downdip.

For steeply inclined coal seams (60°< α < 90°), the failure zones occur not only in the overlying and underlying strata, but also in the steeply inclined coal seam. Rubble originating updip is displaced down-seam by its gravity, causing much larger failure zone in the upseam (refer to Fig. 9.5).

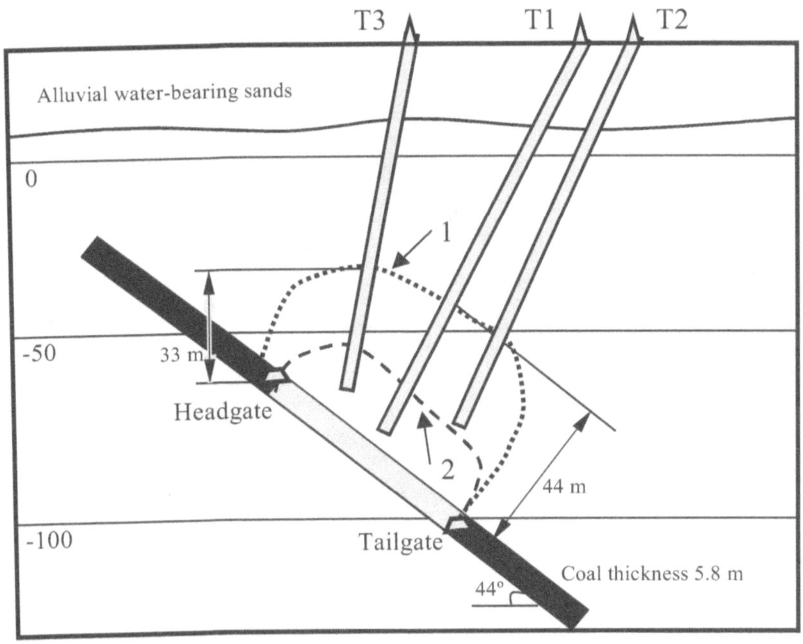

Fig. 9.4. Schematic observation section of hydraulic conductivity changes and rock failure in overburden strata for inclined thick coal seam in Huannan Liyi Colliery, Anhui Province, China.

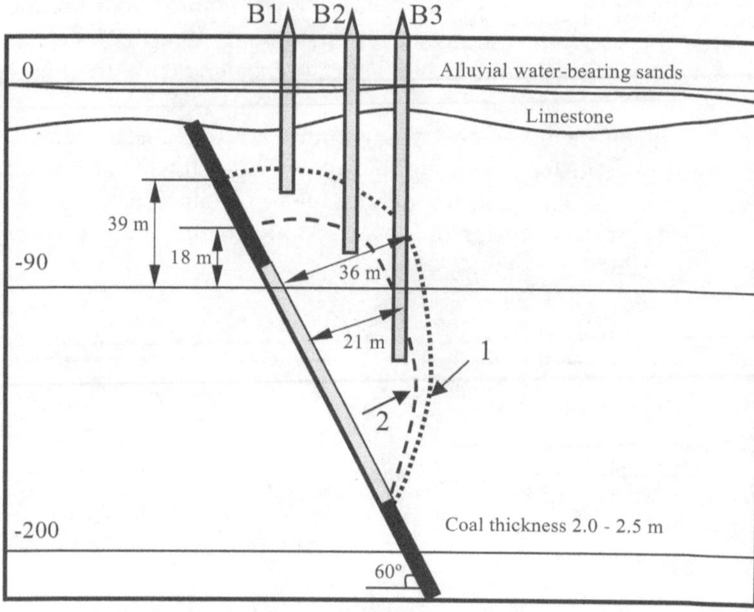

1. Water-conducting failure zone; 2. Caving zone

Fig. 9.5. Schematic observation section of hydraulic conductivity changes and rock failure in overburden strata for steeply inclined thick coal seam in Huannan Lizuizi Colliery, Anhui Province, China.

9.2.2 Empirical prediction of water-conducting height

A considerable number of in-situ observations have shown that heights of strata caved and fractured zones in the overburden formation depend primarily on the lithology and strength of the overlying strata, as well as the inclination of the extracted seam. The following formulae have been obtained according to in-situ observations in thousands of longwall faces (Liu et al. 1981, Bai and Elsworth 1990, Zhang and Shen 2004).

The maximum height of strata fractured zone is given empirically as follows.

For strong strata (UCS > 40 MPa)

$$H_f = \frac{100M}{1.2M + 2} + 8.9 \qquad (9.1)$$

where H_f is the maximum height of the fractured zone starting from the seam immediate roof (m); M is the extracted seam thickness (m); UCS is the uniaxial compressive strength.

For medium strong strata (UCS = 20 - 40 MPa)

$$H_f = \frac{100M}{1.6M + 3.6} + 5.6 \tag{9.2}$$

For weak strata (UCS < 20 MPa)

$$H_f = \frac{100M}{3.1M + 5} + 4 \tag{9.3}$$

For weathered weak strata

$$H_f = \frac{100M}{5M + 8} + 3 \tag{9.4}$$

The maximum height of strata caved zone is given empirically as follows.

For strong strata (UCS > 40 MPa)

$$H_c = \frac{100M}{2.1M + 16} + 2.5 \tag{9.5}$$

where H_c is the maximum height of the caved zone starting from the seam immediate roof (m).

For medium strong strata (UCS = 20 - 40 MPa)

$$H_c = \frac{100M}{4.7M + 19} + 2.2 \tag{9.6}$$

For weak strata (UCS < 20 MPa)

$$H_c = \frac{100M}{6.2M + 32} + 1.5 \tag{9.7}$$

For weathered weak strata

$$H_c = \frac{100M}{7M + 63} + 1.2 \tag{9.8}$$

Figure 9.6 compares the maximum height of the water-conducting fractured zone between the observation data and the empirical formula (Liu et al. 1981). It can be seen that the empirical formulas work well for estimating the height of the water-conducting fractured zone.

For mining under aquifers, it is desirable to avoid the extra expense of strata dewatering. This can only be achieved when aquifers are located outside the water-conducting fractured zone. In this case, water inflow into the mine workings does not increase. When an aquifer lies within the fractured zone, but outside the caved zone, excessive groundwater discharge to the mine occurs (according to the mining experiences in China); however, the sand in the unconsolidated aquifer does not flow into the mining area. When an unconsolidated aquifer is situated within the caved zone, both water and sand can rush into the mining area, and this may even cause disastrous consequences, if the aquifer is very permeable and strongly water-bearing.

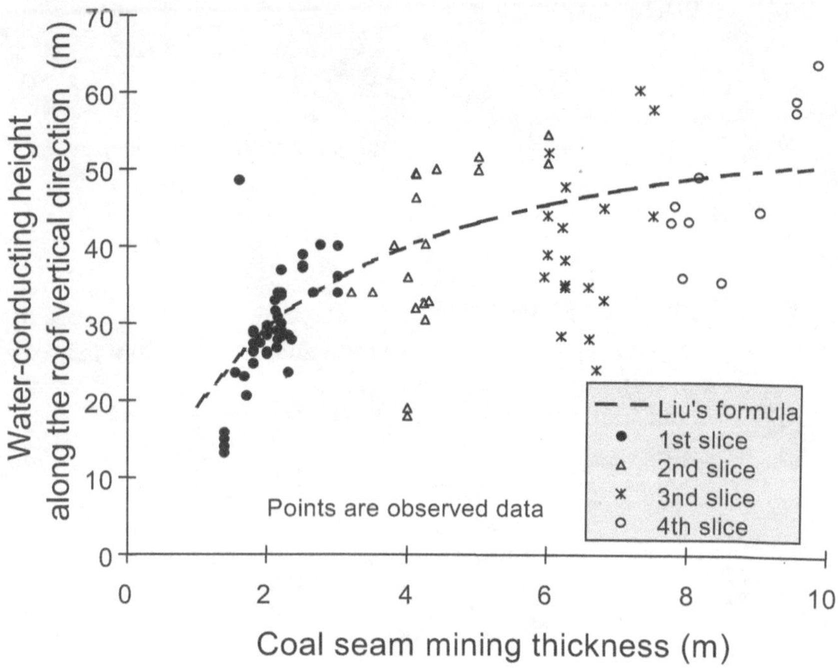

Fig. 9.6. Comparison between the empirical formula (Eq. 9.2) and observed heights of water-conducting fractured zone in overburden strata induced by coal mining with top slicing and full caving.

9.3 Case studies

9.3.1 Introduction

The Daliuta coal mine, affiliated with the Shenhua Group, is located in ShenFu Coalfield, Northern Shaanxi Province and on the southwestern bank of the Yellow River, Northern China (Fig. 9.7). It is one of the major coal mines in China. This coalfield consists of nearly flat-lying beds of Jurassic coal measure. The thickness of the primary coal seam, No. 2, is approximately 4 m with the roof consisting of medium-grained sandstones. The overlying coal measures are 19 to 65 m in thickness, comprising weak, weathered strata in the uppermost reaches. The bedrock is overlain by unconsolidated alluvium comprising mixed impermeable clay layers with water-bearing sands and gravels. The alluvium is generally 38 to 43.4 m in thickness, in which one aquifer underlies lowermost in the unconsolidated overburden. The total depth of cover for seam No. 2 ranges approximately from 20 to 100 m. Comprehensive mechanized longwall mining with full caving is used in the coal extraction.

Fig. 9.7. Location of Daliuta coal mine in Shaanxi Province, Northern China.

The coalfield has a very dry temperate climate and is situated in the southeastern border of the Maowusu Desert. Most of the surface is covered by sand, in which little vegetation exists. The water resource is very precious in this region. Only one perched aquifer in the Quaternary alluvium overlies directly on the coal measure. Therefore, the protection of the water resource and mining safety from groundwater hazards are common concerns of both the mine operator and government.

9.3.2 Water inrush incident in Face 1203

Face 1203 with a retreating longwall full caving system was the first mining panel in the Daliuta coal mine, as shown in Fig. 9.8. The thickness of seam extraction was 3.38 m. The thickness of overburden strata generally was 20 to 25 m and the minimum was 15 m, consisting of strongly to slightly weathered siltstones and mudstones. Above the strata, the formation was an unconsolidated alluvium with medium water-bearing sands and gravels.

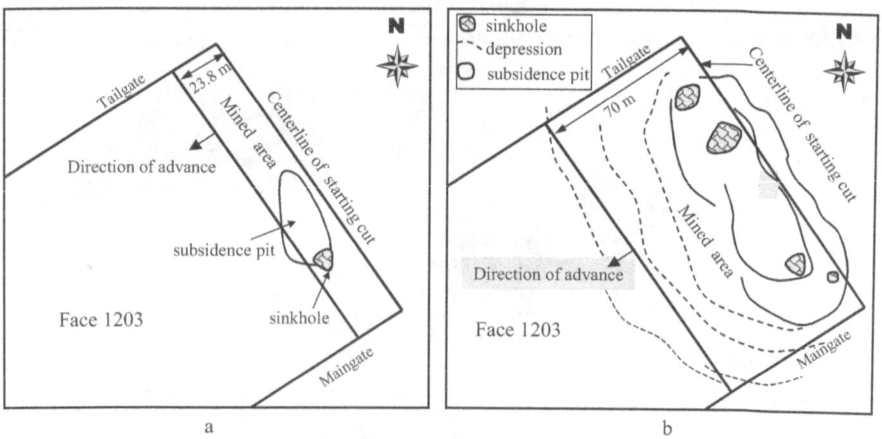

Fig. 9.8. A plan view showing the subsidence pits and sinkholes induced by water inrush in Face 1203, Daliuta coal mine: **a.** After 23.8 m of seam extraction; **b.** After 70 m of seam extraction.

When the mining face proceeded to a distance of 23.8 m, the roof pressures increased dramatically and a small amount of water flowed into the working space from three locations of the roof. Later, a 90 m length of strata in the roof was cut off along the coal seam in the middle of the face, and water with mud and sands rushed suddenly into the workings, with a maximum flow rate of 408 m^3/h. Meantime, a subsidence pit on the surface appeared and at the south corner of the pit, a cone-shaped sinkhole was formed, as shown in Fig. 9.8a. When the face was advanced to 70 m from the starting cut, the longer axis of the subsidence pit reached to 142 m, and 4 sinkholes were formed with a maximum subsidence of 2.95 m (refer to Fig. 9.8b). In investigation and analysis of the water inrush mechanism in Face 1203, the major cause of the water inrush was that the thickness of the overlying strata was too small, i.e. the distance between the aquifer and coal seam was too short. Therefore, determining the critical distance between the aquifer and coal seam is important to maintain mine safety. To find the critical distance, the mining induced failure zone and enhancement of hydraulic conductivity in the overlying strata need to be studied.

9.3.3 Case study in longwall panel Face 102

In this study Face 102 was selected as the experimental panel. The designed mining width was 220 m, the mining length along the strike direction was 2660 m and the thickness of seam extraction was 4 m. The thickness of the overlying strata of the seam was quite different, generally from 39 to 64 m, with a minimum of 2 m. An aquifer mentioned earlier varying in thickness from 2.8 to 29.8 m overlay directly on the strata.

Finite element numerical simulation

The finite element method (FEM) was used to simulate the strata failure and hydraulic conductivity changes associated with coal mining in Panel 102. The finite element calculation was conducted in Computational Lab, China Coal Research Institute by using a finite element software. The stress-dependent hydraulic conductivity derived by Zhang et al. (1997) was incorporated into the finite element method. The hydraulic conductivity and strata failure resulting from seam extraction were calculated for several different mining widths. In this model the depth of cover was 90 m, thickness of extraction was 4 m, thickness of immediate roof was 25 m, thickness of the roof was 30 m, and total thickness of the strata was 180 m. The fracture spacing was assumed as 1 m. Table 9.2 gives the rock mechanical parameters used for the finite element calculation. Due to the symmetric geometry, a half panel and a similar FEM mesh as given in Zhang et al. (2001) were adopted for the ongoing analyses.

Figures 9.9, 9.10, and 9.11 give the Drucker-Prager failure zones for different mining widths. It can be seen that the rock failure zone decreases as the mining width decreases. When the mining width is 220 m, the maximum vertical height of the failure zone in the overlying strata is 42.5 m. That means that the water may flow into the workings when the distance between the coal seam and the aquifer is less than this magnitude.

Table 9.2. Rock mechanical parameters in FEM calculation

Depth (m)	Rock	E (MPa)	ν	K_n (MPa/m)	UCS (MPa)
0 - 35	unconsolidated	540	0.5	2500	-
35 - 65	sandstone	2160	0.32	13000	45
65 - 90	shale	1935	0.38	11000	35
90 - 94	coal	1500	0.36	4500	25
94-112	shale	1935	0.32	11000	35
112-180	sandstone	2160	0.38	13000	45

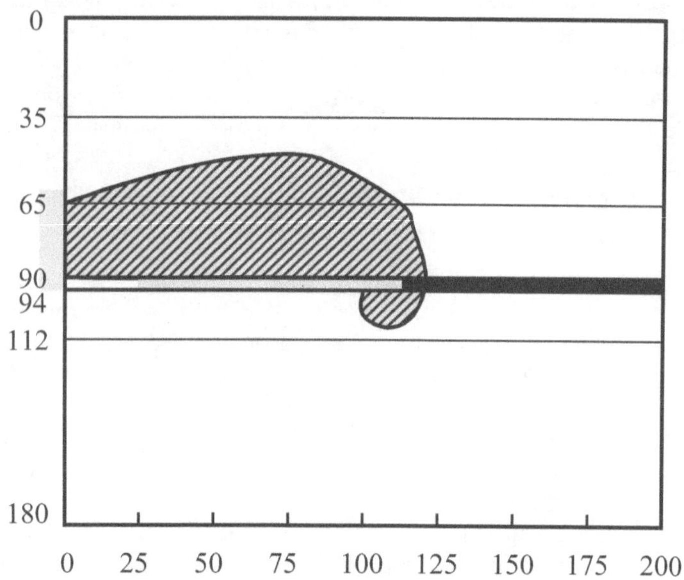

Fig. 9.9. The FEM calculated failure zone for the mining width of 220 m.

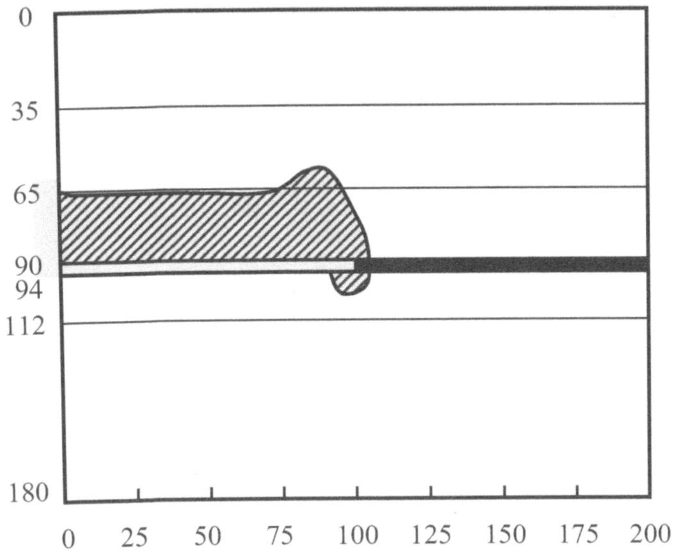

Fig. 9.10. The FEM calculated failure zone for the mining width of 200 m.

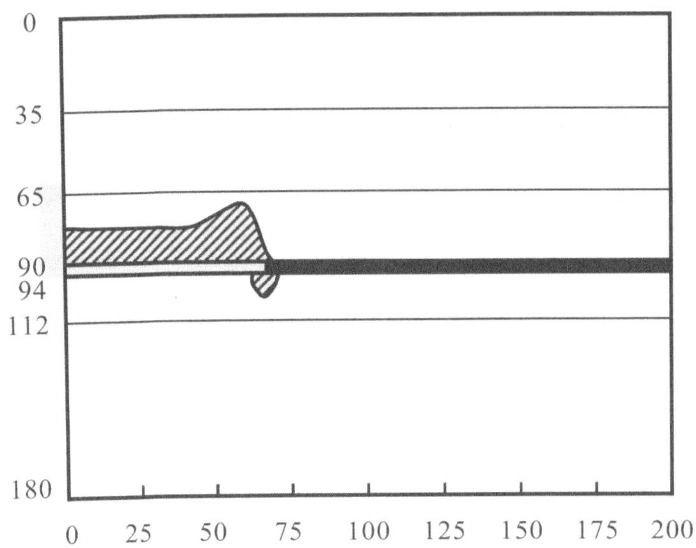

Fig. 9.11. The FEM calculated failure zone for the mining width of 120 m.

Experimental results have shown that hydraulic conductivity is very dependent on the stress changes, and various relationships between the changes of hydraulic conductivity and stresses have been presented. The following relationship was adopted in this FEM analyses to calculate hydraulic conductivity changes due to stress variations (refer to Eq. 8.22 in Chap. 8 Sec. 8.3):

$$k_k = k_{0k}\left\{1-\left(\frac{1}{K_{ni}b_i}+\frac{1}{K_{ni}s_i}+\frac{1}{E_r}\right)\left[\Delta\sigma_i-\nu\left(\Delta\sigma_j+\Delta\sigma_k\right)\right]-\left(\frac{1}{K_{nj}b_j}+\frac{1}{K_{nj}s_j}+\frac{1}{E_r}\right)\left[\Delta\sigma_j-\nu\left(\Delta\sigma_i+\Delta\sigma_k\right)\right]\right\}^3 \tag{9.9}$$

where k_{0k} and k_k are the hydraulic conductivities along the k-direction before and after stress changes, K_{ni} is the fracture normal stiffness in the i-direction, b_i is the initial fracture aperture in the i-direction, s_i is the initial fracture spacing in the i-direction, ν is Passion's ratio, E_r is the Young's modulus of the rock matrix, $\Delta\sigma_i$ is the stress change in the i-directions, $i = x, y, z; j = y, z, x; k = z, x, y; i \neq j \neq k$.

Incorporating Eq. 9.9 into the finite element method and assuming the initial fracture aperture $b_i = b_j = 0.001$ m, hydraulic conductivity due to the stress changes induced by the extraction was calculated. In the following presentation of figures, the contours for the ratio of post- to pre-mining hydraulic conductivities are plotted. Fig. 9.12 presents the vertical hydraulic conductivity contour around the mined area for the mining width of 220 m. It shows that the enhancement of the conductivity concentrated mainly around the mined area, including the strata in the floor. It can be observed that along the vertical direction, the vertical conductivity no longer increases when the strata are located at 46.2 m high from the seam.

Fig. 9.13 shows the hydraulic conductivity in the horizontal direction for the mining width of 220 m. Comparison of Fig. 9.13 and 9.12 shows that the range of the conductivity enhancement in the horizontal direction is much larger than that in the vertical direction. Since the aquifer is located directly above the overlying strata in this case, the conductivity in the vertical direction dominates the water flow.

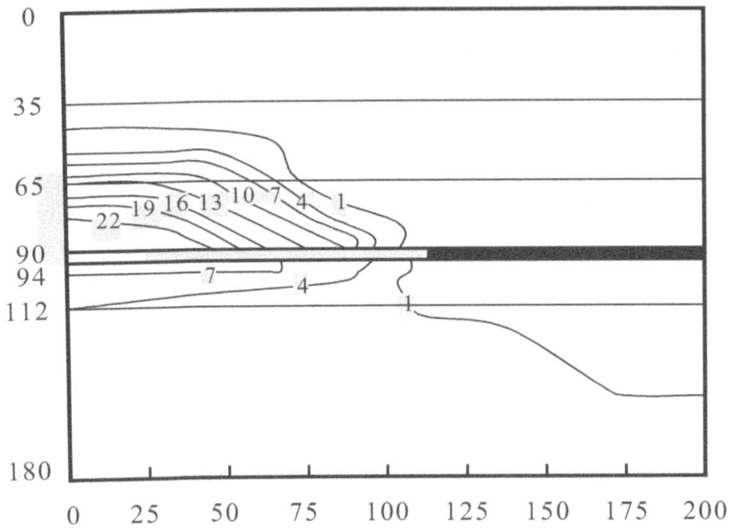

Fig. 9.12. Variations of hydraulic conductivity in the vertical direction for the mining width of 220 m.

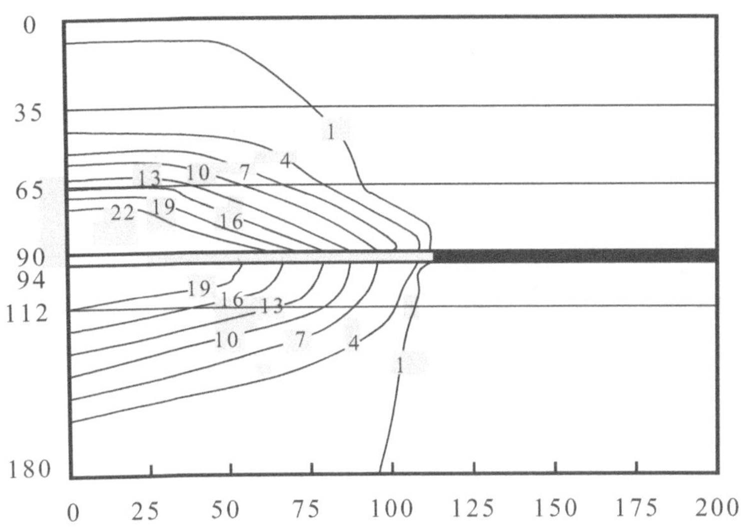

Fig. 9.13. Variations of hydraulic conductivity in the horizontal direction for the mining width of 220 m.

Lab simulation using physical models

Physical simulation by using a similarity model was conducted to verify the foregoing numerical calculations, particularly the height of the strata fractured zone, which is the most critical criterion for design of mining under aquifers. The model was built in layers by using a mixture of water, fine-grained sand, and gypsum, with different mass proportions to simulate the strata with different strengths. The joints between layers and within each layer were made and filled with micas. The model measured 240 cm in length by 156 cm in height and 20 cm in width. The geometry ratio of the simulation model to the in-situ panel was $C = 1:100$, i.e., 1 cm in the model is equal to 1 m in the prototype. The bulk densities of the strata in the prototype and the simulated materials were 2,500 and 1,600 kg/m^3, respectively. Therefore, all the related simulation constants were as follows:

- for bulk density: C_γ=1600/2500 = 0.64;
- for strength and stress: $C_\sigma = CC_\gamma = 0.0064$;
- for loading: $C_F = C^3 C_\gamma = 6.4\times10^{-7}$;
- for time: $C_t = C^{0.5} = 0.1$.

Displacement gauges and stress sensors in both vertical and horizontal directions were installed to study the stress distribution, strata deformation and failure induced by seam extraction. In this physical model, the simulation height of the overburden was 90 m, including 35 m of unconsolidated alluvium and 55 m of roof strata, the thickness of the underburden was 62 m, and the thickness of the extraction seam was 4 m. Three models were conducted and extracted repeatedly for mining with full caving to ensure accuracy and one other model was constructed to model mining with partial sand backfilling. Figure 9.14 presents the model geometry and locations of stress sensors. The six horizontal layers of stress cells along five vertical sections were installed in different heights: four layers in the roof and two layers in the floor.

The extraction was simulated by cutting the seam from left to right (Fig. 9.14) in intervals of 5 cm. During the extraction 60 cm of pillar was left in the left side and 50 cm in the right side to avoid boundary effects. The roof failure, stress and displacement were measured and recorded for each cutting step as advance of seam extraction.

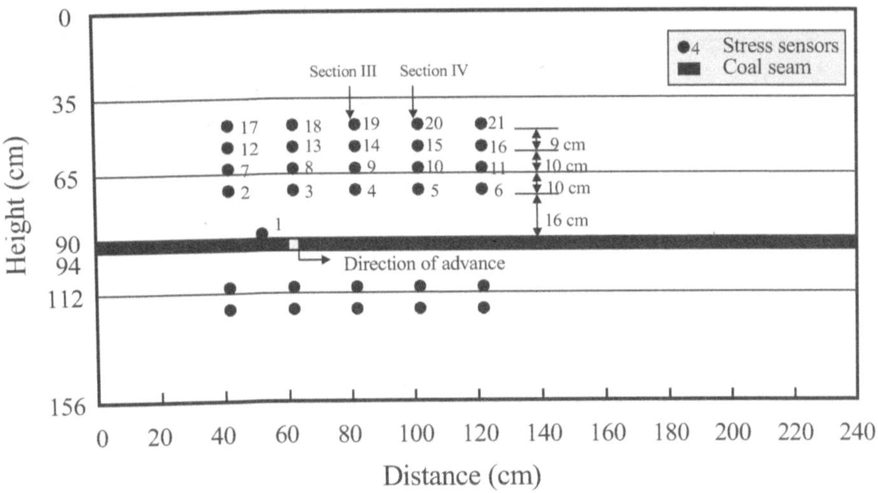

Fig. 9.14. A plan view of the simulation model showing the locations of stress sensors (1 cm in the model = 1 m in the prototype).

Figure 9.15 shows the vertical stress increments (the difference between the induced and origin stresses) in Section III located 20 m away in the direction of advance from the starting cut (refer to Fig. 9.14). It can be seen that only one stress increment in the four sensors increases (so called abutment pressure) before the mining face approaches the sensor located closest to the coal seam. After the face passes the sensors, the stress increments in all four sensors decrease gradually, i.e. the strata experience a state of unloading in which the induced fractures developed easily (Zhang and Shen 2004). This phenomenon is more pronounced for the strata closer to the seam. Figure 9.16 shows the vertical stress increments in Section IV located 40 m in front of the starting cut (see Fig. 9.14). The abutment pressures exist in all four sensors before the face advances through the observed section. These higher stresses are caused by the transfer of the weight of the overburden roof strata in the excavated area to the solid coal in front and on both sides of the face and to the broken strata behind the face. After the strata in the observed section lie in the extracted area, the stress redistribution, as advance of the extraction, experiences three stages, i.e. dramatic decrease and gentle reduction, and gradual recovery to the origin stress.

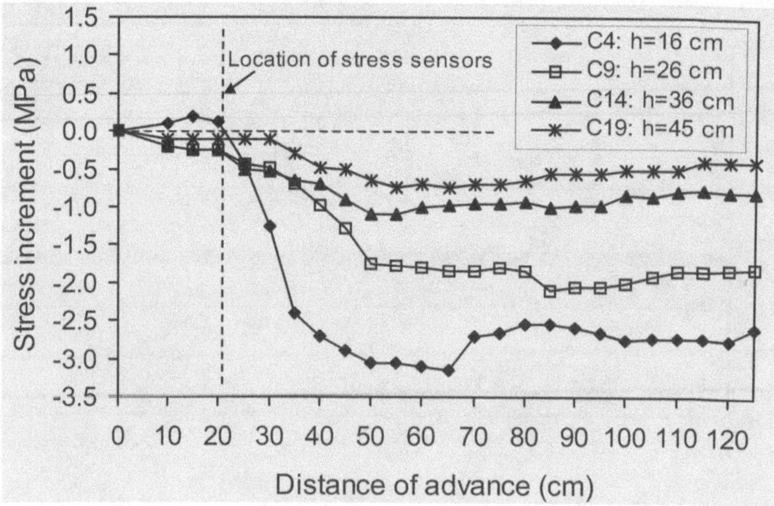

Fig. 9.15. Observed vertical stress increment for a section with the stress sensors located 20 cm from the starting cut (1 cm in the model = 1 m in the prototype).

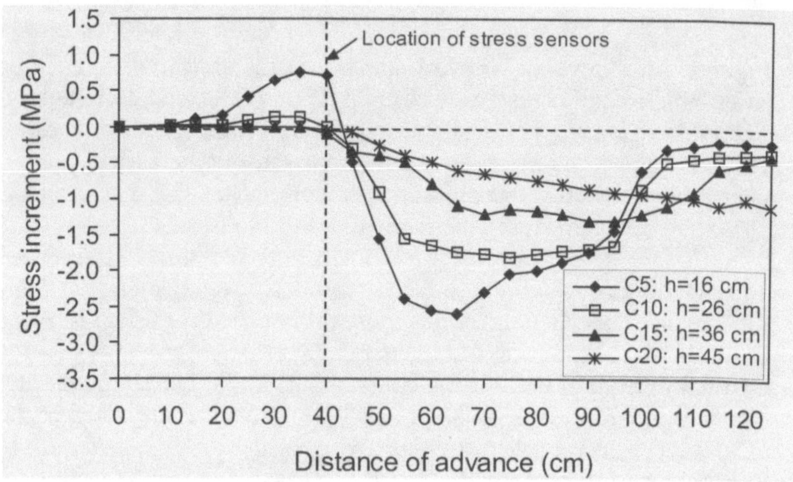

Fig. 9.16. Observed vertical stress increment for a section with the stress sensors located 40 cm from the starting cut (1 cm in the model = 1 m in the prototype).

The strata fractured zone was recorded during the seam extraction. Figure 9.17 plots the observed height of the fractured zone with the distance of advance of the extraction. It shows that the height of the fractured strata increases gradually with extraction until reaching a maximum of 41 m as the face advances 100 m.

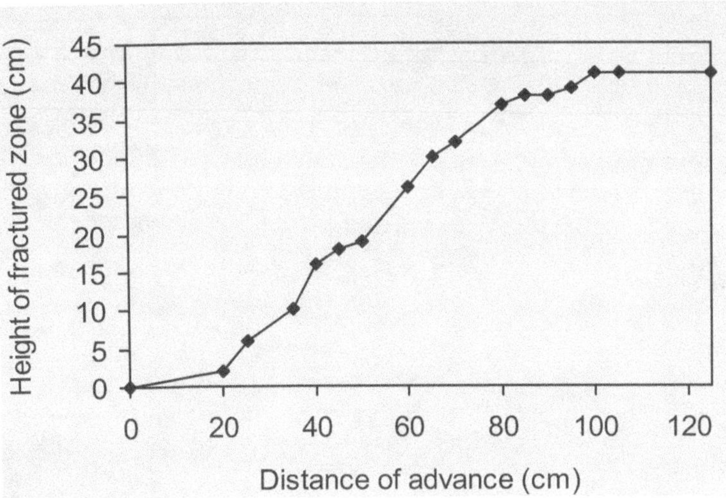

Fig. 9.17. Observed development of the fractured zone with advance of the seam extraction (1 cm in the model = 1 m in the prototype).

The changes of hydraulic conductivity due to stress redistribution were calculated by Eq. 9.9 according to the observed vertical and horizontal stresses in the physical models. Only the vertical conductivity ratio (K/K_0, post- to pre-mining conductivities) was considered in this case. Figure 9.18 presents the conductivity ratio in a section of the strata located 20 m in front of the starting cut (Zhang and Shen 2004). It is observed that the conductivity is basically unchanged before the face reaches the studied section. For the strata located 16 and 26 m above the seam, after the face passes the section (the strata lie in the mined area), the conductivities increase considerably; as the conductivities reach maximum magnitudes (such as K/K_0 = 27.4, when h = 16 m), the increases slow down. For the strata located 36 and 45 m over the seam, after the face passes the section, the conductivities only have very small increases (such as, the maximum K/K_0 = 2.5 for h = 36 m). Figure 10.19 plots the conductivity ratio in a section of the strata located 60 m in front of the starting cut. Comparing Fig. 10.19 to Fig. 9.18 it can be seen that the conductivity changes have same trend except that Fig. 9.19 has a smaller magnitude for the strata located 16 and 26 m above the seam after the strata lie in the mined area. This is due to the fact that the studied strata in Fig. 9.18 are closer to the abutment area where a larger stress change prevails. It is also observed from Figs. 9.18 and 9.19 that the conductivity only has a very small variation when the strata lie 45 m above the extracted seam.

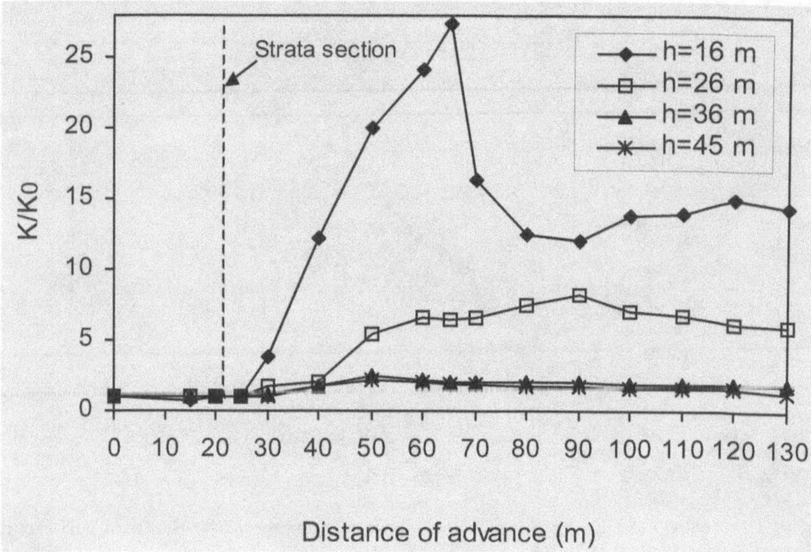

Fig. 9.18. Calculated hydraulic conductivity ratios for a section with the strata located 20 m from the starting cut.

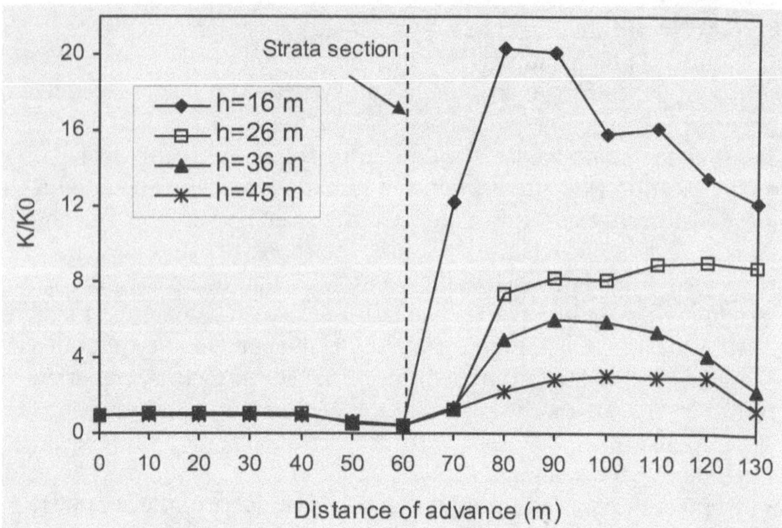

Fig. 9.19. Calculated hydraulic conductivity ratios for a section with the strata located 60 m from the starting cut.

When the mined area is backfilled, the stress changes are greatly reduced. Figure 9.20 gives the conductivity change in a physical model in which the extracted area was backfilled partially 30 m from the cut with sands. Compared to the full caving extraction in Figs. 9.18 and 9.19, the conductivity in the partially backfilled case as shown in Fig. 9.20 has very small changes after mining. This indicates that the mining operation will be much safer for mining with backfilling, when aquifers are near the mining workings.

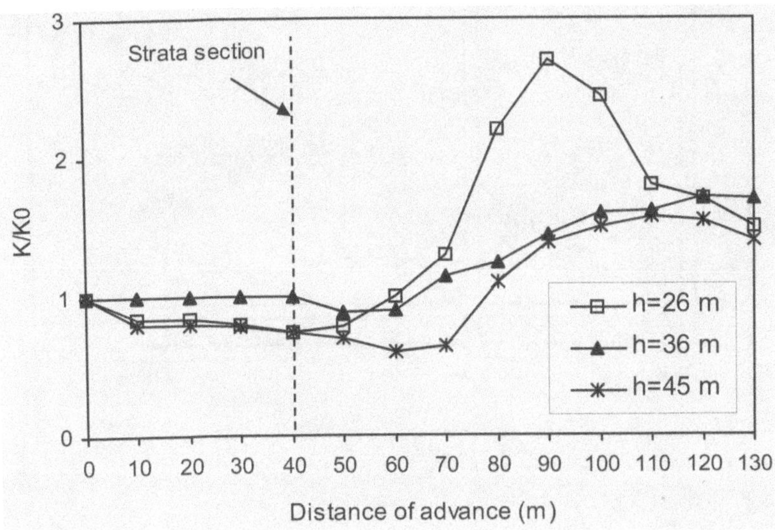

Fig. 9.20. Calculated hydraulic conductivity ratios for a section with the strata located 40 m from the starting cut, for the extraction with partial backfilling.

Field monitoring in Face 102

Summarizing the numerical calculation and physical simulation, the height of the strata failure zone ranges from 41 to 42.5 m, and the height of the conductivity enhancement zone in the vertical direction varies from 45 to 46.2 m. Furthermore, from the empirical formula (Eq. 9.2) it can be obtained that for the strata with medium strength the maximum height of the water-conducting fractured zone is 45.6 m. According to the above results, it can be concluded that disastrous water inrush would not occur when the distance between the aquifer and coal seam is greater than 46.2 m. Therefore, the old design for Face 102 was revised and the location of the starting cut was changed, as shown in Fig. 9.21, since the strata thickness was less than 46.2 m around the old location.

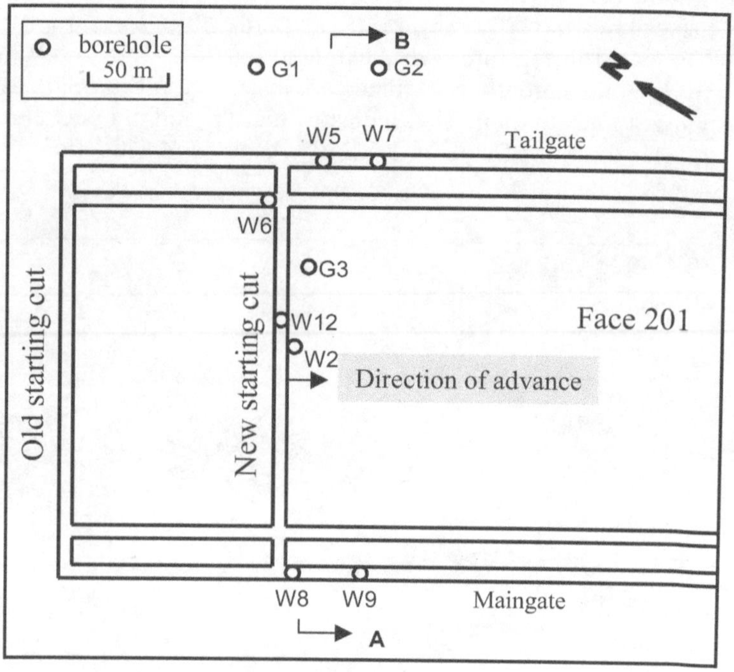

Fig. 9.21. A plan view for mining panel Face 102 that shows the location of observing boreholes.

In order to secure the mining operation, several boreholes (refer to Fig. 9.21) were drilled to monitor the changes of the water table in pre- and post-mining. The boreholes were located near the new starting cut, since the thickness of the covering strata in this area was the smallest in this panel (Zhang and Shen 2004). Figure 9.22 shows the geologic section of the face around the new cut. Note that the shortest distance between the aquifer and the mining workings was 46.73 m, located in borehole W2. Therefore, this was the key location needed to be monitored. Figure 9.23 records the changes of water tables in boreholes W2 and G3 pre- and post-mining. The starting time of seam extraction in Face 201 was July 10. Before the extraction, the water tables in boreholes W2 and G3 lowered slightly, as shown in Fig. 9.24. On July 16 when Face 102 advanced 10 m (the face just passed borehole W2), the water table in this hole lowered moderately, illustrating that the strata conductivity increased in this area. On August 11 when Face 102 advanced 60 m (the face passed borehole W2 50 m), the water table in W2 dropped dramatically. The water table in

W2 decreased 11.18 m, relative to that 2.5 hrs earlier. However, the water table in the adjacent borehole G3 and other holes only had small fluctuations (refer to Figs. 9.22 and 9.23). These indicate that the strata conductivity had a great enhancement in the strata located 46.73m high (hole W2) from the seam, but no obvious increase in the strata located 51.95 m away from the seam (hole G3). Later, the water table in W2 increased gradually. However, it could not recover to the pre-mining level. The water table variation in W2 induced by the hydraulic conductivity change verifies the previous calculated results shown in Figs. 9.18 and 9.19. That is, the post-mining conductivity increases gradually, then reaches a maximum after a certain distance (20 to 60 m), and finally decreases to a certain value. Since the minimum strata thickness in Face 102 was strictly controlled and was not allowed to be less than the calculated value, 46.2 m, the seam was successfully extracted and 2,825 thousand tons of coal reserve was recovered with neither water inrush nor even excessive groundwater discharge to the workings.

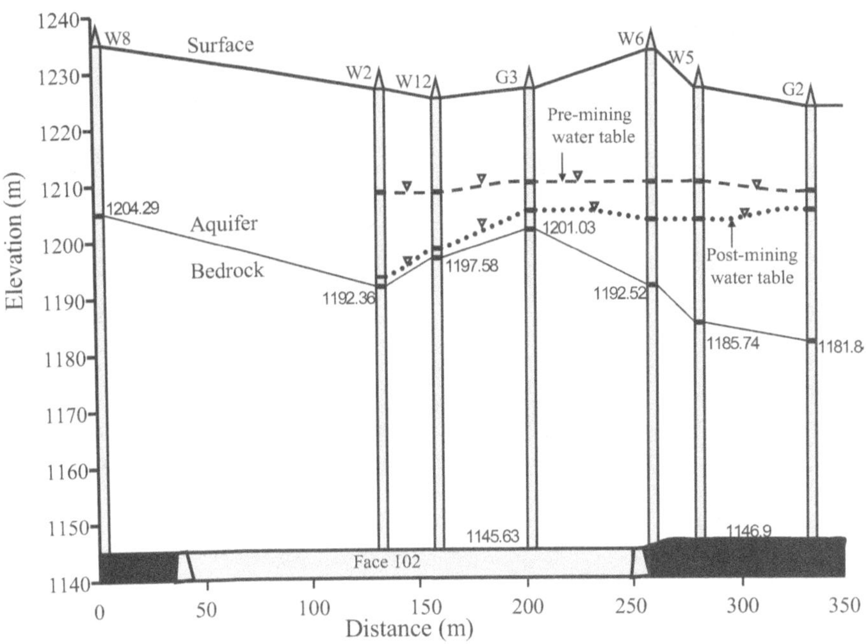

Fig. 9.22. The A-B section in Fig. 9.21 illustrating the aquifer location, pre-mining water table and the maximum water table drop induced by mining.

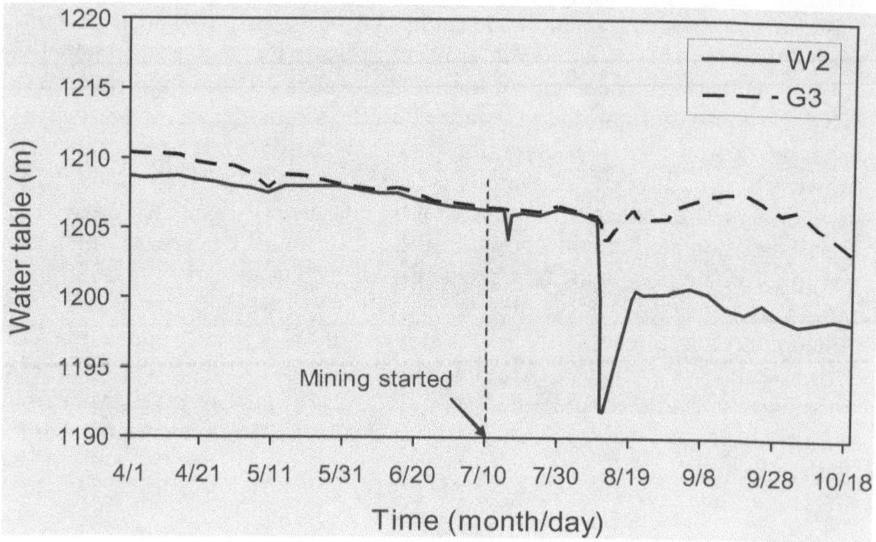

Fig. 9.23. Changes of water tables (elevations from the sea level) pre- and post-mining in observing boreholes of W2 and G3.

9.4 Mining under bodies of water at shallow depths

9.4.1 Introduction

To enhance recovery, it is often necessary to extract coal seams in very difficult geological and hydrogeological conditions and at shallow and deep depths (Peng et al. 2002b). However, in some cases coal production, environment, and lives of miners have suffered serious threats from water inrushes, methane explosions and roof collapses. In a considerable number of coalfields in China the coal beds are covered by unconsolidated Cenozoic alluvium, such as those in the Yellow river and the Huai river alluvial plain areas. The alluvium is comprised of mixed impermeable clay layers with water-bearing sands and gravels. Usually, the lowermost layer in the alluvium is permeable water-bearing sands mixed with gravels. Underlying the aquifer is the bedrock of coal beds. The water-bearing layers, particularly the lowermost aquifers, are potential threats to safe mining at shallow depth. In China, water inrush incidents happen frequently during mining at shallow depth. Statistical results show that water inrushes take place about 125 times annually, resulting in serious economic loss and personnel casualties. Figure 9.24 shows the schematic relationship between the alluvium, coal seam and surrounding strata. The size of the outcrop coal pillar is defined as the vertical distance between the lowermost layer in the alluvium and the uppermost part of the mining face (usually called mining tailgate). If the mining face is very close to the alluvium aquifer, water inrush from the aquifer will be a serious threat. Therefore, the optimal design of the pillar size is of crucial importance for safe production and groundwater protection. If the size is too large, it will lead in loss of coal resources. Liu (1998) made statistical analyses of 20 coal mines with conservative designs of coal pillars in China. Total reserves in those mines were as high as 780 million tons, about 40 million tons for each mine. However, if the size is too small, then the water inrush will cause serious consequences, such as mine submersion, surface sinkhole collapse, and destruction of groundwater resources. There is a paradox between enhancing recovery and environmental protection, and the best solution should be safe mining with maximum coal recovery.

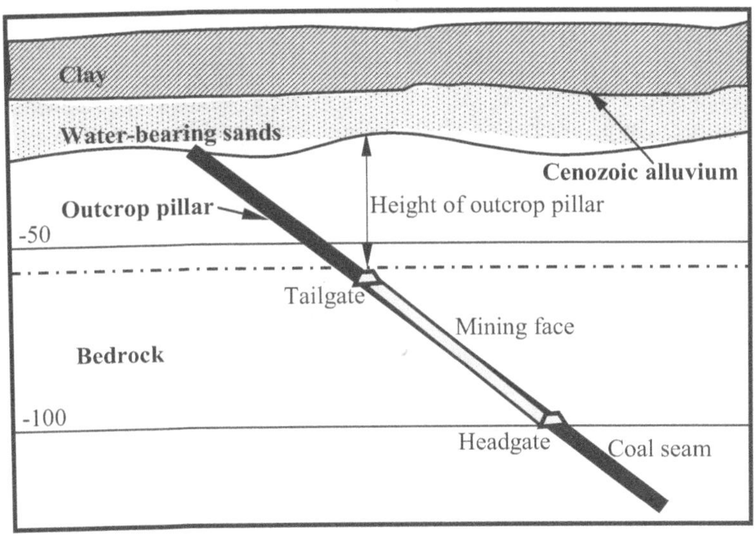

Fig. 9.24. A typical longitudinal section to illustrate the relation between the alluvium and its underlying coal measures strata.

In the USA and other countries, groundwater, domestic water supply, and mining safety are also affected by shallow coal seam mining (Booth et al. 1998). For example, most alleged domestic water supply cases in 73 investigations were impacted by shallow underground mining in Virginia (Zipper et al. 1997). An extensive hydrological monitoring program was conducted at a longwall coal mine in West Virginia. In-situ experiments showed that when the monitoring wells were 36 m vertically away from the mining seam, the water wells went dry after mining and did not recover (Hasenfus et al. 1990, Liu and Elsworth 1997). Even though aquifers were not affected by mine drainage, they still exhibited changes in groundwater chemistry induced by mine subsidence (Booth and Bertsch 1999).

9.4.2 Incidents of mining at the shallow depth

Many incidents have happened in China during mining at shallow depths, causing serious loss of personnel casualties and in environment quality. The accidents have primarily been caused by the following aspects.

Small Outcrop pillar size

When the outcrop pillar size is not large enough to prevent mining induced fractures from reaching the lowermost aquifer in the alluvium, water, or even sand in the aquifer, can intrude into the mining space, causing excessive water drainage and destructive surface subsidence. For example, in the Daliuta coal mine of the Shenhua Corporation, a water inrush caused a serious impact on groundwater resource and farm land in 1995, as shown in Fig. 9.8.

Influence of faults

At shallow depth, when tunneling and mining expose faults, collapses of the coal seam roof with water and sand inrushes may occur. The strata near a fault have a much lower strength and are sometimes poorly consolidated. Furthermore, if the fault is permeable, water inrush occurs when the fault is unveiled. Many shallow mining incidents of water inrushes were caused by faults. Figure 9.25 shows a typical example of water and sand inrush from the alluvium into the mining face. The incident took place in the Lujiatuo coal mine of the Kailuan coalfield in Hebei Province. The thickness of the coal seam was 4 m. The hydraulic mining method was used in Face # 3371. The rocks in the roof were very fissured, water-bearing, and poorly consolidated. A faulted zone existed nearby. When the mining face was 15 m away from the fault, the roof collapsed, and rocks in the faulted zone rapidly fell into the mining space. Sand and mud with water from the alluvium intruded into the mining face. Consequently a cone-shaped collapse pit with a diameter of 14 m and depth of 4.2 m was formed on the surface.

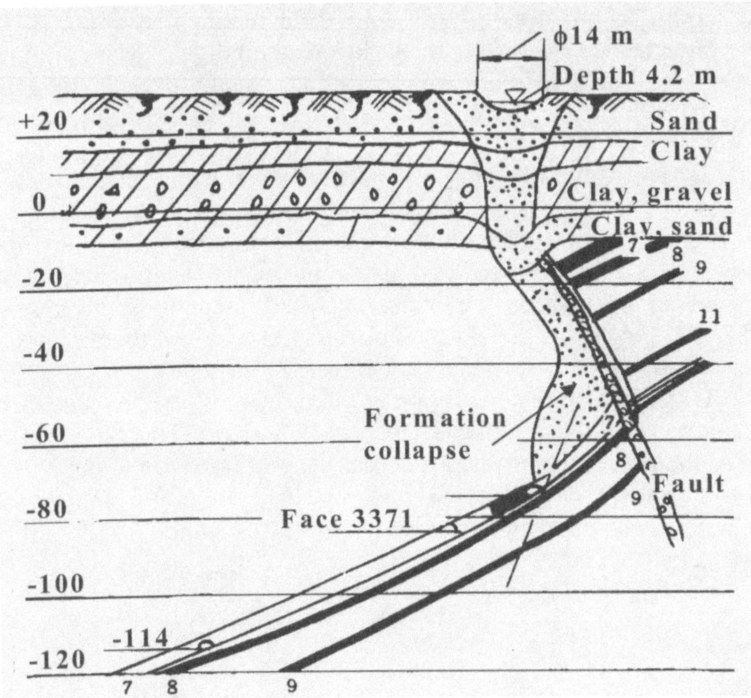

Fig. 9.25. A longitudinal section showing water and sand inrush incident and surface collapse occurred in the Lujiatuo coal mine due to fault effect (Liu 1998).

Steeply inclined seam

When extracting the steeply inclined seam from the high strength of the surrounding strata, the outcrop coal pillar is more likely to collapse into the mining space, allowing water and sand inrushes, alluvium collapses, and sinkholes (Liu et al. 1981, Zhou 1997).

9.4.3 Overburden strata failure induced by the shallow mining

Generally two failure zones that affect strata hydraulic conductivity are formed in the strata overlying the mined area: caved zones and fractured zones. On the surface, both subsidence troughs and tensile fissures are created due to shallow mining. Table 9.3 lists observed surface fissures induced by underground longwall mining in alluvial areas.

Figure 9.26 presents a typical observed section for strata failures at a deep depth. The profiles of the caved and fractured zones are broad in section with extended lobes over the headgate and tailgate. Maximum heights of caved and fractured zones are in the strata directly over the tailgate. Observed data show that the failure zones decrease as the strata become increasely softer. For shallow mining, because the strata are very close to the weathered rocks, the strata become much weaker. Consequently the failure zones have lower maximum heights and more specific features than those in normal conditions (Liu 2000 personal communication). Figure 9.27 shows the in-situ measured failured zone of the shallow depth in the Yangzhuang coal mine, Feicheng coalfield, Shandong Province (Liu 1998). The heights of the caved and fractured zones in the upseam of the strata are much smaller, which is opposite from the deeper mining (Fig. 9.26). This is due to that the strata in the upseam in the shallow strata are much softer than those in the downseam (Fig. 9.28). The experimental results of the core samples obtained from the X-2 borehole in the Xuzhuang coal mine show that the uniaxial compressive strength of the rock sample (σ_c) increases as the depth of cover (h) increases, which has the following relationship (Zhang and Peng 2005):

$$\sigma_c = 0.058e^{0.04h} \tag{9.10}$$

Table 9.3. Observed depths of surface fissures induced by longwall mining in some China's coal mines

Mine, coalfield	Depth of surface fracture (m)
Qinghemen, Fuxin	0.4 – 0.6
Tangjiazhuang, Kailuan	5 – 6
Fangezhuang, Kailuan	1.76
Shengli, Liaoyuan	5
Shengli, Fushun	7 – 8
Suncun, Xinwen	2.5 – 3.0
Chaili, Zaozhuang	6 – 10
Zilia Nuoer, Xinjiang	1.9 – 2.0
Bijiagang, Huainan	2.8 – 3.0

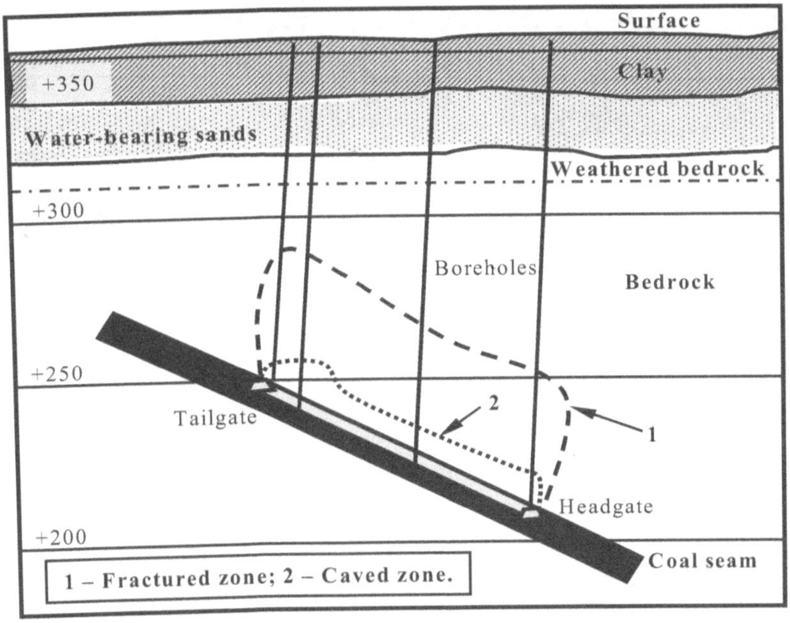

Fig. 9.26. The observed strata failures induced by mining in a coal mine.

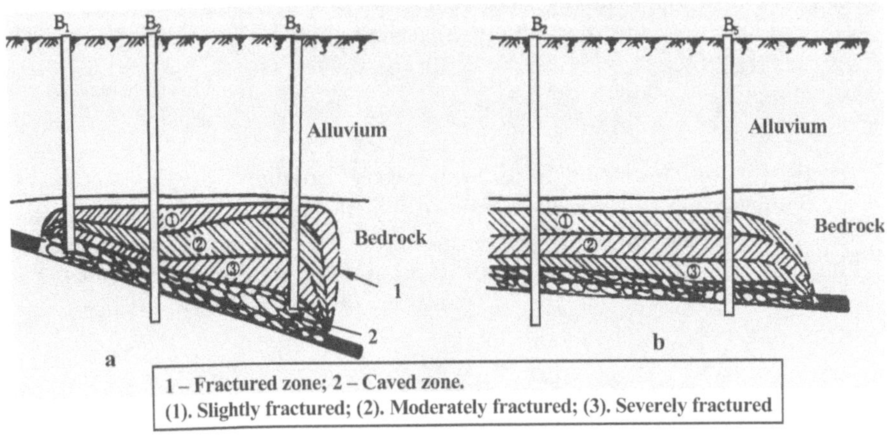

Fig. 9.27. Downhole observed strata failures induced by shallow mining in a coal mine. **a.** Along the inclined direction; **b.** Along the striking direction.

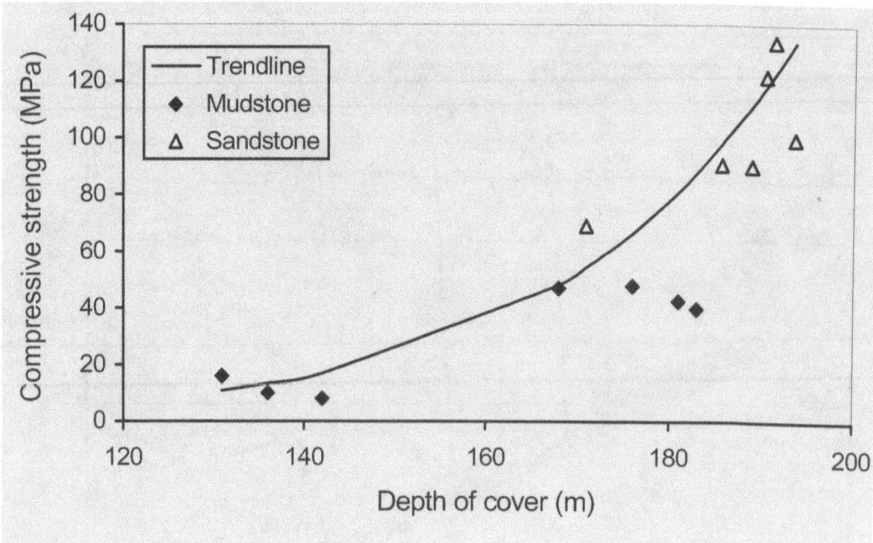

Fig. 9.28. Rock compressive strength versus the depth of cover (mining depth) in X-2 borehole in the Xuzhuang coal mine, the Datun coalfield in Jiangsu Province.

The in-situ observation demonstrates that as the mining depth decreases, the height of the fractured zone decreases (Fig. 9.29). The observations from the Shuangyang coal mine, Shuangyashan coalfield in Heilongjiang Province have shown that the height of the fractured zone increases (H_f) linearly with mining depth (h), i.e.:

$$H_f = 0.32h + 1.63 \tag{9.11}$$

In-situ measurements from 19 boreholes in the Chaili coal mine in Shandong Province show the height of the fractured zone decreases as the height of the outcrop pillar decreases (Fig. 9.30). The relationship between the ratio of the height of the fractured zone to the mining thickness (H_f/M) and the outcrop pillar height (h_p) can be expressed as follows (Zhang and Peng 2005):

$$H_f / M = 0.156h_p + 2.61 \tag{9.12}$$

Figure 9.31 illustrates the relationship between the fractured zone shape and mining depth. As the mining depth or outcrop pillar decreases, the maximum height of the fractures zone decreases, which is advantageous for the shallow mining.

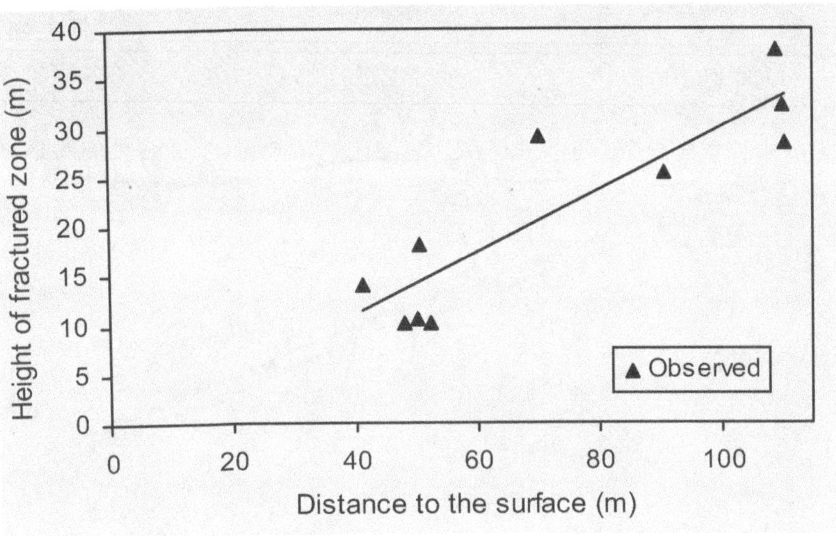

Fig. 9.29. The observed height of fractured zone versus the distance from mining face to the surface in Shuangyang coal mine.

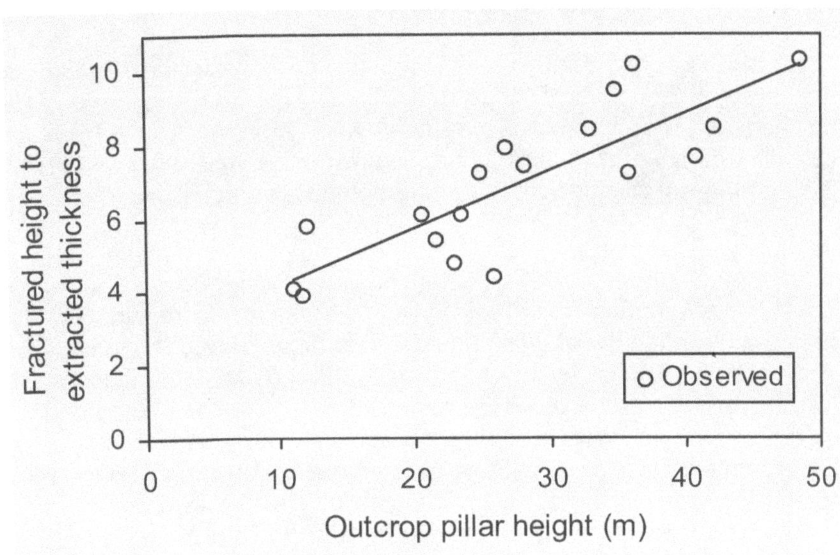

Fig. 9.30. The observed height of the fractured zone to mining thickness versus the outcrop pillar height in the Chaili coal mine.

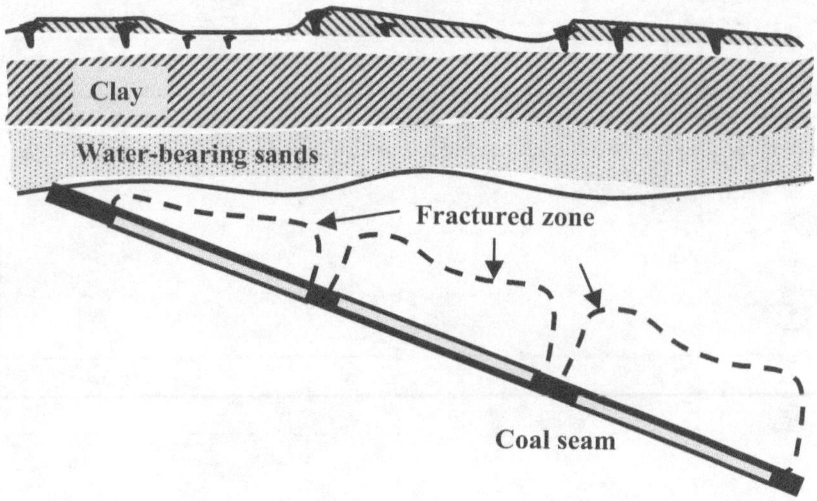

Fig. 9.31. A schematic diagram showing the shapes of the fractured zone versus the outcrop pillar size or the mining depth.

9.4.4 Design of outcrop coal pillars

There are three types of outcrop coal pillars, i.e., the water-proof pillar, the sand-proof pillar and the caving-proof pillar (Zhang and Peng 2005). Their functions, sizes and applicable conditions and response to hydrogeology vary. The design of the pillars in the shallow mining are based primarily on the hydrological characteristics, strata lithology and mining conditions.

The water-proof pillar

The water-proof pillar is used for preventing groundwater or surface water to flow into mining workings through the mining-induced fractured zone. The water-proof pillar guarantees successful seam extraction with neither water inrush nor even excessive groundwater discharge to the mine. It requires that the fractured zone does not penetrate upwards into the overlying aquifers. In this case, both water resources and mining safety can be achieved. When the alluvial aquifer, particularly the lowermost water-bearing layer, belongs to a strong aquifer (unit flowrate $q > 1.0$ l/m·s), or medium strong aquifer (unit flowrate $q = 0.1 - 1.0$ l/m·s), or any aquifer used as a water resource and needed to be protected, the water-proof pillar is applied. Normally, the minimum height of the water-proof pillar must not be less than the maximum height of the fractured zone and the height of a protective layer, as shown in Figs. 9.32a and 9.32b, i.e.

$$H_w \geq H_f + H_p \tag{9.13}$$

where H_w is the vertical height of the water-proof pillar; H_f is the maximum vertical height of the fractured zone, which can be obtained from the empirical formulae given in the previous section (Eqs. 9.1 - 9.4); H_p is the height of the protective layer (refer to Table 9.4, Liu 1998).

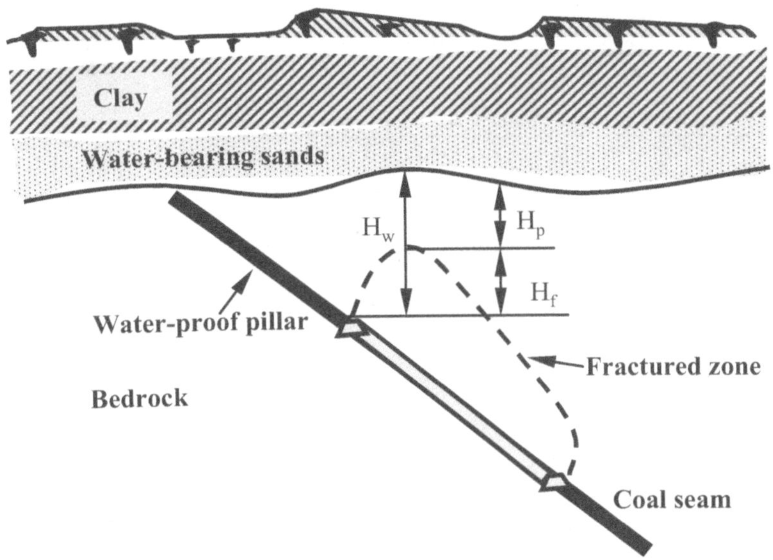

Fig. 9.32 a. A schematic diagram showing the water-proof pillar consisting of the fractured zone and the protective layer in slightly inclined strata.

If there is no alluvium covered overlying the coal measures or the mining depth is very small, the maximum depth of the surface fissures induced by mining needs to be considered. In this case the height of the water-proof pillar should be as follows (see Fig. 9.32c):

$$H_w \geq H_f + H_p + H_e \tag{9.14}$$

where H_e is the maximum vertical depth of the surface fissure, as demonstrated in Table 9.3.

If the lowermost layer in the alluvium is strong or medium strong aquifer, and the weathered zone in the bedrock is permeable, then the maximum thickness of the weathered zone needs to be considered. In this case the height of the water-proof pillar can be expressed as follows (refer to Fig. 9.32d):

$$H_w \geq H_f + H_p + H_b \tag{9.15}$$

where H_b is the maximum vertical height of the weathered bedrock.

Table 9.4. The height of the protective layer for the water-proof pillar

Strata strength σ_c (MPa)	a	b	c	d
>40	5A	7A	6A	4A
20-40	4A	6A	5A	3A
20-10	3A	5A	4A	2A
<10	2A	4A	3A	2A

Note, $A = M/n$; M = total thickness of the extraction (m); n = number of the lift mining.
a – the thickness of clay in lowermost alluvium is less than the total thickness of extraction; b – it has no clay in the lowermost alluvium; c – the alluvium consists of clay, and its total thickness is less than the total thickness of extraction; d – the thickness of the clay in the lowermost alluvium is greater than the total thickness of extraction.

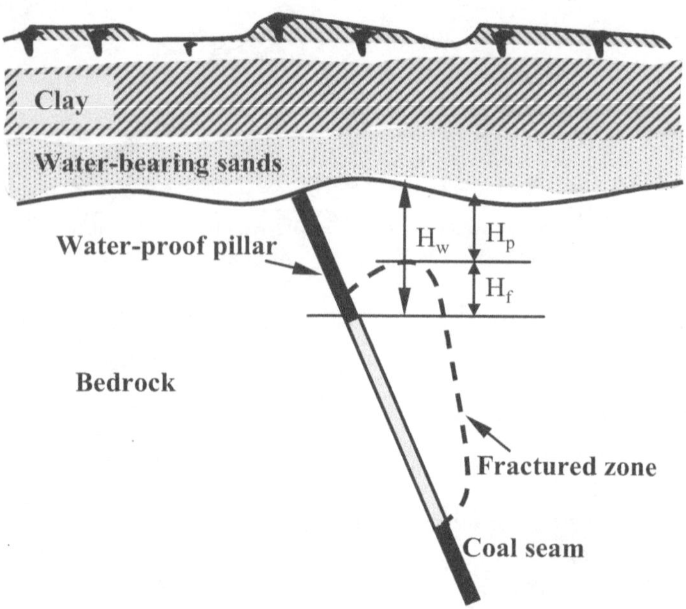

Fig. 9.32 b. A schematic diagram showing the water-proof pillar consisting of the fractured zone and the protective layer in steeply inclined strata.

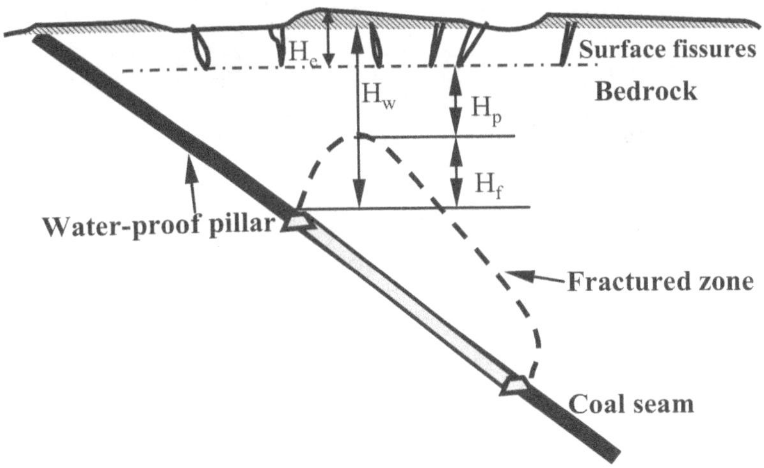

Fig. 9.32 c. A schematic diagram showing the water-proof pillar consisting of the fractured zone, the protective layer and surface fissures without alluvium covered on the bedrock.

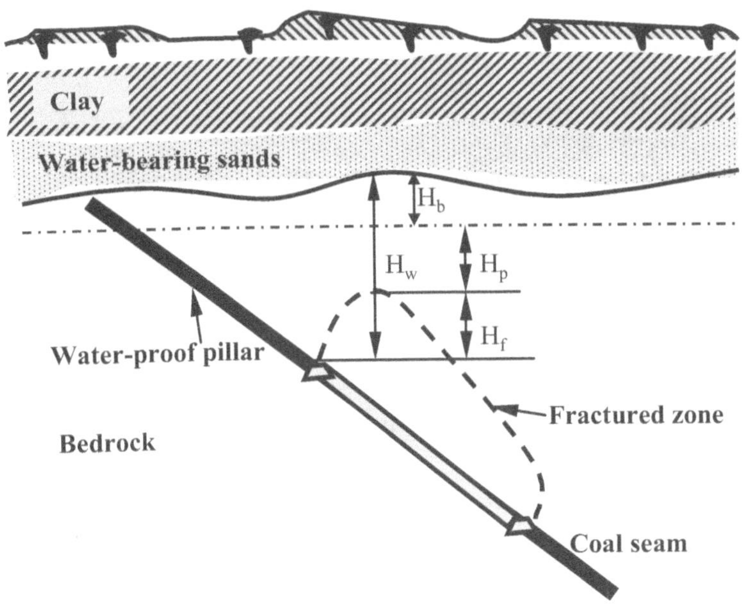

Fig. 9.32 d. A schematic diagram showing the water-proof pillar consisting of the fractured zone, the protective layer and the weathered permeable zone in slightly inclined strata.

The sand-proof pillar

The sand-proof pillar is designed for preventing sand in the alluvium from flowing into mining workings through mining-induced caving. It allows the fractured zone to penetrate partially into the aquifer. The function of the sand-proof pillar is to make successful coal seam extraction without sand inrush, but with a slight increase of groundwater discharge to the mine. When the lowermost water-bearing layer in the alluvium belongs to a weak aquifer, and the aquifer does not need to be protected as water resource, the sand-proof pillar can be applied (Zhang and Peng 2005). The minimum height of the water-proof pillar must not be less than the maximum height of the caved zone and the height of a protective layer, as shown in Fig. 9.33, i.e.

$$H_s \geq H_c + H_{ps} \tag{9.16}$$

where H_s is the vertical height of the sand-proof pillar; H_c is the maximum vertical height of the caved zone, which can be obtained from the empirical formulae given in the previous section (Eqs. 9.6 - 9.8); H_{ps} is the height of the protective layer (Table 9.5, Liu 1998).

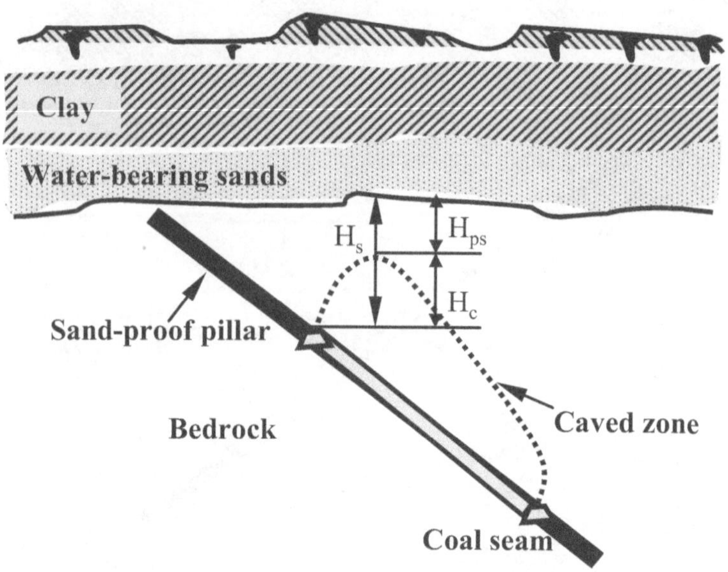

Fig. 9.33. The sand-proof pillar design.

Table 9.5. The height of the protective layer for the sand-proof pillar

Strata strength σ_c (MPa)	e	f
>40	4A	2A
20-40	3A	2A
20-10	2A	2A
<10	2A	2A

Note, A is the same as that in Table 9.4.
e – the thickness of the weak aquifer in lowermost alluvium is greater than the total thickness of the extraction; f – the total thickness of clay layers in the alluvium is greater than the total thickness of extraction.

The caving-proof pillar

When the lowermost layer in the alluvium belongs to a thick layer of clay or a drainable weak aquifer, the caving-proof pillar can be applied. The minimum height of the caving-proof pillar is equal to the maximum height of the caved zone, as shown in Fig. 9.34, i.e.

$$H_t = H_c \tag{9.17}$$

where H_t is the vertical height of the caving-proof pillar.

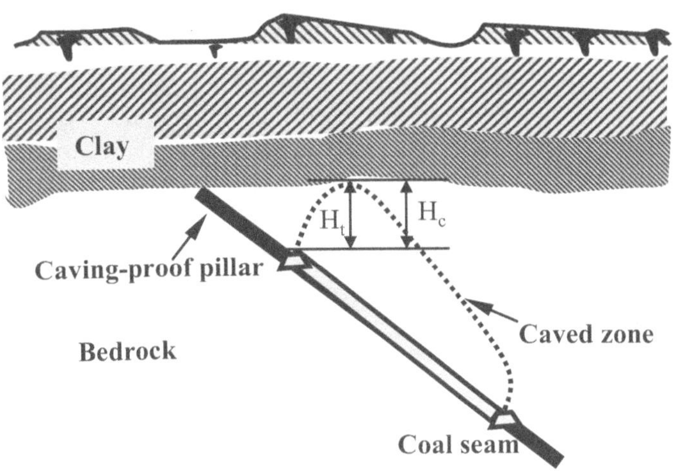

Fig. 9.34. The caving-proof pillar design.

9.4.5 Technical measures for the shallow mining

Several case studies and in-situ experiments were conducted in shallow coal mining areas in China, such as the Xingtai mine and the Dongpang mine in Hebei Province, the Xuzhuang mine and the Longdong mine in Jiangsu Province, the Daliuta mine in Shaanxi Province, the Xinglongzhuang mine and the Yangzhuang mine in Shangdong Province, and the Zhuxianzhuang mine in Anhui Province, etc. Table 9.6 lists some mines with successful reduction of outcrop coal pillar sizes without any safety incidents.

Summarizing the experiences obtained from the case studies, the following mining techniques are of crucial importance for safe mining at shallow depths:

(1). Design optimal coal pillar size according to the hydrogeological conditions of the alluvium, mining parameters and the empirical formulae given in the previous sections.

(2). Apply lift mining for thick seams. The lift mining offers the substantial benefit of reducing water and sand inflows by minimizing total heights of the fractured and caved zones resulting from extraction. Reduction of the thickness of extraction within the first or second lift is extremely beneficial in minimizing the height of the failure zones.

(3). Adopt short-wall mining. Reducing mining width can decrease permeability enhancement and reduce mining-induced fractured height (Zhang 2005).

(4). Detect faults, coal pillar size and aquifers prior to active mining reaching the area in question. The form and magnitude of the potential hazard can be diagnosed, and appropriate remedial measures need to be taken.

(5). Grout permeable faults or leave certain coal seams intact for maintaining distance between the faults and the mining face.

Table 9.6. Outcrop coal pillar reduction in some coal mines in China

Coal mine	Thickness of alluvium (m)	Original pillar height (m)	Reserves in the pillar (Mt)	Height Reduction of pillar (m)	Pillar type for reduction
Xingtai	80-290	60	28	35-45	Sand-proof
Longdong	154-213	45	21	27	Sand-proof
Xinglongzhuang	180-200	80	73	20	Water-roof
Baishan	114-117	40	14.2	20	Sand-proof

References

Bai M, Elsworth D (1990) Some aspects of mining under aquifers in China. Min Sci Tech 10:81-91

Booth CJ, Bertsch LP (1999) Groundwater geochemistry in shallow aquifers above longwall mines in Illinois, USA. Hydrogeo J 7(6):561-575

Booth CJ, Spande ED, Pattee CT, Miller JD, Bertsch LP (1998) Positive and negative impacts of longwall mine subsidence on a sandstone aquifer. Environ Geol 34(2/3):223-233

Hasenfus GJ, Johnson KL, Su DHW (1990) A hydromechanical study of overburn aquifer response to longwall mining. In Peng SS (ed) Proc 7th Int Conf on Ground Contr in Min pp149-162

Liu J, Elsworth D (1997) Three-dimensional effects of hydraulic conductivity enhancement and desaturation around mined panels. Int J Rock Mech Min Sci 34(8):1139-1152

Liu T et al. (1981) Coal mine ground movement and strata failure. Coal Industry Press Beijing (in Chinese)

Liu T (1998) Theory and techniques for optimal design of outcrop coal pillar, Coal Industry Press Beijing (in Chinese)

Peng S (1997) Investigation of engineering disasters induced by coal mining at deep depth in China. 4th Ann meeting, Geology Branch, China Coal Industry Tech Committee, Qingdao (in Chinese)

Peng S (1998) Establishment and implementation of geological support system of high production and high efficiency in China coal mines. Symp 31st Young scientists, China Sci Tech Soc, Coal Industry Press Beijing (in Chinese)

Peng S (1999) The state-of-art and the future work of engineering disasters induced by coal mining at deep depth in China. Symp Engineering and Disaster, Dept Material Science and Engineering, China NSF, Mt Jiuhua, Jianxi (in Chinese)

Peng S, Ling B, Liu D (2002a) Application seismic CT detection technique into roof coal caving comprehensive mechanical longwall mining. Chinese Rock Mech Eng 21(12):1786-1790 (in Chinese)

Peng S, Sun L, Ma C (2002b) Comprehensive analyses of rockmass deformation due to underground mining and their application. J China Min Tech Univ (Natural Sci Ed) 31(1):80-83 (in Chinese)

Peng S, Meng Z (2002) Theory and practice of mining engineering geology. Geological Press (in Chinese)

Zhang J (2005) Investigations of water inrushes from aquifers under coal seams. Int J Rock Mech Min Sci 42(3):350-360

Zhang J, Liu T, Zhang Y (1997) The study of permeability behaviors for fractured rock masses. J China Coal Soc 22:481-485 (in Chinese)

Zhang J, Peng S (2005) Water inrush and environmental impact of shallow seam mining. Environ Geol 48:1068-1076

Zhang J, Roegiers JC, Bai M, Zhang Y (2001) Stress-dependent permeability variation and mine subsidence, In: 38th U.S. Rock Mechanics Symposium, Washington DC Rotterdam Balkema

Zhang J, Shen B (2004) Coal mining under aquifers in China: a case study. Int J Rock Mech Min Sci 41(4):629-639

Zhang J, Zhang Y, Liu T (1997) Rock mass permeability and coal mine water inrush. Geological Publ House Beijing (in Chinese)

Zhou WF (1997) The formation of sinkholes in karst mining areas in China and some methods of prevention. Environ Geol 31(1/2):50-58

Zipper C, Balfour W, Roth R, Randolph J (1997) Domestic water supply impacts by underground coal mining in Virginia, USA. Environ Geol 29(1/2):84-93

10 Water inrush and mining above confined aquifers

10.1 Introduction

In many coal mines the limestone confined aquifers underlie coal seams. During coal extraction from these mines, water inrushes occur frequently with disastrous consequences. The most serious one of possible water disasters affecting the safe operation of coal mines in China is water inrushes from the Ordovician limestone under the Permo-Carboniferous coal seams in Northern China (Zhang and Shen 2004). The Ordovician limestone is a confined karst aquifer containing an abundant supply of water and a very high water pressure. Furthermore, the strata between coal seams and the aquifer are relatively thin varying in thickness from 30 to 60 m. Due to these characteristics of the aquifer, plus mining-induced strata failure and inherent geological structures (such as water-conducting faults, fractures) high pressure groundwater can break through seam floors and burst into mining workings. Therefore, water inrushes from the aquifer occur frequently, and coal mines often suffer from serious water disasters during coal extractions (Peng 1997, 1999). Water inrush incidents have shown that the maximum water inflow in a coal mine has reached as much as 2,053 m^3/min, which submerged the mine in a very short time.

The total coal reserves threatened by bodies of water are estimated at 25 billion tons. For example in Northern China, the yearly coal production from the Permo-Carboniferous coal-bearing formations is more than 200 million tons. However, the coal extraction has been threatened by frequent water inrushes from the Ordovician aquifer. In this region, the lower level seams which have more than half of the total coal reserves are much more difficult to mine due to this threat of water inrushes (Table 10.1).

Table 10.1. Coal reserves threatened by the Ordovician aquifer in China's coal mines

Coalfield	Total reserves (100 million tons)	Reserves threatened (100 million tons)	Ratio of threatened to total (%)
Jiaozhuo	5.65	4.95	87.6
Fengfeng, Handan, Xingtai	70	35	50
Hancheng	12.7	7.83	61.7
Chenghe	3.26	2.2	67.5
Feicheng	4	2.5	63
Huozhou	1.37	0.81	59.1
Total	96.98	53.29	55

For example, from 1950 to 1990 a total of 222 serious water inrush incidents took place in China causing collieries to be submerged by water intrusions from the confined karst aquifers. More recently, total water inrushes at nationalized key coal mines occur about 125 times annually resulting in an economic loss of about 180 million US dollars. In addition, local coal mines run by provinces, counties, and private business have a larger annual economic loss induced by water inrushes.

The main coalfields threatened by the Ordovician aquifer are Jiaozhuo in Henan Province, Fengfeng, Handan and Xingtai in Hebei Province, Zibo and Feicheng in Shandong Province, and Hancheng and Chenghe in Shaanxi Province. If the problems of safe mining over the aquifer cannot be solved efficiently, some coal mines in the mentioned coalfields will be faced with gradual reduction of production or even abandonment of the mines. In general, there are two different ways to solve the problems of mining over confined aquifers. One is to drain the aquifer before mining operation, the other is to mine without drainage. Geological investigation and mining practice have unveiled that there are many environmental problems induced by water drainage from limestone aquifers, such as the Ordovician limestone aquifer in Northern China, and the Maokuo limestone aquifer in Southern China. Fissures in these karstified limestones are well developed and interconnected within the aquifer such that when water is drained from one particular region, the drainage has an extensive influence on the whole aquifer. For example, a dewatering in the Ordovician limestone aquifer was conducted in Wangfen colliery, Hebei Province. The pumping flowrate was 96 m^3/min. However, the drawdown in the central well was only 2.8 m, and the radius of the cone of depression extended to 10 km which caused the loss of many drinking water wells. This resulted in a shortage of water supply for 100,000 people. Therefore water drainage is infeasible. The only solution for coal extraction over the limestone aquifers is to mine with technical measures and without drainage. In order to do so, it is of vital importance to study strata failure characteristics and hydraulic conductivity changes due to mining and thereafter find a way to predict and prevent water inrushes. Various research has been conducted in this area (e.g. Zhang 1989, Gao and Li 1992, Zheng et al. 2000, Peng and Wang 2001, Peng and Meng 2002, Li and Guan 2002, Wang et al. 2002, Li and Gao 2003, Yang et al. 2003, Zhang 2005, Yin and Zhang 2005). However, the mechanism of water inrushes is still not well understood.

10.2 Underlying strata failure due to mining

10.2.1 Determination of the water-conducting failure zone in the seam floor

Coal extraction causes strata deformation and failure which may enhance hydraulic conductivity in the surrounding strata. Therefore, it is desirable to accurately determine pre- and post-mining hydraulic conductivities in the overburden and underlying strata of the coal seam. To measure the hydraulic conductivity in the underlying strata, boreholes are drilled pre-mining in underground roadways for observation. In each borehole, water injection and a number of well logging techniques (such as electric resistivity, sonic log, acoustic emission, and hole televiewer, etc.) are used to determine rock strength, borehole fissure, and changes in hydraulic conductivity. Figure 10.1 gives a schematic diagram of a water injection instrument (Zhang and Zhu 1994). The key technique during measurements is to control the injection pressure. The pressure should not be high enough to create new fractures in the strata, since the experiment is to determine the changes in hydraulic conductivity induced by mining. Therefore, the injection pressure should not exceed the least principal stress of the surrounding strata.

Figure 10.2 shows the observing borehole locations and layout in Xingtai coal mine. In-situ stresses in this area were: $\sigma_v = 7.4$ MPa, $\sigma_h = 4.5$ MPa. The roadway in Fig. 10.2 was located 36 m below the mining face and four boreholes were drilled at different angles. The water injection instrument described in Fig. 10.1 was applied to measure the flowrate of water injection pre- and post-mining, using an injection pressure of 0.35 - 0.5 MPa. The water injection along each borehole was conducted by pumping water into the instrument, then into the borehole. The measurements were taken in each hole at different sections throughout the borehole and at different times.

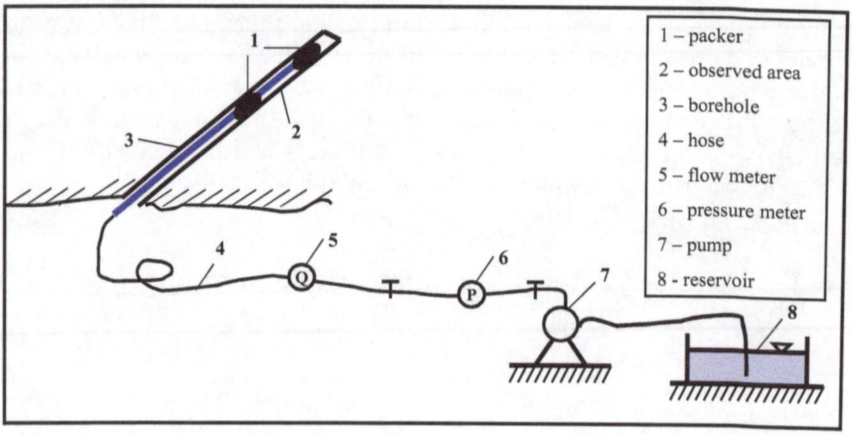

Fig. 10.1. The instrumentation for water injection observation in a borehole.

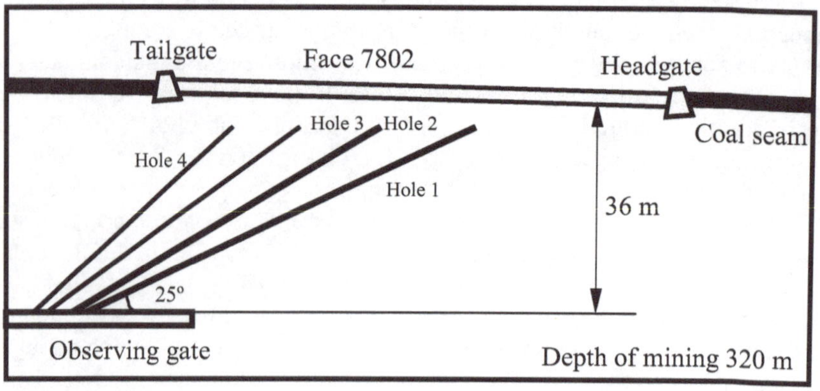

Fig. 10.2. Observing borehole layout for water injection measurements in Xingtai coal mine, Hebei Province.

Figure 10.3 presents the measured pre-mining and post-mining flowrate of water injection in Hole 1 (refer to Fig. 10.2). Note that in this context pre-mining corresponds to a state before the mining face passes the borehole, and post-mining means after the mining face passes the borehole. Before the mining face passed Hole 1 (in pre-mining the mining face was 32 m away from the borehole), the injection rate was zero from 53 to 68 m in the inclined borehole. This means that the strata in this area were impermeable. However, when the mining face passed the borehole, the injection rate (refer to Fig. 10.3 for post-mining at 19 and 63 m) increased dramatically, and the strata in some areas changed from being impermeable to permeable. Since the borehole wall collapsed by mining when the mining face passed 63 m from the borehole, water injection data could not be obtained after 60 m from the borehole opening. The borehole collapse in post-mining illustrates that the borehole was seriously damaged, and that rocks around the borehole failed due to mining.

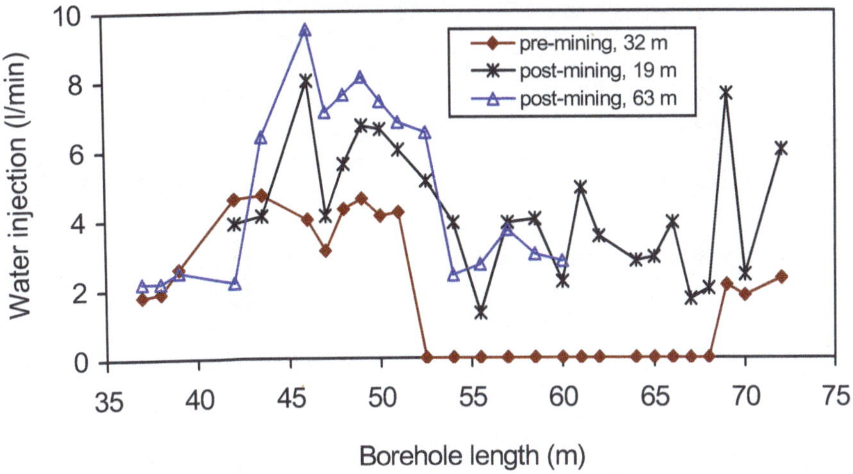

Fig. 10.3. Flowrate of water injection along a borehole (Hole 1 in Fig. 10.2.) pre- and post-mining in Xingtai coal mine, Hebei Province.

Figure 10.4 plots the increments of water injection rates after mining, which were obtained by subtracting the pre-mining injection rates from those of the post-mining (Zhang 2005). These increments represent injection rates caused by permeability changes induced by coal extraction. Figure 10.4 shows that along the inclined borehole from 43 m to 72 m (the borehole end), the injection rate increases compared to the pre-mining (in-situ) state. Therefore, the strata in this area were fissured by mining, and this fractured area is defined as the water-conducting failure zone. Using the same method to analyze the observed data from all boreholes, the mining-induced water-conducting failure zone can be obtained. This failure zone is of critical importance for mine design and water inrush prevention for mining over aquifers.

Figure 10.5 shows the changes of the average water injection rate in the seam floor of 10 m deep from the seam in the Xingtai coal mine. The injection rate clearly increases compared to the unmined area. The injection rate decreases inside the abutment. This decrease is due to stress concentration and high abutment pressure occurring in this area, causing the fractures to close.

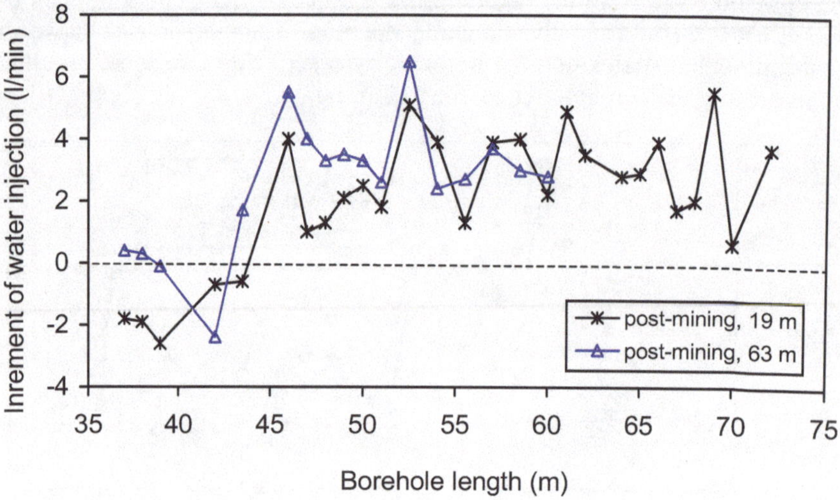

Fig. 10.4. Flowrate increment of water injection along a borehole (Hole 1 in Fig. 2) after mining in Xingtai coal mine, Hebei Province.

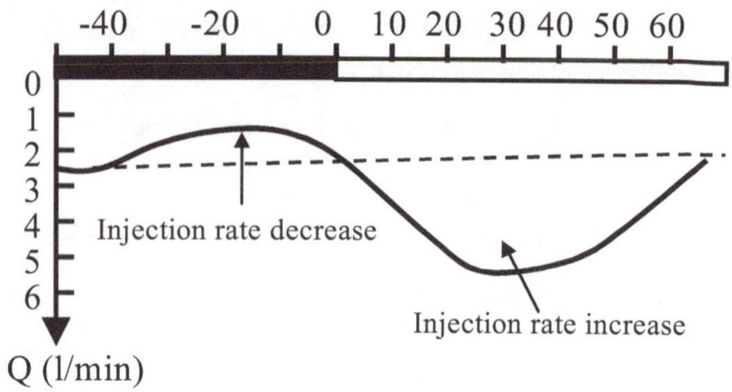

Fig. 10.5. Water injection rate versus the advance of the mining face in the Xingtai coal mine. In the figure the negative distance represents pre-mining state, and the positive means post-mining.

Figure 10.6 displays the changes of the injection rates with distance of mining advance for two different depths beneath the coal seam in Wangfeng coal mine, Hebei Province. It clearly shows that in the mined area the injection rate increases significantly compared to the unmined area. It also can be seen that due to coal extraction the water-conducting capacity increases in the floor strata, and this water-conducting capacity decreases as the distance from coal seam to the floor strata increases. This indicates that the closer the strata is to the extracted seam, the higher the permeability in the seam floor. It is also noticeable that the injection rate decreases inside the abutment. This decrease is due to the fact that stress concentration and high abutment pressure occurred in this area cause the fractures to be closed.

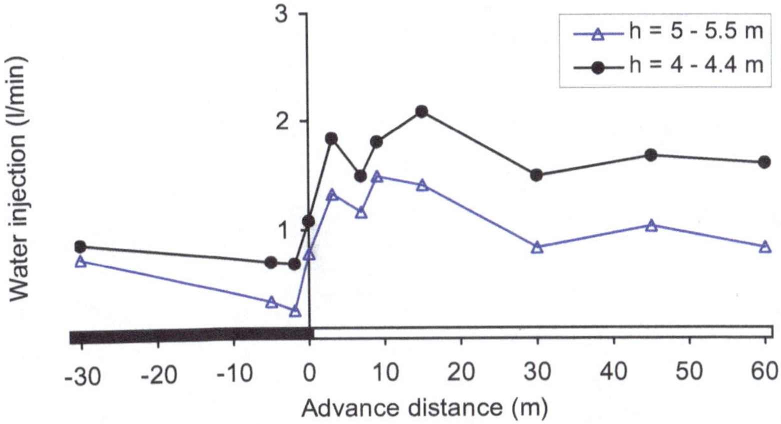

Fig. 10.6. Flowrate of water injection versus mining distance for two different depths beneath coal seam (the negative distance represents pre-mining state, and the positive means post-mining).

Field observations have shown that characteristics of failures in the floor strata are considerably different for different inclinations of the extracted seams. For flat or slightly inclined seams (dip angle, $\alpha < 25°$), the profile of the water-conducting failure zone is broad in section with extended lobes, and the maximum failure depth occurs beneath the headgate and tailgate, respectively shown in Fig. 10.7 (Fengfeng Mining Bureau et al. 1985). For inclined seams ($25° < \alpha < 60°$), the failure zone propagates downwards in an asymmetric manner in the dip direction, as shown in Fig. 10.8 (Huainan Mining Bureau et al. 1983). The extent of the failure zone increases gradually from updip to downdip, and the maximum failure depth appears in the floor strata beneath the area around the lower gate. For steeply inclined seams ($60° < \alpha < 90°$), the failure zones in the floor strata are opposite to the inclined seams, i.e., the maximum failure depth appears in the strata beneath the area around the upper gate (Zhang 2005).

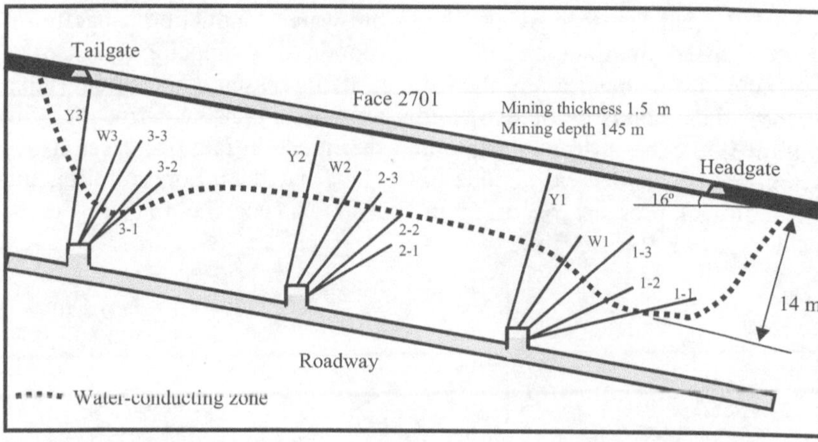

Fig. 10.7. Observing borehole layout and observation section of the water-conducting failure zone in the underlying strata for slightly inclined coal seam in Fengfeng coal mines, Hebei Province.

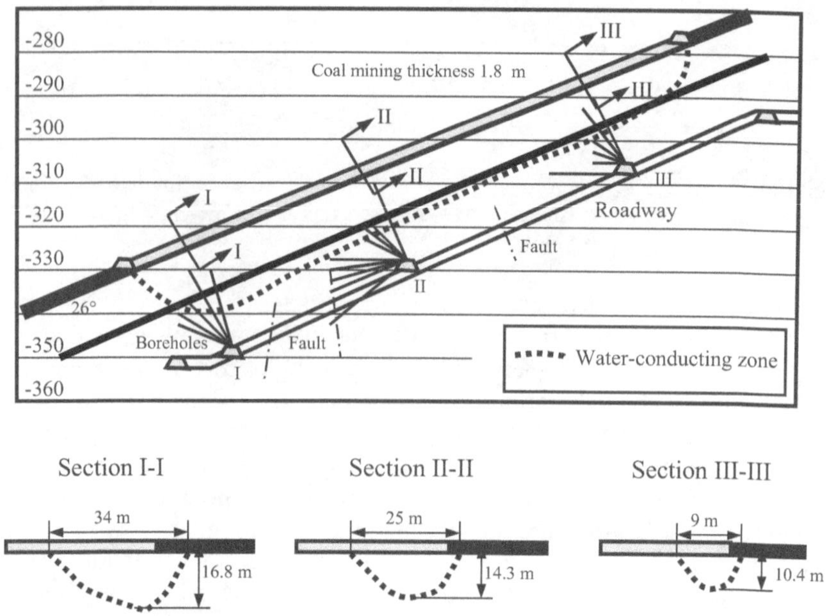

Fig. 10.8. Observing borehole layout and observation section of the water-conducting failure zone in the underlying strata for inclined coal seam in Huainan coal mines, Anhui Province along both strike and dip directions failure zones increase from the upstream to downstream.

10.2.2 Empirical prediction of the depth of the water-conducting failure zone

According to field observations, a number of parameters affect the development and depth of the water-conducting failure zone. Mining width of the working face and uniaxial compressive strength of the strata are the most important in all parameters. An empirical formula for predicting the depth of the water-conducting failure zone was developed from field test results in long-wall and short-wall mining faces (Zhang et al. 1997). The formula is expressed as (refer to Fig. 10.9):

$$h_1 = 0.303 L_x^{0.8} \tag{10.1}$$

where h_1 is the depth of the water-conducting failure zone starting from the immediate floor of the seam (m); L_x is the mining width of the mining face (m). Note that the measured data were obtained from coal mines in Northern China, with mining depths ranging from 103 to 560 m, and uniaxial compressive strengths from 20 to 40 MPa.

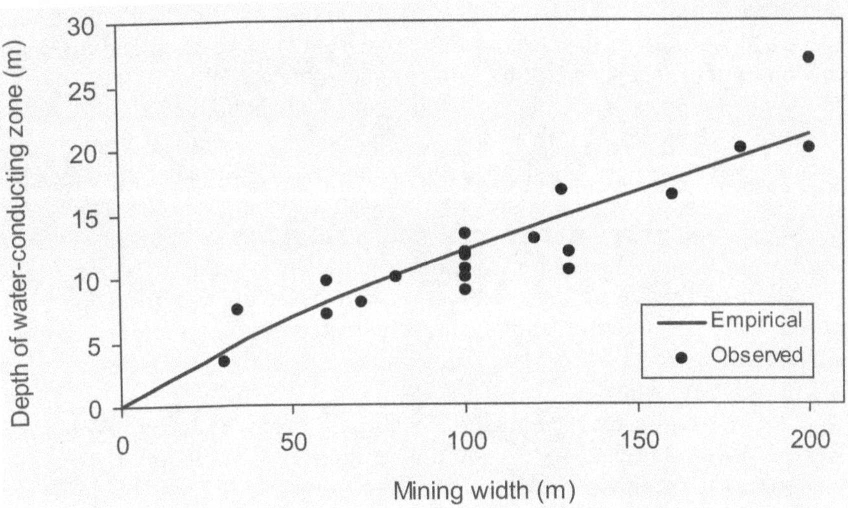

Fig. 10.9. Observed maximum depth of the water-conducting failure zone in the seam floor strata for different mining widths in China.

For mining above aquifers, it is desirable to avoid water inrushes and the extra expense of strata dewatering. This can be achieved only when aquifers are located a certain distance outside the water-conducting failure zone. If an aquifer which is confined, very permeable and with abundant water, lies within the failure zone, water with high pressure will rush into the mining area, and may cause a disastrous consequence.

10.2.3 Theoretical and numerical analyses of strata failure and hydraulic conductivity change around a mining panel

Many efforts have been made to predict the mining-induced failure area. These include theoretical analysis as well as numerical simulation. According to the plastic sliding theory, Zhang and Liu (1990) proposed the following equation to calculate the maximum depth of failure zone induced by mining in the seam floor:

$$h_1 = \frac{x_a \cos\varphi_0}{2\cos(\frac{\pi}{4}+\frac{\varphi_0}{2})} \exp[(\frac{\pi}{4}+\frac{\varphi_0}{2})\tan\varphi_0] \tag{10.2}$$

where x_a is the length of failure zone of the coal seam around the mining face, which can be calculated from Zhang et al. (1997); φ_0 is the angle of the internal friction of the strata. Using the above equation combined with empirical formula in Eq. 10.1, some good results have been achieved for predicting the water-conducting failure zone (Zhang et al. 1997).

For mining near aquifers, the finite element method (FEM) can analyze not only the strata failure, but also hydraulic conductivity changes associated with mining. The latter is more important for predicting water inflow into the mining panel. For this purpose, the stress-dependent hydraulic conductivity was incorporated into the FEM model (refer to Sect. 9.3.3 of Chap. 9). Therefore, hydraulic conductivity variations and strata failures resulting from seam extraction can be calculated from the proposed FEM. In the following finite element analysis, a typical mining panel is selected from Yanzhou Coal Mines, Shandong Province (see Sec. 8.5 for this case study).

Figure 10.10 displays the vertical hydraulic conductivity contour around the mining panel for a mining width of 100 m. In the figure, the contours of the ratios of post- to pre-mining hydraulic conductivities are plotted for a half of the panel, which is geometrically symmetric. The area having the ratio of hydraulic conductivity greater than 1 is defined as the enhancement of hydraulic conductivity; while the area with hydraulic conductivity ratio less than 1 is called the decrease of conductivity. Figure 10.10 shows that the enhancement of conductivity concentrates mainly on the mined area both in the roof and floor. Compared to the pre-mining state, hydraulic conductivity of the floor strata increases in the mined area and decreases in the abutment area (Fig. 10.10). These results are coincident with the field measurements as shown in Fig. 10.6. It is also noted from Fig. 10.10 that there is a small area with a conductivity increase beneath the coal wall.

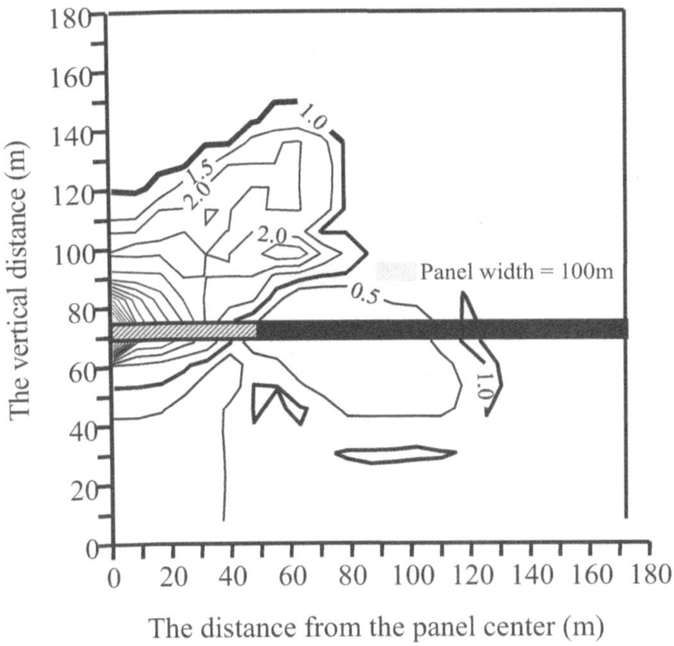

Fig. 10.10. Calculated contours of hydraulic conductivity ratios induced by mining in the vertical direction for a mining width of 100 m.

Figure 10.11 presents the vertical hydraulic conductivity enhancement areas, where the conductivity ratios are all greater than 1, for different mining widths. It is obvious that the area of the conductivity enhancement in the roof as well as floor increases as the mining width increases. It can be seen that decreasing mining width can reduce the extent of the water-conducting zone.

10.3 Water inrush mechanism

There are many factors that control and influence the occurrences of water inrushes. The main factors are strata pressure or stress, water pressure, geologic structures and mining parameters (such as mining width and advancing distance, etc.).

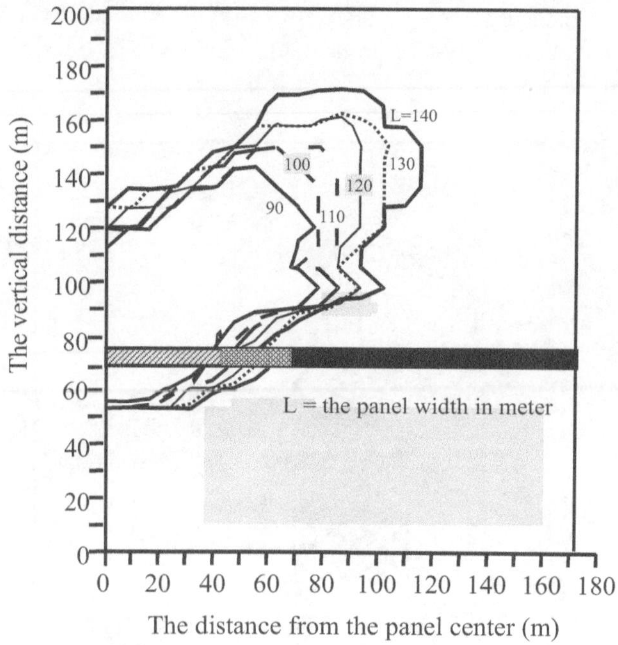

Fig. 10.11. Calculated enhancement zone of hydraulic conductivity induced by mining in the vertical direction for different mining widths.

10.3.1 Strata stress

Strata stress is caused by mining-induced in-situ stress redistributions. The strata stress distribution and behavior in the overburden strata have been extensively studied. However, the strata stress in the floor strata is not fully understood. Some of field measurements and laboratory physical modeling have given good insights on stress characteristics in the floor strata.

Figure 10.12 shows the vertical stress increments (the difference between the induced and original stresses) in the floor strata located 5 m below the seam in a section located 60 m away in the direction of advance from the starting cut (refer to Fig. 9.14 in Chap. 9 for the model setup). It can be seen that the stress increment increases (the so called abutment pressure) before the mining face approaches the sensor. After the face passes the sensor, the floor strata in the observed section lie in the extracted area, and the stress increment decreases gradually, i.e. the strata experience a state of unloading. In this stage stress relaxation and confinement reduction occur in the floor strata, which significantly reduce the inherent stability of the rock mass and induce mining fractures. As the extraction advances, stress in the floor strata experiences three stages. The stages are the increase of stress in pre-mining, dramatic decrease then gentle reduction of stress after mining, and gradual recovery to the original stress level. Corresponding to the stress redistribution, displacement in the floor strata also experiences three stages, which include compression before mining, expansion after mining, and gradual recovery, as shown in Fig. 10.13. During floor expansion or unloading stage (stress relaxation in Fig. 10.12), the strata are more prone to create tensile fractures. In the intersection between compression and expansion located beneath the area around the coal wall of the mining face, the strata easily produce shear fractures. These fractures induced in the floor cause an increase in permeability in floor strata after mining, as shown in Fig. 10.6 where the water injection rate increases considerably, when inject water into the strata. This is also the reason that the floor failure zone is larger in the strata right beneath the area around the coal wall, as shown in Figs. 10.7 and 10.8.

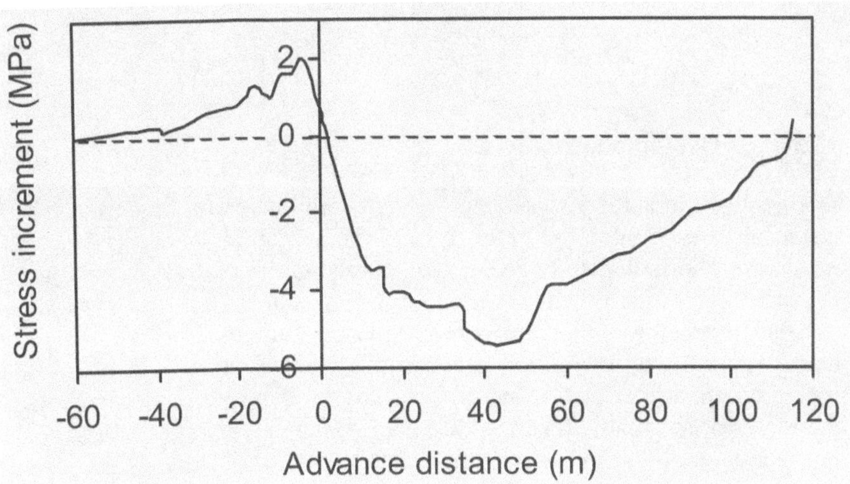

Fig. 10.12. The vertical stress increment in the floor strata versus mining advance distance in a physical model (the negative distance represents pre-mining state, and the positive means post-mining).

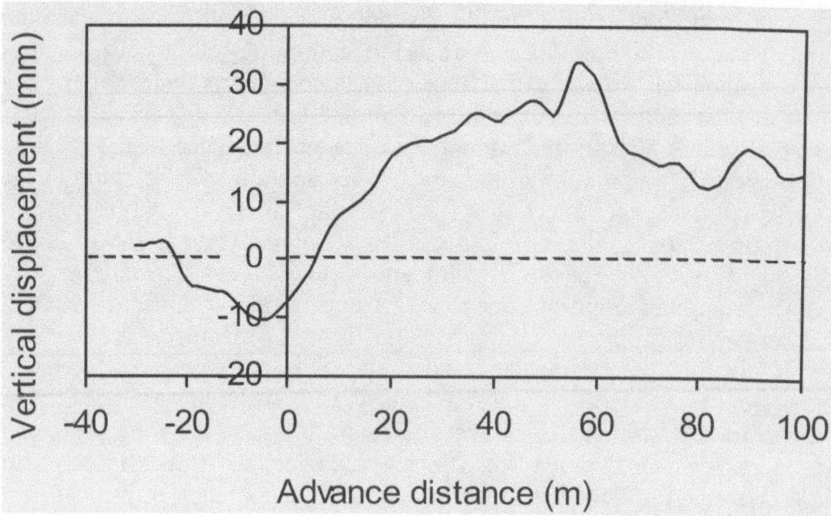

Fig. 10.13. The vertical displacement in the floor strata versus mining advance distance in a physical model (the negative displacement represents strata compression and the positive means expansion).

10.3.2 Geologic structures

The geologic structures, particularly faults, are one of the main causes of water inrushes. Table 10.2 gives the statistical relations between water inrush times and faults, and other geologic structures for 163 long-wall mining faces in Northern China. It is obvious from Table 10.2 that most water inrushes from underlying aquifers were related to faults. The following explanations may be applied to interpret this phenomenon. First of all, the rock strength in the fault zone is much lower than that of normal rocks, therefore rocks in faulted zones fail easier, and create a larger failure zone. For example, in-situ measurements in Face 4303 of Xinzhuangzi colliery, Huainan coal mines showed that the maximum depth of the water-conducting zone in the floor was 16.8 m for the normal case (refer to Fig. 10.8). However, it was 29.6 m in the fault zone. Secondly, existence of a fault may reduce the distance between the seam and aquifer. Finally, when the faulted zone is permeable and reaches to the aquifer, it will cause water inrush once the fault is exposed by mining.

Table 10.2. Statistical results of times of water inrushes from aquifers under coal seams versus geologic structures in Northern China

Coalfield	Times of water inrushes	Times with fault effects	Times with other structure effects	fault effects (%)	Total structure effects (%)
Fengfeng	11	9	1	81.8	90.9
Jiaozhuo	24	13	1	54.2	58.3
Zibo	55	31	2	56.4	60
Feicheng	73	53	0	72.6	72.6
Total	163	106	4	65	67.5

10.3.3 Mining size

Mining width is a critical parameter for mining over aquifers. In-situ observations of water injection in the floor strata have shown that the depth of the failure zone increases as mining width increases, as shown in Fig. 10.9. The reason is that as the exposure of the mining area increases, the floor strata have more space to deform and expand. This results in confinement reduction, inducing more fractures. The distance of advancement from the starting cut is also an important parameter for long-wall full caving mining. Analyses of water inrush incidents indicate that most of water inrushes occurred when the mining faces advanced 20 to 30 m from the beginning of mining operations (Fig. 10.14). At this distance the roof collapses for the first time. The first collapse of the roof causes significant stress concentration and relaxation in some areas of both the seam roof and floor. The consequences include more fractures and a larger failure zone in the floor. Thus, if an aquifer is located under the seam floor, it is more likely for water inrush to take place. It also can be seen from Fig. 10.14 that after the mining faces advance 110 m, it is very rare to have water inrush incidents.

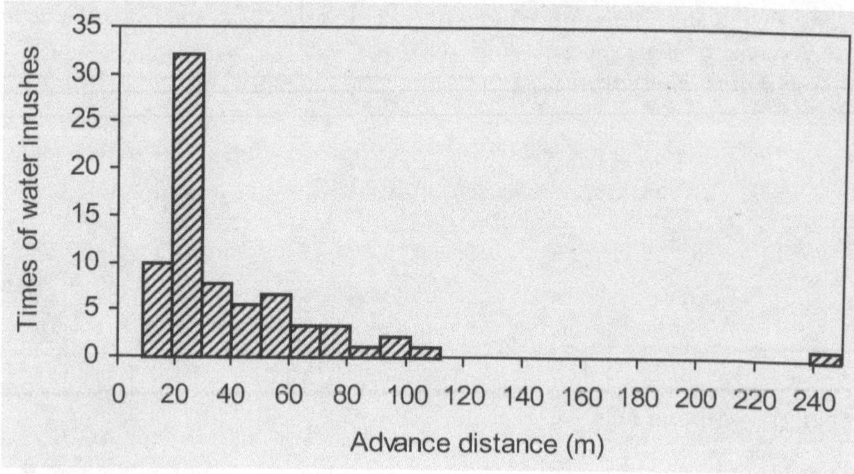

Fig. 10.14. Times of water inrushes for mining over aquifers versus mining advance distance for 72 water inrushes in long-wall mining faces in coal mines in Northern China.

10.3.4 Water pressure

Water pressure in the aquifer that underlies the coal seam plays an important role in water inrushes during mining. Acting as a kind of stress on the bottom of the floor, water pressure makes the floor strata much more apt to expand into the exposed space of the mined area. Therefore, the higher the water pressure is, the easier water inrush occurs. That is the reason why some mines use water pressure as one of the important parameters to estimate potential water inrushes. In China the water inrush index or coefficient is applied, as a rule of thumb, for preliminary evaluation of water inrushes in some coal mines. The water inrush index is defined to be the water pressure bearing capacity per unit thickness of the floor strata. Table 10.3 lists the water inrush indexes in some coal mine areas in China.

Table 10.3. Critical water inrush index in some coal mine areas in China

Coalfield, province	Critical water inrush index (MPa/m)
Fengfeng, Hebei	0.05 – 0.065
Jiaozhuo, Henan	0.05 – 0.10
Zibo, Shandong	0.05 – 0.12
Jingjing, Hebei	0.05 – 0.13
Kailuan, Hebei	0.067 – 0.10

It can be seen that water inrushes can occur when the ratio of water pressure to thickness of the floor strata in a study area exceeds the critical water inrush index of 0.05 - 0.13 MPa/m. This index varies for different mines (Table 10.3).

10.4 Predictive methods of water inrushes

10.4.1 Empirical method of water inrush prediction

According to the field data of water inrushes from coal mining faces, there are two parameters affecting water inrushes from an underlying confined aquifer. They are the water pressure of the confined aquifer, and the thickness between the coal seam and the aquifer. By analyzing the field data of water inrushes in different coal mines (data from a coalfield are shown in Fig. 10.15), an empirical relationship between the critical water pressure and the thickness of the strata from the coal seam to its underlying aquifer can be obtained. It can be expressed as the following generalized form:

$$p_c = Ah^2 + Bh + C \tag{10.3}$$

where p_c is the critical water pressure (MPa); h is the thickness of the strata between the coal seam and its underlying aquifer (m); A, B, and C are the coefficients depending upon the strata lithology.

The parameters in Eq. 10.3 for different coal mines are listed in Table 10.4 (Zhang 2005, Zibo Mining Bureau 1979). These empirical formulas have played an important role for water inrush prediction in coal mines with similar geological and mining conditions.

Table 10.4. Parameters in Eq. 10.3 for predicting the critical water pressure p_c in some coal mines in China

Coalfield	A	B	C
Fengfeng	0.0006	0.026	0
Jiaozhuo	0.0018	-0.016	0.16
Lianshao	0.014	-0.0836	0
Heishan colliery, Zibo	0.00177	0.015	-0.43
Shigu and Xiazhuang collieries, Zibo	0.0016	0.015	-0.3
Hongshan and Zhaili collieries, Zibo	0.002	0.015	-0.158
Shuangshan and Fucun collieries, Zibo	0.00084	0.015	-0.168

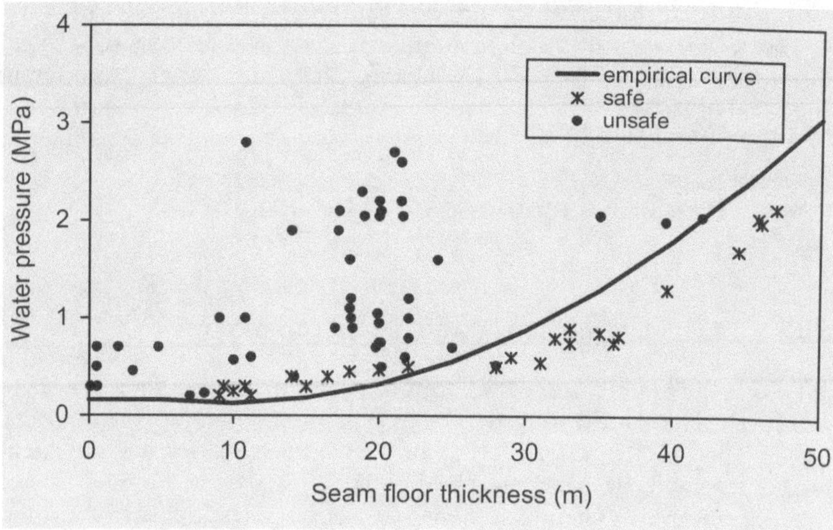

Fig. 10.15. Field data showing the relationship between water pressure and seam floor thickness in mining faces with and without water inrushes in Jiaozhuo Coal Mine Area, Henan Province.

10.4.2 Theoretical method of water inrush prediction

A theoretical model for analyzing the floor strata was given by Zhang (1989, 1990). In this model, the strata between the coal seam and the underlying aquifer were classified into two zones; one is the water-conducting failure zone with a thickness of h_1, the other is the intact strata zone with a thickness of $h-h_1$, as shown in Fig.10.16. The elastoplastic plate model was applied to analyze the stresses and failures of the intact strata. The following relationships among the critical water pressure, strata mechanical parameters and mining size were derived (Zhang 2005):

$$p_1 = A_1(h-h_1)^2\tau_0 + \gamma h \tag{10.4}$$

$$p_2 = A_2(h-h_1)^2 S_t + \gamma h \tag{10.5}$$

where p_1, p_2 are the critical or maximum water pressures that strata can bear; h is the distance between the seam and the underlying aquifer; h_1 is the depth of water-conducting failure zone in the floor, which can be obtained from Eq. 10.1; γ is the unit weight of the floor strata; τ_0, S_t are the strata shear and tensile strengths, respectively; and

$$A_1 = \frac{\pi^2[3(L_x^4 + L_y^4) + 2L_x^2 L_y^2)]}{6L_x^2 L_y^2 (L_x^2 + \nu L_y^2)}, \quad A_2 = \frac{12L_x^2}{L_y^2 (\sqrt{L_y^2 + 3L_x^2} - L_y)^2}$$

where ν is Poisson's ratio; L_x, L_y are the mining width and length, respectively.

When the critical pressure, calculated in Eqs. 10.4 and 10.5 respectively, is greater than the real water pressure in the aquifer, then, no water inrush occurs during mining. The criteria for water inrush prediction can be obtained from Eqs. 10.4 and 10.5, and may be expressed as:

$$A_1(h-h_1)^2 \tau_0 + \gamma h \geq p \qquad (10.6)$$

$$A_2(h-h_1)^2 S_t + \gamma h \geq p \qquad (10.7)$$

where p is the real water pressure in the underlying aquifer.

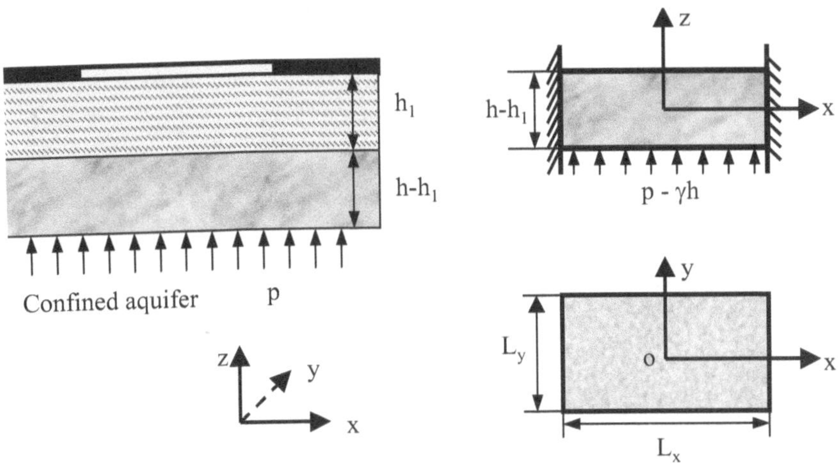

Fig. 10.16. Mechanical model of the plate for analyzing relationship between the strata failure and water pressure of the confined aquifer in seam floor after mining.

The above equations have been applied to predict water inrushes in some coal mines in Northern China. By analyzing the water inrush data, strata strength parameters, and by combining the empirical formulas for different coal mines (Eq. 10.3 and Table 10.4), the prediction equations (Eqs. 10.6 and 10.7) have a good prospect to predict water inrushes and guide mining design. These equations were also validated by 111 mining faces extracted over confined aquifers (72 water inrush cases of the 111 faces) in Northern China.

A case study was conducted in Shuanggou coal mine, Zibo Coal Mine Area, Shangdong Province. The case study included in-situ stress and water injection measurements for determining the water-conducting failure zone (Zhang et al. 1997), as well as theoretical calculations based on the parameters obtained from in-situ measurements. Figure 10.17 exhibits the calculated results by applying Eq. 10.5 with S_t = 1.1 MPa, γ = 0.023 MPa/m. The old mining design in Shuanggou mine was long-wall mining with a mining width of L_x = 140 m. Since the distance from the coal seam to the Ordovician aquifer was 52 m, it can be seen from Fig. 10.17 that the critical water pressure is 2.67 MPa for L_x = 140 m and h = 52 m. However, the real water pressure for the Ordovician aquifer was 3 MPa. This means that this mining design has the possibility for a water inrush. Therefore, the design needs to be modified either by using a shorter mining width (L_x < 120 m from Fig. 10.17) or taking other technical measurements. In fact, double mining faces with mining widths of 60 m and 70 m were adopted, and the seam was successful extracted without any water inrush.

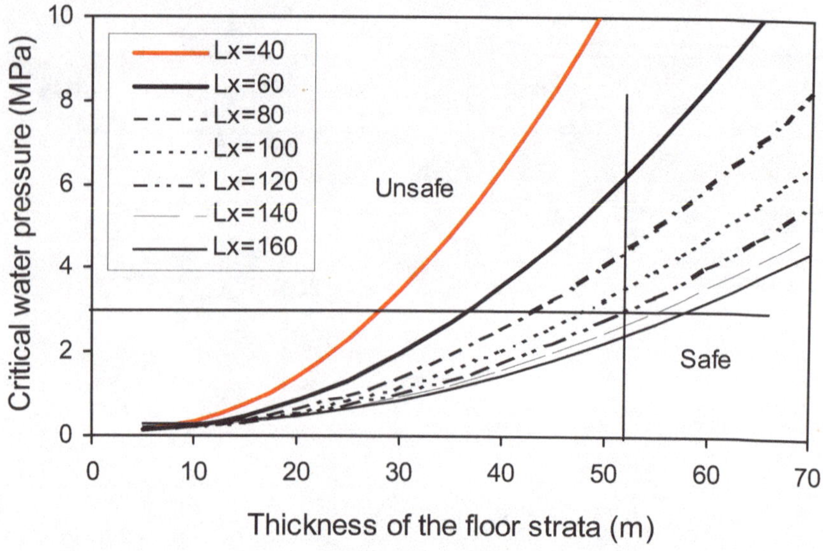

Fig. 10.17. Calculated relationships between the critical water pressure strata can bear and the floor strata thickness with different mining widths for Shuanggou coal mine, Zibo Coal Mine Area, Shangdong Province. In the figure, the upper region of each curve is the unsafe zone, and the lower region is the safe zone. In the case study the strata thickness between the seam and its underlying aquifer was 52 m, and water pressure in the aquifer was 3 MPa.

10.5 Suggested technical measures for mining over aquifers

10.5.1 Coal seam floor grouting

When faults and fractures exist in the seam floor, they can penetrate into the aquifer and cause water intrusion into the mining area. Grouting can seal these faults and fractures and reduce the possibility of water inrushes. Statistical data in Table 10.2 indicates that many water inrushes were caused by faults. Therefore, grouting faults and fractures before mining will be of vital importance for safe coal production. In some coalfields in Northern China there are several thin limestone layers laid in the seam floor which belong to aquitards or weak aquifers. Grouting these weak aquifers can not only change the weak aquifer into an impermeable layer, but also increase the mechanical strength of the floor strata, which makes the strata less likely to fail. For example, Caozhuang coal mine of Feicheng Coal Mine Area employed grouting techniques to grout the No. 5 limestone weak aquifer interbedded in the floor strata. Five mining faces were grouted and 500 thousand tons of coal reserves were successfully extracted over the Ordovician limestone aquifer with neither water inrush nor excessive groundwater discharge.

10.5.2 Mining method modification

Empirical results indicate that reducing mining width can greatly reduce the depth of the failure zone in the floor strata. This is considerably beneficial in minimizing water intrusions. For example, Face 1002 with a mining width of 50 m and Face 1081 with a mining width of 100 m in Xiazhuang coal mine of Zibo Coal Mine Area had similar geologic conditions; however, only the face with the larger mining width had a water inrush incident. The same result occurred for Face 110 with a mining width of 60 m and Face 116 with a mining width of 100 m in Zhaili coal mine, Zibo Coal Mine Area. Therefore, short-wall and room-and-pillar mining, which have much smaller mining widths than long-wall mining, are appropriate mining methods for preventing possible water disasters. For short-wall mining, the mining width should be kept at 40 to 60 m in Northern China (Zhang 1991).

Double-face mining with a reasonable distance between the two mining faces is also a mining measure that reduces floor strata failure. Zibo,

Huaibei, and Kailuan Coal Mine Areas have applied this technique to extract coal reserves threatened by confined aquifers (Peng et al. 2003, Luo et al. 2002, Wang et al. 2002). The advance distance between double faces should be 18 to 60 m (Zhang 2005).

When the mined area is backfilled, the mining-induced stress is greatly reduced in the strata (Zhang and Shen 2004). This implies that the difference between the maximum and minimum principal stresses around mining panel is reduced due to backfilling. Therefore, the strata based on the Mohr-Coulomb failure theory are less likely to fail. When aquifers are located near the mining workings, the mining operation will be much safer with backfilling. For example, Xingwen Coal Mine Area has used sand with water to backfill the mined areas. It has never had any water inrush incidents although it has similar hydrogeological conditions as the adjacent coalfields in Zibo and Feicheng Coal Mine Areas. In the latter two areas, many water inrush incidents have occurred. Zibo had more than 180 water inrushes, and Feicheng more than 80. Therefore, mining with backfilling is one of the technical measures for mining over confined aquifers. Its drawback is related to higher cost and more complicated mining technology.

10.6 Mining above confined aquifers with paleo-sinkhole effects

10.6.1 Introduction

The karst paleo-sinkhole is an ancient karst cavity collapsed and filled with falling rocks from the overburden formations. There are many paleo-sinkholes formed in the Permo-Carboniferous coalfields in northern China. Due to their specific structures and hydrogeological characteristics, the paleo-sinkholes normally act as pathways to connect groundwater from the Ordovician limestone aquifer (simplified as O_2) to the strata of coal beds. Water disasters caused by the paleo-sinkholes usually are catastrophic and unpredictable. When the paleo-sinkhole connects to the mining space through mining induced fractures, the water influx is often rapid and very significant. Many water inrush incidents have happened in some coal mines in China, resulting in considerable damages. For example, in the Fangezhuang coal mine (with an annual coal production of 3.1 million tons) of the Kailuan area, a disastrous water inrush took place on June 2, 1984 at mining face No. 2171. The maximum water flowrate reached as high as 2,053 m^3/min, which submerged the colliery and other three nearby collieries in less than 21 hours. This not only caused a serious economic loss, but also had a tremendous environmental impact causing a local water supply shortage and surface collapse of sinkholes (affecting distances as far as 3.5 km) due to the drawdown of the water level. Similar incidents have happened in other coal mines in northern China. Figure 10.18 shows a seismic section of a typical paleo-sinkhole in China's coal measures. Figure 10.19 presents the plane view of paleo-sinkholes and coal mining panel. The existence of paleo-sinkholes affects the mining panel design and mining operation (Yang et al. 2002).

10.6.2 Impact of the karst paleo-sinkholes on coal mining

Paleo-sinkholes were primarily formed in carbonate rocks. The two factors are necessary for the formation of paleo-sinkholes: karstified cavities in carbonate rocks, such as limestones, dolomites and gypsums, and flowing groundwater in the karstified rocks. When overburden formation falls down into the cavities, a vertical collapsed column is formed within, which is the paleo-sinkhole. In northern China, paleo-sinkholes are widely distributed in 45 coal mining areas within 20 coalfields coalfields (Yin et al. 2003). The paleo-sinkholes are mainly situated in the highlands of Shanxi Province, Taihangshan Mountains, Luliangshan Mountains, and Shandong, Henan, Jiangsu, Qinghai and Shaanxi Provinces. Statistical results show that more than 3,000 paleo-sinkhohles have been found in these coalfields (Li and Zhou 1988). The paleo-sinkoles are much more densely distributed in the coalfields of the Xishan and onshore of the Fenhe River in Shanxi Province. It was discovered that there were 1300 paleo-sinkholes in an area of 70 km^2 in the Xishan coalfield. (Li and Zhou 1988). The maximum distributing density of paleo-sinkholes in coal seam No. 2, the Fenxi coalfield, reached 100 /km^2 (Yin 2002).

The paleo-sinkhole is a potential geological disaster when it either contains water or connects to the karst aquifer as a pathway for groundwater. Many water inrushes in coal mines have taken place because of these paleo-sinkholes. Table 10.5 presents water inrush disasters occurred in some coalfields in northern China (Yin and Zhang 2005, Yin and Wu 2004). In these tragic incidents both economic loss and personnel casualties were very significant.

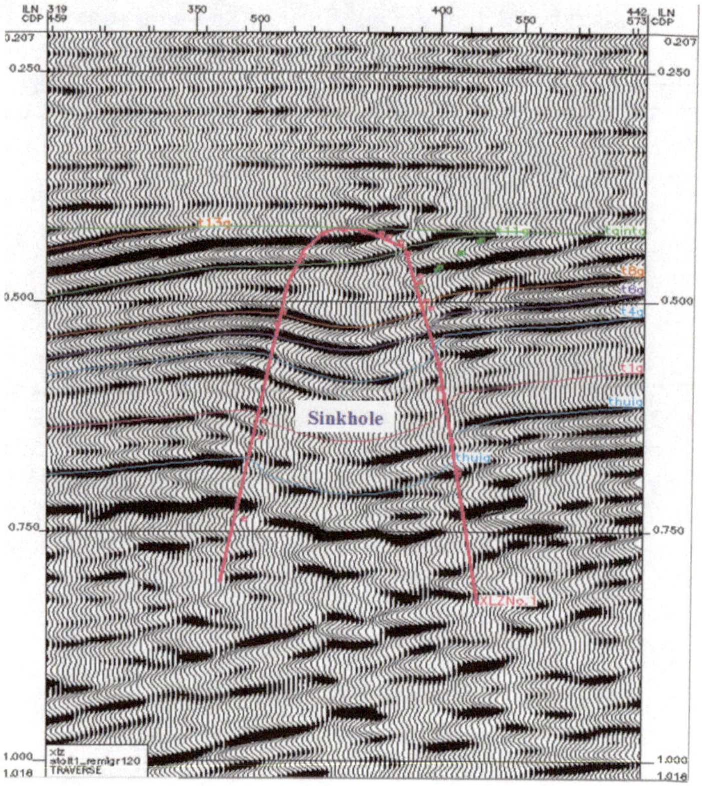

Fig. 10.18. A seismic section showing a paleo-sinkhole in coal measures in Huainan Coalfield of China.

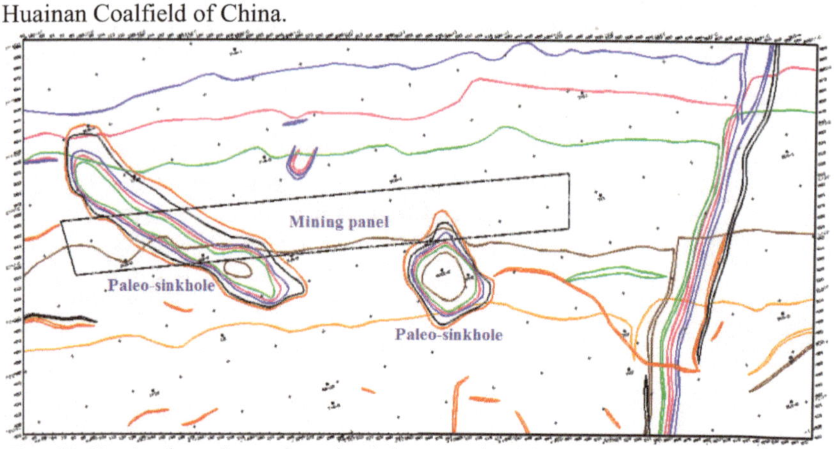

Fig. 10.19. A plan view of coal mining panel with paleo-sinkholes in Huainan Coalfield of China.

Table 10.5. Water inrush disasters induced by the paleo-sinkholes in some coal mines

Mining face No.	Mine, coalfield	Max. water inflow (m^3/min)	Date of water inrush (month-day-year)	Loss
2171	Fangezhuang, Kailuan	2053	6-2-1984	Four coal mines submerged
2903	Dongpang, Xingtai	1167	4-12-2003	Mine submerged, Lost US $38 millions
$7_2$22	Renlou, Wanbei	576	3-4-1996	Mine submerged, Lost US $44 millions
Roadway	Guojiazhuang, Feicheng	550	1-5-1993	Mine submerged
-300 level roadway	Zhangji, Xuzhou	402	2-18-1997	Mine submerged
Dong 18	Lifeng, Jiaozhuo	120	3-29-1967	Mine submerged
32031	Wucun, Huixian	40	11-25-1999	Panel submerged
103	Tongye, Anyang	23	8-25-1965	Mine submerged

10.6.3 Mechanism of water inrushes from karst paleo-sinkholes

Figure 10.20 shows a section for typical paleo-sinkholes (Yin and Zhang 2005). When the paleo-sinkholes are permeable, the effective distance between the aquifer and the coal seam is reduced greatly, which makes the water inrush more likely, as shown in Fig. 10.21 (Yin and Zhang 2005).

The paleo-sinkhole also changes the structure of the floor strata by becoming a discontinuity. This in turn increases the strata failure in the seam floor, as shown in Fig. 10.21. Figure 10.22 presents the mining induced strata failure zone in the seam floor in the Huainan coalfield. The maximum depth of the floor strata failure occurs in the strata directly beneath the area around the coal wall where water inrush is most likely to take place. The in-situ measurements in the Huainan coalfield also show that the maximum depth of the failure zone in the floor was 16.8 m for the normal case (refer to Fig. 10.22a). However, it was 29.6 m in the fault zone (Fig. 10.22b). This indicates that a geological structure existed in the seam floor makes the strata more susceptible to fail.

Also, with the effect of water pressure, the floor strata above the top of the paleo-sinkhole may induce tensile fractures by hydraulic fracturing. Furthermore, an existing fissure zone of 3 to 10 m in the surrounding rocks of the paleo-sinkhole makes rocks easier for groundwater to reach the mining face.

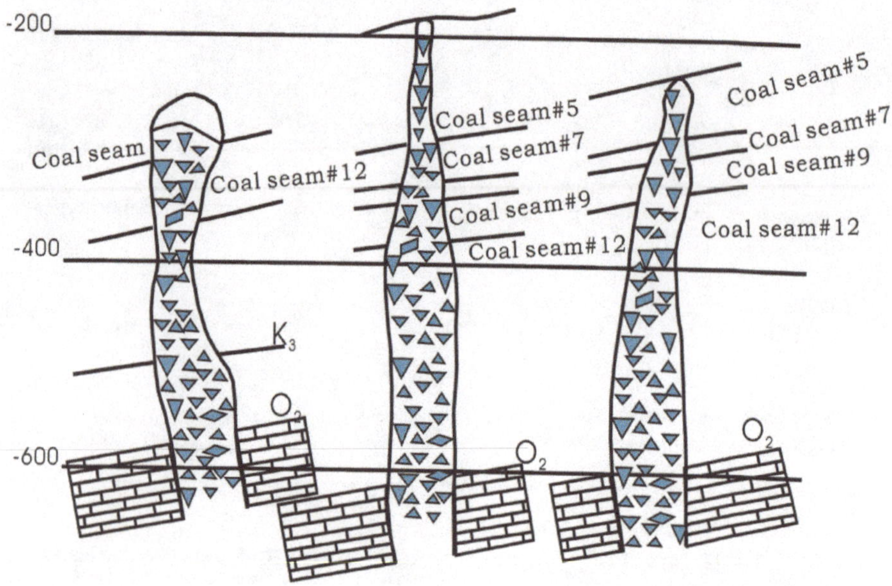

Fig. 10.20. Longitudinal cross sections of paleo-sinkholes No. 9 (highly permeable), No. 2 (impermeable) and No. 1 (poorly permeable) in the Fangezhuang coal mine.

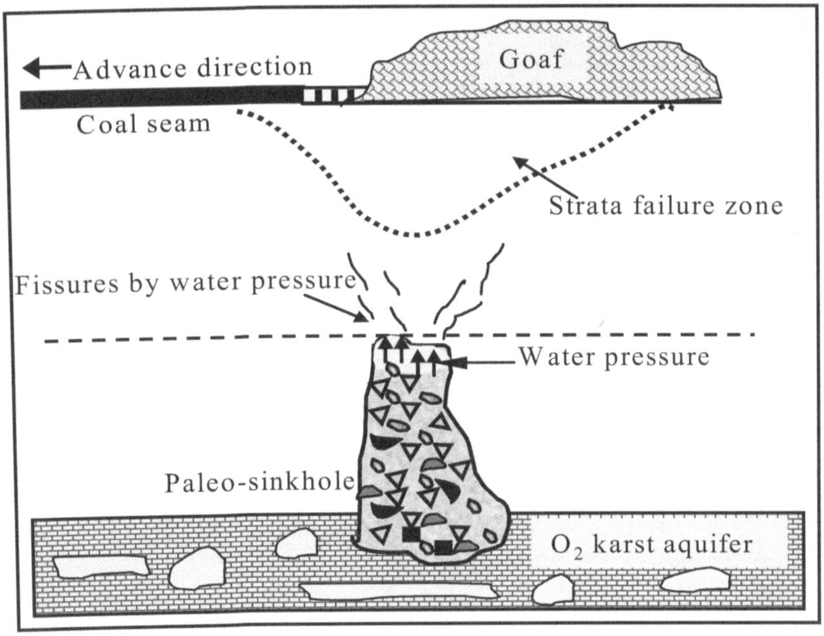

Fig. 10.21. Sketch of the strata failure in the seam floor caused by the paleo-sinkholes and coal mining.

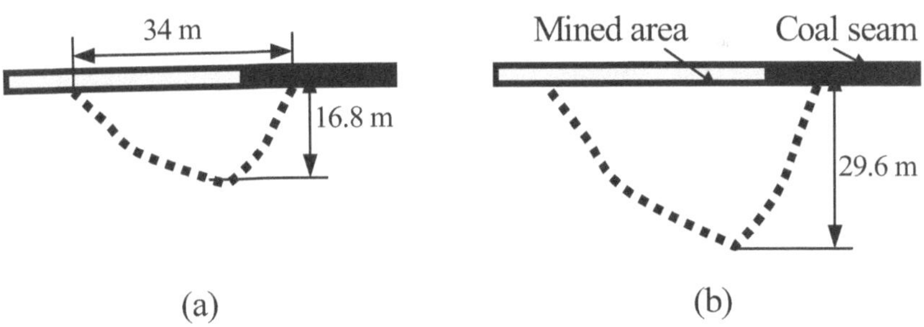

Fig. 10.22. Observed floor strata failure zone induced by mining along mining direction in the Huainan coalfield. a. Normal geological condition. b. With geological structure in the floor strata.

10.7 Case study

10.7.1 Introduction

The Yangzhuang coal mine is located in Huaibei Coalfield, Anhui Province, Eastern China. This coalfield consists of slightly inclined coal seams of Permian and Carboniferous coal measure. The thickness of the primary coal seam, No. 6, was approximately 2.9 m with the roof consisting of sandy shales and fine sandstones. The seam floor was comprised of sandy shales and sandstones. There was a strong limestone karst aquifer in the Taiyuan limestone, which was about 44 m underneath the coal seam. The thick limestone was highly fractured and contained abundant water with high water pressure. Water inrush from this aquifer was the potential disaster of safe mining. The total depth of cover for seam No. 6 was approximately 330 m. Comprehensive mechanized longwall mining with full caving was used in the coal extraction.

10.7.2 Water inrush incidents

Several water inrushes from the Taiyuan limestone aquifer took place during coal mining. Figure 10.23 shows that there were several water inrush incidents occurred in longwall mining faces (Peng and Wang 2001). Longwall face II 617 was located in the second production level. The length of the face along the strike direction was 680 m, the length of the width was 105 m, and the thickness of the mining seam was about 2.86 m. The Taiyuan limestone aquifer (K_1 in Fig. 10.24) was located 29.4 to 44.3 m below the coal seam. The water pressure in this aquifer was about 3.2 MPa (Peng and Meng 2002).

The production in Face II 617 began on September 1, 1988. When the mining distance reached 42 m on October 2, water at a flow rate of 59.9 m^3/h intruded into the working space from inside of mined area. Later on October 24, when the face was advanced to 60 m from the starting cut, water inflow increased dramatically into the workings, with a maximum flow rate of 3,153 m^3/h. Several hours later the second production level was submerged totally, causing serious economic losses. Meantime, the observing boreholes in various locations on the surface showed that the water level in the Taiyuan limestone was lowered around 10.25 to 61.38 m, which verified that the water in the incident came from the Taiyuan limestone (K_1) underlying the coal seam. The similar water inrushes happened in the nearby mining faces in the past with much smaller intensity. For example, Face II 611 had a water inrush with a maximum flow rate of 350 m^3/h (refer to Fig. 10.23).

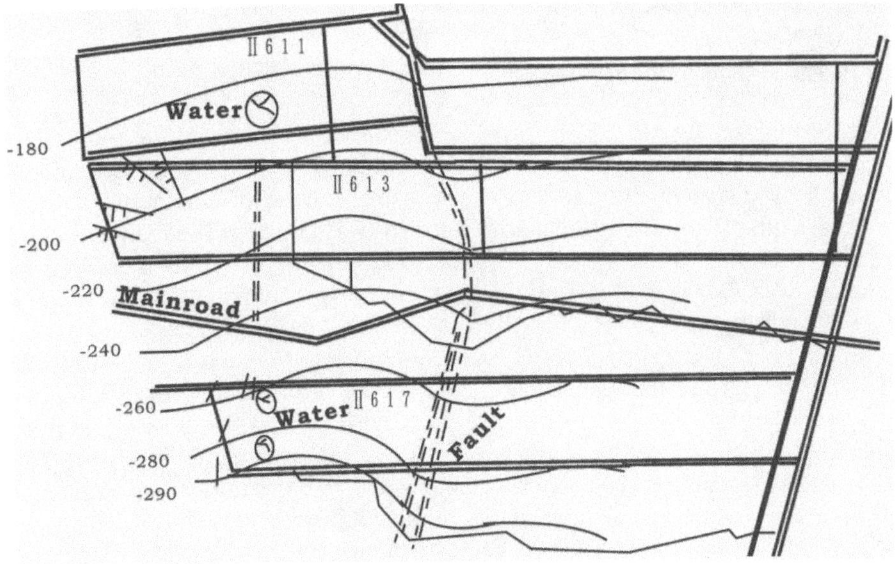

Fig. 10.23. Sketch of the plan view of mining panel and water inrush from a karst aquifer in the seam floor.

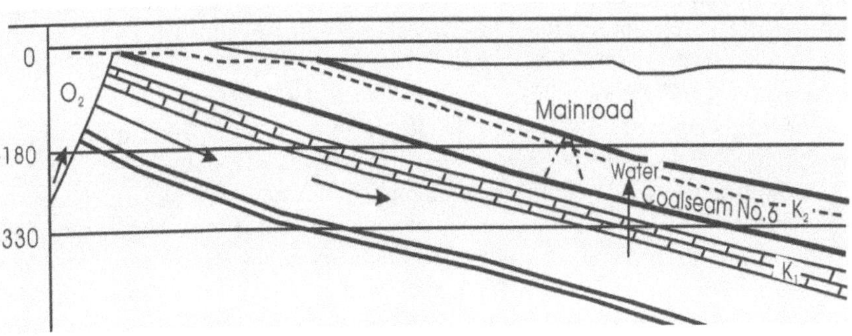

Fig. 10.24. Sketch of the plane view of mining panel and water inrush from a karst aquifer in the seam floor.

In investigation and analysis of the water inrush mechanism in Face II 617, the major cause of the water inrush was that the thickness of the underlying strata was too small, i.e. the distance between the aquifer and coal seam was small. Furthermore, these strata were partially fractured due to mining. Therefore, detecting the failure zone of the strata induced by mining and determining the critical distance between the aquifer and coal seam are important to maintain mine safety.

10.7.3 Field observations

The comprehensive underground observation was conducted in double mining face II 625. The observations included in-situ stress measurement, roadway displacement and support pressure, and water-conducting failure zone in the floor strata (Peng and Wang 2001). The face II 625 was a double longwall mining system with the face interval of 10 m. The mining seam was 2.6 m in thickness, with mining depth of 289 m. The mining width of the face A was 140 m and the face B was 80 m. The length of the mining panel was about 600 m. The seam dip angle ranged from 17 to 21°.

To measure the water-conducting failure in the underlying strata, inclined boreholes (59 mm of diameter) are drilled into the floor strata premining from the conveyor roadway for observation, as show in Fig. 10.25. In each borehole, water injection was used to determine rock strength, borehole fissure, and changes in hydraulic conductivity.

The water injection along each borehole was conducted by pumping into the borehole. The measurements were taken in each hole at different time. It should be noted that mining-induced fractures might cause the conductivity of the observing strata to increase, inducing water inflow from the borehole. In this case the observation was to be changed as water inflow measurements instead of water injection. Figure 10.26 shows the water injection/inflow and strata stress changes with the distance of the mining face and the borehole. It can be seen that prior to mining both the water influx and strata stress are small. In other words, the mining impact is trivial. The floor strata experienced stress increase/compression, then reduction/extension after mining, causing floor strata fractured. Therefore, the water influx increases post-mining, as shown in Fig. 10.26. According to the field observation the water-conducting depth in the floor strata in this panel was about 15 m, which was verified by the physical simulation (Fig. 10.27) and numerical models (Peng and Wang 2001).

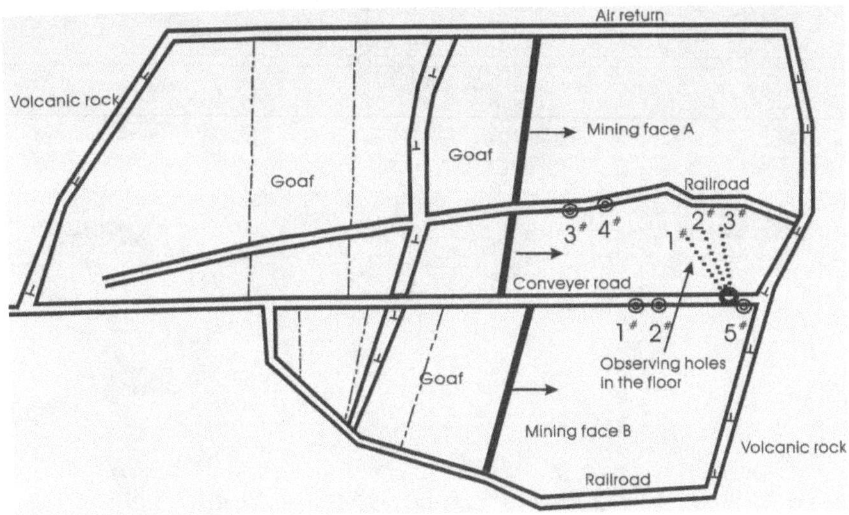

Fig. 10.25. Sketch of the plane view of the double mining faces II 635 and locations of the observation holes.

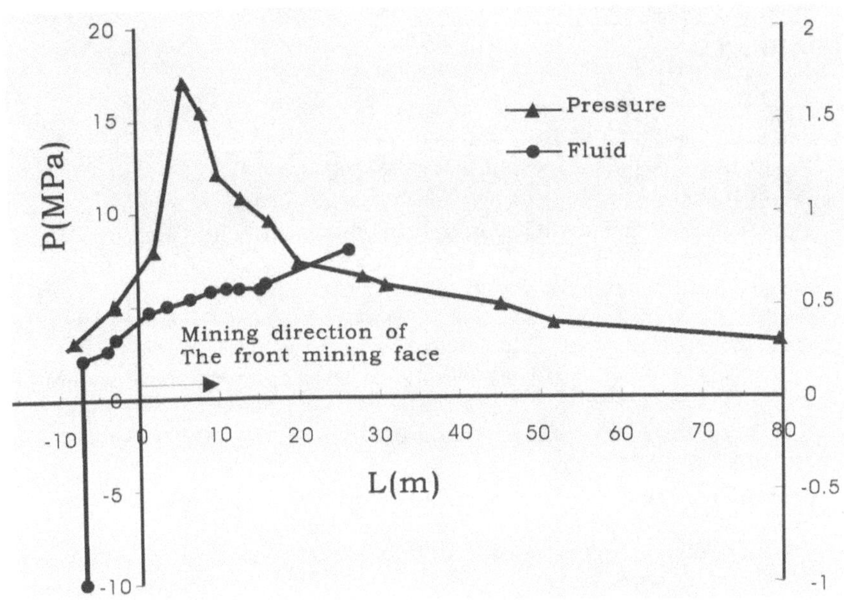

Fig. 10.26. Observed water injection into and water inflow from the observed hole before/after the mining face passing the observing hole. In the figure the negative distance represents pre-mining state, and the positive means post-mining. Also the negative flow rate represents injection, and the positive is influx.

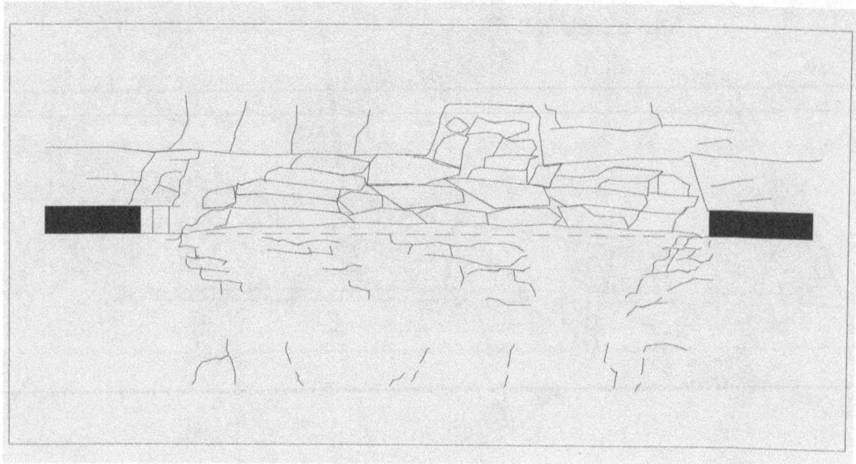

Fig. 10.27. Rock failure observed from the physical simulation model, showing that both the overburden strata and floor rocks generate mining-induced fractures.

References

Fengfeng Mining Bureau, Shangdong University of Science and Technology. In-situ measurement of the floor strata failure in the No. 2 coal mine of Fengfeng Coalfield. Internal research report, 1985 (in Chinese)

Gao Y, Li B Y (1992) Study on coal seam floor deformation and failure induced by mining faces threatened by Ordovician limestone confined aquifer. J China Coal Society 17(2)

Huainan Mining Bureau, Xi'an Branch of China Coal Research Institute 1983 Hydrogeological conditions and control methods of the karst aquifer under # A coal seam in Huainan coalfield. Internal research report (in Chinese)

Li J, Zhou W (1988) Karst groundwater inrush and its prevention and control in coal mines in China. Karst hydrogeology and karst environment protection, Yuan D (ed.), Proc 21st IAH Conf Beijing, Geological Publ House 176(2):1075-1082

Li M, Guan Y (2002) Coal seam floor failure depth of fully-mechanized mining face. Ground Pressure Strata Control 19:52-54 (in Chinese)

Li X, Gao Y (2003) Damage analysis of floor strata. Chinese J Rock Mech Eng 22:35-39 (in Chinese)

Luo L, Peng S, Wang J (2002) Measurements on mine pressure in double mining face for mining above confined aquifers. Mine pressure and roof management 19(1):83-84

Peng S (1997) Investigation of engineering disasters induced by coal mining at deep depth in China. 4th Ann meeting, Geology Branch, China Coal Industry Tech Committee, Qingdao (in Chinese)
Peng S (1999) The state-of-art and the future work of engineering disasters induced by coal mining at deep depth in China. Symp Engineering and Disaster, Dept Material Science and Engineering, China NSF, Mt Jiuhua, Jianxi (in Chinese)
Peng S, Wang J (2001) Safe mining above confined aquifers. Coal Industry Press Beijing (in Chinese)
Peng S, Meng Z (2002) Theory and practice of mining engineering geology. Geological Press (in Chinese)
Peng S, Luo L, Wang J (2003) Reasonable distance of double mining face for mining above confined aquifers. Chinese J Rock Mech Eng 22(1):48-52 (in Chinese)
Wang J, Peng S, Meng Z (2002) Strata failure of the coal floor in double mining face for mining above confined aquifers. J Beijing Univ Sci Tech 24(3):243-247
Wang L, Song Y, Miao X (2002) Lyapunov exponent of chaos feature in the processing of deformation failure for coal floor. Chinese J Geotechnical Eng 24:356-359 (in Chinese)
Yang D, Peng S, et al. (2002) Specific geologic section and its application in study of karst paleo-sinkholes. Coal Geology Exploration 30(6):47-49
Yang T, Tang C, Liu H, Zhu W, Feng Q (2003) Numerical model of the instability-failure process of the coal-bed floor due to confined water inrush. J Geomechanics 9:281-288 (in Chinese)
Yin SX (2002) Analysis, simulation and applications of water inrush system in mine area. PhD dissertation. China Univ of Min Tech (in Chinese)
Yin SX, Wang SX, Wu Q (2003) Characteristics of Karstic collapse columns in north coalfields of china and mechanism of water inrush from coal seam floor. Rock Stress. 3rd Int Symp Rock Mech, Kumamoto, Japan. Rotterdam, A A Balkema: 517-522
Yin S, Wu Q (2004) Simulation and mechanism analysis of water inrush from karstic collapse columns in coal floor. Chinese J Rock Mech Eng 23:2551-2557 (in Chinese)
Yin S, Zhang J (2005) Impacts of karst paleo-sinkholes on mining and environmental in northern China. Environ Geol 48:1077-1083
Zibo Mining Bureau 1979 Data analyses and application of coal seam water inrushes in Zibo Coalfield. Zibo Coal Sci Tech (2) (in Chinese)
Zhang J (2005) Investigations of water inrushes from aquifers under coal seams. Int J Rock Mech Min Sci 42(3):350-360
Zhang J (1990) Theoretical criterion of water inrushes from confined aquifers under coal seams and its application. Mechanics and Practice 12:35-38 (in Chinese)
Zhang J (1989) Theory and practice on prediction of water inrushes from coal seam floor. Coal Geology and Exploration (4):38-41 (in Chinese)

Zhang J, Shen B (2004) Coal mining under aquifers in China: a case study. Int J Rock Mech Min Sci 41(4):629-639

Zhang J, Zhang Y, Liu T (1997) Rock mass permeability and coal mine water inrush. Geological Publ House Beijing (in Chinese)

Zhang J, Liu T 1990 On depth of fissured zone in seam floor resulted from coal extraction and its distribution characteristics. J China Coal Society 15(2):46-55 (in Chinese)

Zhang J, Wang J (2006) Coupled behavior of stress and permeability and its engineering applications. Chinese J Rock Mech Eng in press

Zhang J, Zhu Z (1994) In situ observation of water-conducting failure zone in Xingtai coal mine. China Coal Research Inst and Xingtai Coal Mine, Internal research report (in Chinese)

Zheng S, Zhu W, Wang S (2000) Study on the coupling problem between flow and solid of mine in confined aquifer. Chinese J Rock Mech Eng 19: 421-424 (in Chinese)

11 Stability of underground excavations

11.1 Analytical solutions of underground excavations

11.1.1 Circular excavation

The closed-form elastic solution for a circle excavation can be obtained in two-dimensional geometry. A circular cross section of a long excavation in a formation subject to biaxial stress is examined, as shown in Fig. 11.1. The biaxial stresses in the far-field of the excavation are assumed as σ_0 and $k\sigma_0$. The analytical solutions for stress and displacement distributions around the circular opening can be expressed as follows (Brady and Brown 1985):

Total radial, tangential, and shear stresses after excavation:

$$\begin{cases} \sigma_r = \dfrac{\sigma_0}{2}\left[(1+k)\left(1-\dfrac{a^2}{r^2}\right)-(1-k)\left(1-\dfrac{4a^2}{r^2}+\dfrac{3a^4}{r^4}\right)\cos 2\theta\right] \\ \sigma_\theta = \dfrac{\sigma_0}{2}\left[(1+k)\left(1+\dfrac{a^2}{r^2}\right)+(1-k)\left(1+\dfrac{3a^4}{r^4}\right)\cos 2\theta\right] \\ \tau_{r\theta} = \dfrac{\sigma_0}{2}\left[(1-k)\left(1+\dfrac{2a^2}{r^2}-\dfrac{3a^4}{r^4}\right)\sin 2\theta\right] \end{cases} \tag{11.1}$$

where a is the excavation radius; r is the distance from the center of the excavation; σ_r, σ_θ are the radial and tangential stress components, respectively; $\tau_{r\theta}$ is the shear stress component; θ is the angle as shown in Fig. 11.1; σ_0 and $k\sigma_0$ are the vertical and horizontal stresses in the far-field stress field, respectively.

Radial and tangential displacements induced by excavation can be expressed as follows:

$$\begin{cases} u_r = -\dfrac{\sigma_0 a^2}{4Gr}\left[(1+k)-(1-k)\left(2(1-2\nu)+\dfrac{a^2}{r^2}\right)\cos 2\theta\right] \\ u_\theta = -\dfrac{\sigma_0 a^2}{4Gr}\left[(1-k)\left(2(1-2\nu)+\dfrac{a^2}{r^2}\right)\sin 2\theta\right] \end{cases} \tag{11.2}$$

where u_r, u_θ are the radial and tangential displacements, respectively; ν is the Poisson's ratio; and G is the shear modulus.

The induced stresses on the excavation wall can be obtained by putting $r = a$ in Eq. 11.1:

$$\begin{cases} \sigma_r = 0 \\ \sigma_\theta = \sigma_0\left[(1+k)+2(1-k)\cos 2\theta\right] \\ \tau_{r\theta} = 0 \end{cases} \tag{11.3}$$

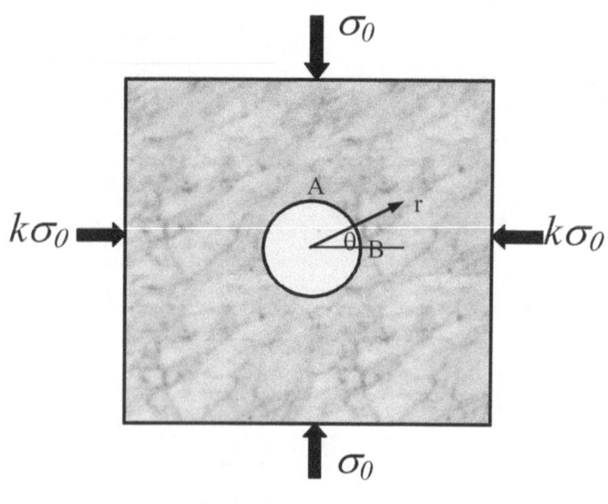

Fig. 11.1. Two-dimensional circular excavation under biaxial stresses.

At the crown (point A) and sidewall (point B) (Fig. 11.1) the excavation induced stresses are:

At point A:

$$\sigma_\theta = (3k-1)\sigma_0 \tag{11.3}$$

At point B:

$$\sigma_\theta = (3-k)\sigma_0 \tag{11.4}$$

For a horizontal tunnel, the vertical stress component (σ_0) is the overburden stress, which is $\sigma_0 = \gamma H$, and the constant k can be estimated by $k = \nu/(1-\nu)$. The stresses at the crown and sidewall can be rewritten as:

At point A:

$$\sigma_\theta = \left(\frac{4\nu - 1}{1-\nu}\right)\gamma H \tag{11.5}$$

At point B:

$$\sigma_\theta = \left(\frac{3 - 4\nu}{1-\nu}\right)\gamma H \tag{11.6}$$

where H is the depth of the excavation; and γ is the average specific weight of the rocks.

11.1.2 Elastoplastic solution of a circular excavation

With a circular opening excavated in a formation subject to hydrostatic stress (σ_0) as shown in Fig. 11.2, Ladanyi (1974) gave an elastoplastic analysis on stress distributions and applied support pressure (p_i). It is assumed that the rock strength is described by the Mohr-Coulomb criterion (Brady and Brown 1985), i.e.:

$$\sigma_\theta = q\sigma_r + UCS \tag{11.7}$$

Where σ_θ and σ_r are the tangential and radial stresses, respectively; UCS is the uniaxial compressive strength of the rock; $q = (1+\sin\phi)/(1-\sin\phi)$; and ϕ is the internal friction angle of the rock.

The strength of fractured rock (failure zone) as shown in Fig. 11.2 is taken to be purely frictional, with the limiting state of stress within the fractured rock mass defined by (Brady and Brown 1985):

$$\sigma_\theta = q_f\sigma_r \tag{11.8}$$

where $q_f = (1+\sin\phi_f)/(1-\sin\phi_f)$; and ϕ_f is the internal friction angle of the fractured rock.

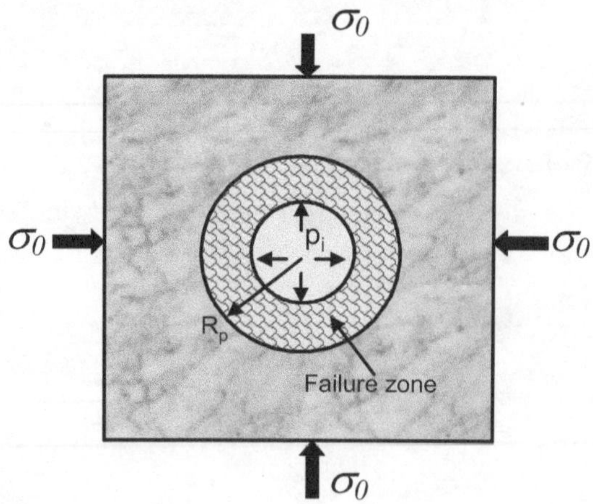

Fig. 11.2. Two-dimensional circular excavation under hydrostatic stresses and support pressure.

Applying Eq. 11.8 the stress distributions in the failure zone can be expressed as follows (Brady and Brown 1985):

$$\begin{cases} \sigma_r = p_i \left(\dfrac{r}{a} \right)^{q_f - 1} \\ \sigma_\theta = q_f p_i \left(\dfrac{r}{a} \right)^{q_f - 1} \end{cases} \tag{11.9}$$

The stress distributions in the elastic zone can be written as (Brady and Brown 1985):

$$\begin{cases} \sigma_r = \sigma_0 \left(1 - \dfrac{R_p^2}{r^2} \right) + p_i \dfrac{R_p^2}{r^2} \\ \sigma_\theta = \sigma_0 \left(1 + \dfrac{R_p^2}{r^2} \right) - p_i \dfrac{R_p^2}{r^2} \end{cases} \tag{11.10}$$

The radius of the fractured zone induced by excavation can be expressed as:

$$R_p = a \left[\frac{2\sigma_o - UCS}{(1+q)p_i} \right]^{1/(q_f - 1)} \tag{11.11}$$

It can be seen from Eq. 11.11 that the excavation has a smaller fractured zone with a larger support pressure and higher rock strength.

Most underground openings are non-circular, and it is difficult to obtain closed form solutions. The numerical methods, such as the finite element method, can then be used to analyze excavation-induced stress and displacement distributions and determine support pressure.

11.2 Roadway stability and support in underground mining

11.2.1 Controlling factors of roadway stability

Roadways in underground mining are very important underground structures. The stability of a roadway is directly related to safety of production and miners' lives. There are many factors that control and influence roadway stability. The main factors are lithology of the surrounding strata, geologic structures, in-situ stress, and groundwater.

Lithology of surrounding rocks

The lithology and strength of rocks are key factors to affect roadway stability. If a roadway is excavated in the strong rocks, such as limestone and sandstone, the required support pressure to control roadway stability may be very small. In this case the roadway may even not need any support. In contrast, if a roadway is excavated in shale and other weak rocks, the high deformation rate and failure may occur in response to stress redistributions. In this case multiple support systems are needed to keep the roadway stable. In China's coal mining industry about 67% of roadways excavated in shales and mudstones are instable. The roadways excavated in siltstone and sandy shale belong to medium-stable ones. Most roadways in sandstones and limestone are stable.

Geologic structures

Various discontinuities exist in the natural rock formations, such as faults, joints, fractures, beddings. When discontinuities exist in the rock, rock strength is much lower; sometimes it is only one tenth of normal rock strength. Figure 11.3 illustrates the relationship of the strengths of intact rock and the rock with discontinuities (Peng and Meng 2002). The statistical data in China coal industry show that in instable roadways more than 75% roof strata contained densely distributed fractures, and the spacing of the fracture set was less than 0.3 m. These fractured rocks in the roof are hence prone to fall down. Rock bolts with grouting or with metal mesh are necessary to support the roof. If larger fractures existed in the roof of the roadway, additional rock bolts are needed to keep the roof stable. The rock bolts should be installed perpendicularly to the fracture strikes, as shown in Fig. 11.4.

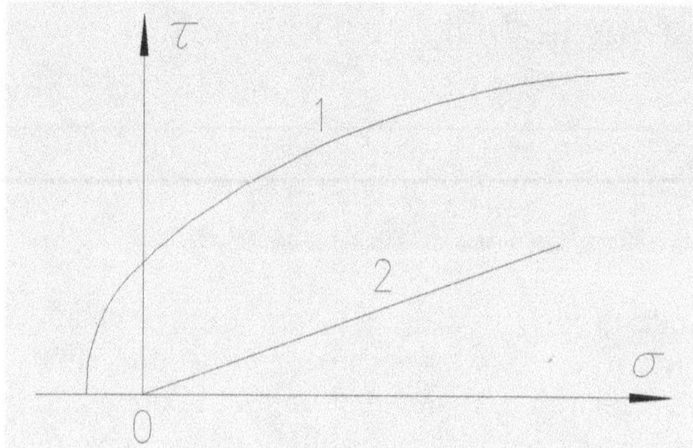

Fig. 11.3. Schematic plot showing that the rock with discontinuities has a much lower strength envelope. 1- for the intact rock; 2-for the rocks with discontinuities.

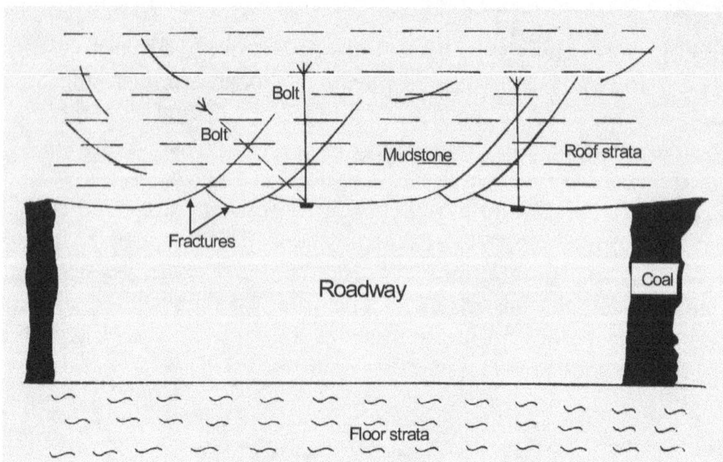

Fig. 11.4. Additional rock bolts needed when the roof strata contain large fractures. The bolts should be perpendicular to the strike of the fractures and penetrated the fractures.

In-situ stress

In-situ stress magnitudes and directions depend on the stress redistributions and roadway stability, as shown in Eq. 11.1. For example, in deep formations because of the high in-situ stress, it needs a larger support pressure to keep the roadway stable. The stress directions and stress regimes also have important influences on roadways stability. In the normal faulting stress regime, the horizontal roadway excavated in the minimum horizontal stress direction may be more stable than in other directions. However, in a strike slip faulting or thrust faulting stress regime a horizontal roadway excavated parallel to the maximum horizontal stress direction has a smaller shear failure area, hence is more stable, refer to Fig. 7.22 (Zhang et al. 2006).

Groundwater

Groundwater affects both stress state and rock strength. The water in some rocks reduces rock strength (as discussed in Chap 5, Sec. 5.6) and makes roof strata more likely to fail. For water-saturated rocks, the effective stress needs to be considered in the stress distribution and support pressure calculation. The stress calculation with consideration of pore pressure effects can refer to Chap. 7, Sec. 7.2.4 and Sec. 7.3.

11.2.2 Prediction of roadway stability

Roadway stability criterion

Roadway stability is a dynamic representation of a complex system in strata and support interaction. It is influenced not only by geological conditions, but also by the size and shape of the roadway cross section, excavation advance speed, support pressure and parameters, etc. Therefore, prediction of roadway stability is a complicated process (Peng et al. 2001).

Displacement criterion

Roadway stability can be estimated according to the critical displacement in the roof as expressed as follows (Peng and Meng 2002):

If $U \leq U_0$, then roadway stable;

If $U > U_0$, then roadway instable.

where U is the measured displacement of roadway roof; U_0 is the calculated critical displacement.

Displacement rate criterion

Preliminary support is needed after excavation, then displacement monitoring is needed to determine time and parameters of the secondary support. According to the experiences in roadway support in China, the following displacement rates may be used to determine roadway stability:

When the net displacement of the strata is greater than 1 mm/day, the roadway is at a stage of serious deformation. If this displacement rate lasts for several days, then additional support is needed.

When the net displacement of the strata varies from 0.2 to 1 mm/day, the roadway is at a stable stage of slow deformation. The roadway is becoming stable.

When the net displacement of the strata is less than 0.2 mm/day, the roadway is at a stage of slow deformation. Only when the displacement satisfies this condition, the secondary/permanent support can be used.

Construction experiences

The roadway instability normally occurs when the following incidents takes place:

- Rocks partially collapse and a large area of the shotcrete layer or liner is fractured;
- Accumulated displacement has reached 2/3 of the critical displacement;
- Daily displacement is excess of 10% of the critical displacement;
- Roadway deformation increase abnormally.

When the conditions of roadway stability, such as displacement, cannot be satisfied, the following techniques need to be used:

- Change the shape of the cross section;
- Modify the excavation order;
- Increase the thickness of the shotcrete, or use the concrete consisted of the cement with a high early strength;
- Increase bolt density or length.

Figure 11.5 shows displacement variations in an observing section of a roadway after preliminary support. After 10 days the displacement rate was still greater than 1 mm/day. Therefore, an inverted arch was built in the floor strata of the roadway to seal the bottom rocks. Consequently, the displacement slowed down and became stable (refer to Fig. 11.5).

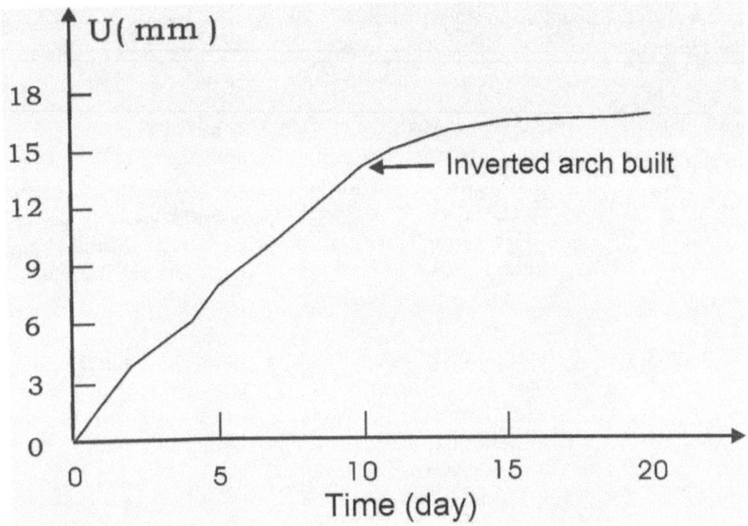

Fig. 11.5. Observed displacements in a section of a roadway after excavation. The displacement rate slowed down after the shape of the roadway cross-section was modified.

11.2.3 Strata classifications for roadway supports

Various classifications for surrounding rocks of roadways/tunnels were proposed to determine roadway stability. Table 11.1 lists strata classifications presented by China coal industry (Peng and Meng 2002). Five types of rock masses were classified according to geological conditions, lithology, rock strength, and stability of the strata in which the roadway was excavated.

Another strata classification was based on the roadway fractured zone (relaxation and failure zone) (Dong et al. 1994). A fractured zone exists due to excavation, and the roadway support is to balance the gravity weight of the rocks in this zone and partial stress induced by elastoplastic deformations of the fractured zone and the deformation in the rocks beyond this zone. The thickness of the fractured zone depends on rock strength, in-situ stress, and excavation-induced stress redistributions. Table 11.2 gives the strata classification according to the thickness of the fractured zone. It can be seen that the larger the fractured zone is, the greater the deformation of the roadway, and more instable the rocks.

Table 11.1. Rock Classifications for roadway stability in China's coal industry

Type	Classification	Description	Stability	Rock
I	Stable	Intact, strong rock mass and uneasy to weather. Single layer thickness of stratified rocks is not less than 0.6 m, and with strong cementation between layers and without weak layer interbedded. UCS > 98 MPa.	Rock mass stable. Do not need support for long term and without fragment falling.	Granite, basalt, strong sandstone, limestone.
II	Fairly stable	1. Intact, fairly strong rock mass. Single layer thickness of stratified rocks is not less than 0.3 m, and with fairly strong cementation between layers. UCS = 78 - 98 MPa. 2. Strong blocky rock mass with closed interface and without fillings of muddy rocks.	Basically stable. It may have small fragments falling without support for a long time.	Fairly strong sandstone and limestone, strong conglomerate
III	Medium-stable	1. Intact, moderately strong rock mass. Single layer thickness of stratified rocks is not less than 0.2 m. Mostly strong layers with minor weak rock layers. UCS = 59 - 78 MPa. 2. Fairly strong blocky rock mass.	Basically stable during excavation for a small cross-section roadway. For a large cross-section one, rock fallings take place without support.	Common sandstone, limestone, sandy shale.
IV	Poorly stable	1. Intact, weak rock mass. Stratified rocks with moderate strong layers interbedded. UCS = 29 - 59 MPa. 2. Moderately strong blocky rock mass.	Keep stable for a very short time without support.	Shale, mudstone, weak sandstone and limestone, hard marlstone.

V	Instable	1. Loose, weak, easily falling, and weathered rock mass. UCS = 9.8 - 29 MPa. 2. Any fractured and fissured rocks.	Immediate support needed, otherwise roof and sidewall collapse occur.	Weak shale, coal, gravel, rubble.

The intact rock mass represents that the fracture spacing or distance between beddings is greater than 1.5 m. The blocky rock mass means that the fracture spacing or distance between beddings is from 0.3 to 1.5m. Note that the groundwater effect does not consider in this classification.

Table 11.2. Rock classification and roadway support using fractured zone concept

Type	Classification	Thickness of fractured zone (m)	Support method	Remark
I	Stable	0 - 0.4	Shotcrete	May not need support for intact strong rock mass
II	Relatively stable	0.4 - 1	Suspension rock bolt and local grout	
III	Normal stable	1 - 1.5	Suspension rock bolt and local shotcrete	May partially damage rigid support
IV	General instable (weak rock)	1.5 - 2	Combined bolt of the voussoir arch and shotcrete, with metal mesh partial support	Rigid support largely destroyed and contractible support recommended
V	Instable (weaker rock)	2 - 3	Combined arch rock bolt and shotcrete, with metal mesh partial support	Can reach stable deformation
VI	Extremely instable	> 3	Comprehensive support	Cannot reach stable deformation

11.3 Mining-induced stress and overburden failure in longwall mining

11.3.1 Mining-induced stress distribution

In longwall caving mining, the roof strata behind the longwall face are abandoned and allowed to collapse. When the face is advanced far enough, the immediate roof collapses behind at a distance depending on geological conditions (Yavuz 2004). The roof above the immediate roof, usually stronger and thicker sandstone, collapses behind the face at a longer distance. When this roof collapses, it generates a strong load to the support of the face. Failure of the roof continues until the roof and caved rocks are in contact. The in-situ stress distribution in the rock strata is disturbed by the excavation; high stress zones are created in the adjacent coal because of the transfer of stresses, as shown in Fig. 11.6. The stress redistributions in the mining panel follow the following rule: i.e. stress increases (higher than the cover stress) in the abutment (pillar and un-mined area) and stress decreases (lower than the in-situ stress) in the goaf (Peng et al. 1996). When the extraction of the panel is completed, the next panel is developed a sufficient distance from the over-stressed area. The width of the pillar left between the panels depends on the stress concentration over the abutment, in addition to the stress induced by the weight of overburden.

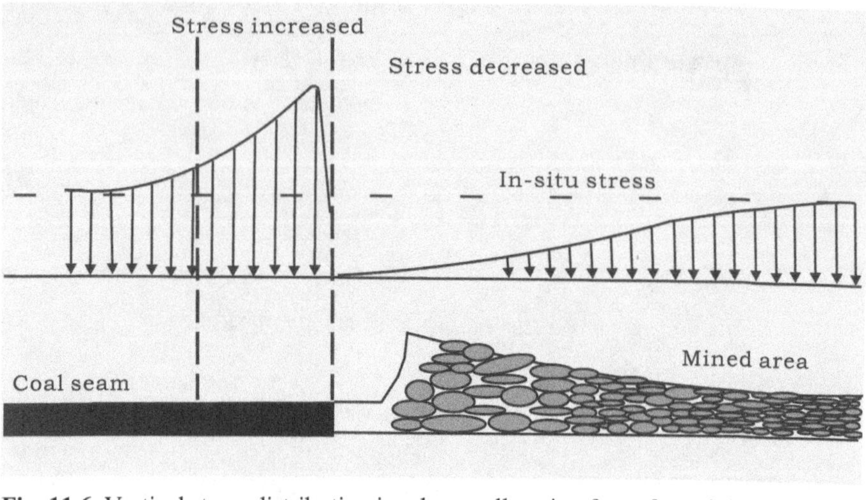

Fig. 11.6. Vertical stress distribution in a longwall caving face after mining.

Wilson (1997, 1981) proposed a methodology to calculate the vertical stress redistributions in the coal seam for a longwall mining. Table 11.3 lists equations for calculating the vertical stress, abutment pressure (yield stress of the coal seam), and the stress in the yield zone (Zhang et al. 1997). It should be noted that the yield zone is the area from the abutment to the location where the coal seam has the maximum vertical stress (yield stress σ_Y), as shown in Fig. 11.7.

Table 11.3. Equations used to calculate vertical stress distribution and yield zone in the coal seam in longwall mining (Wilson 1977)

	Weak coal seam in strong roof and floor rocks	Similar strength in the coal seam and its roof and floor rocks
Vertical stress (MPa)	$\sigma_{zz} = bp^* \exp(xF/m)$	$\sigma_{zz} = bp^*(1+2x/m)$
Abutment pressure or yield stress (MPa)	$\sigma_Y = C_0 + bp$	$\sigma_Y = C_0 + bp$
Width of yield zone (m)	$x_b = \dfrac{m}{F}\ln(\dfrac{p}{p^*})$	$x_b = \dfrac{m}{2}[(\dfrac{p}{p^*})^{1/(b-1)} - 1]$
Vertical stress in yield zone (MN/m)	$A_b = \dfrac{m}{F}b(p - p^*)$	$A_b = \dfrac{m}{2}p^*[(\dfrac{p}{p^*})^{b/(b-1)} - 1]$

In the table, σ_{zz} is the vertical stress (MPa); σ_Y is the abutment stress or yield stress (MPa); p is the in-situ vertical stress, and $p = \gamma h$; h is the mining depth; γ is the rock specific weight; C_0 is the uniaxial compressive strength (MPa); φ is the angle of internal friction of the coal seam; b is the constant, and $b = (1+\sin\varphi)/(1-\sin\varphi)$; m is the height of roadway or mining thickness (m); x is the distance from the abutment/wall of the coal pillar (m); x_b is the width of yield zone, p^* is the sum of the support pressure p_i and uniaxial compressive strength of the failed rock of the wall of the coal pillar, it can be assumed p^* = 0.1MPa; and

$$F = \frac{b-1}{\sqrt{b}}(1+\frac{b-1}{\sqrt{b}}\tan^{-1}\sqrt{b})$$

where $\tan^{-1}\sqrt{b}$ is in radians.

In the region inside the coal seam after the vertical stress reaches peak stress (abutment pressure), the stress decays and approaches gradually the virgin overburden stress, as shown in Fig. 11.7. The vertical stress in the decay stage can be expressed as the follows:

$$(\sigma_{zz} - p) = (\sigma_y - p)\exp(\frac{x_b - x}{c}) \tag{11.12}$$

where c is a constant with a unit of the distance (m).

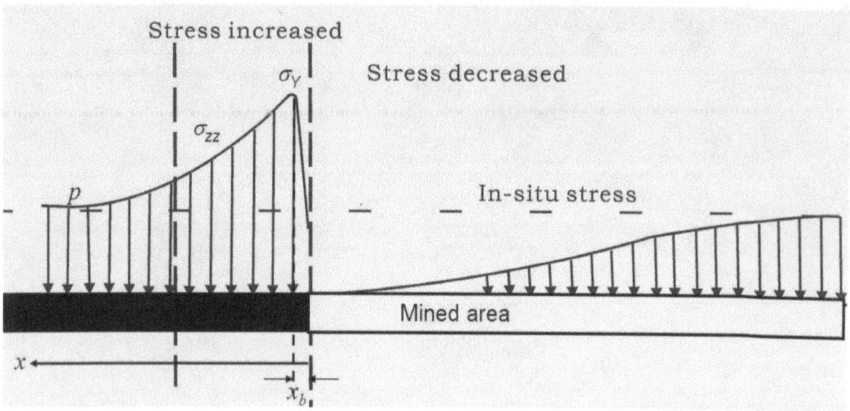

Fig. 11.7. Vertical stress distribution in the coal seam after longwall caving mining.

Numerical simulation results of longwall mining suggest that the stress in the coal seam in front working face of 80-90 m begins to increase slowly. The stress in the coal seam in front working face of 40-50 m increases significantly. The abutment stress usually is 2 to 3 times of the cover stress. A numerical analysis was given for a longwall mining panel, Face #1552 of the coal seam # 13-1 in Huainan Pansan colliery. The buried depth of the seam is 620 to 670 m. The mining thickness of the seam is 3 m. Figure 11.8 presents the vertical stress distribution after advance of 40 m in longwall mining in Face #1552. It can be seen that the stress increases in the seam (un-mined area) compared to the in-situ stress of 15 MPa. The stress decreases in the mined area, due to stress relaxation caused by roof collapse. The numerical analysis in another face with a cover depth of 510 to 560 m in the same colliery shows a similar result, as show in Fig. 11.9 (Peng and Meng 2002).

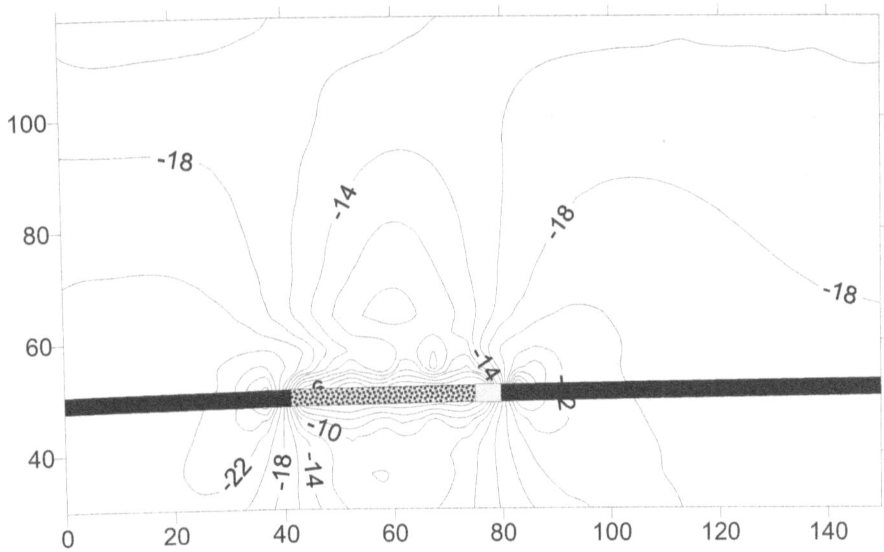

Fig. 11.8. Vertical stress distributions after advance of 40 m in longwall mining face #1552. The stress is in MPa, and negative sign represents compressive stress. The x- and y-coordinates are the horizontal and vertical distances in meters.

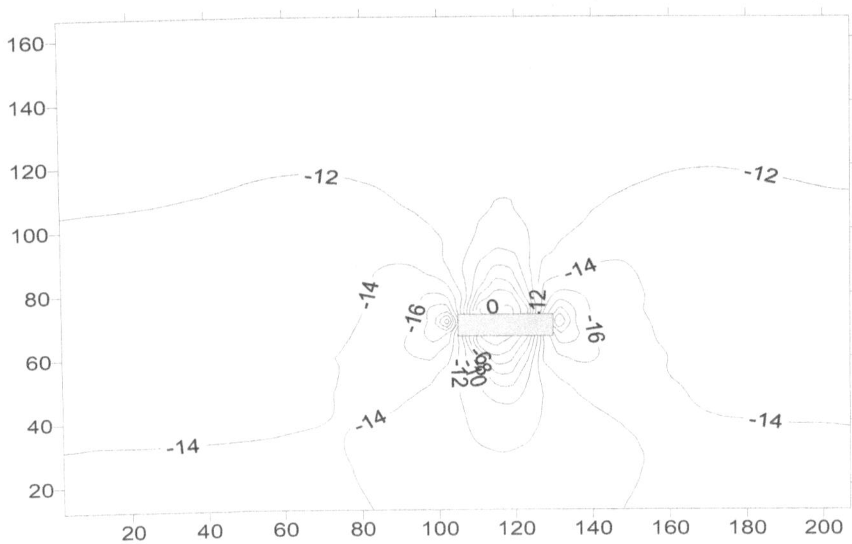

Fig. 11.9. Vertical stress distributions after advance of 24 m in longwall mining face #1221. The stress is in MPa, and negative sign represents compressive stress. The x- and y-coordinates are the horizontal and vertical distances in meters.

11.3.2 Mining-induced overburden failure

Due to longwall mining and rock decompression, roof rocks move downward to deform, and the caved zone, fractured zone, bent and sunk zone yield from bottom rock layer to the surface above mined area (Peng et al. 2002).

Mining engineering theory of strata movement and field observations indicate that the area immediately above the mined panel caves into the void created by the extraction of the coal. This completely caved rubble zone (the caved zone) extends above the mined panel as much as 4 to 6 times of the extracted thickness (Minns et al. 1995). Figure 11.10 shows subsidence zone and mining induced failure zones (Coe and Stowe 1984, Zhang and Shen 2004).

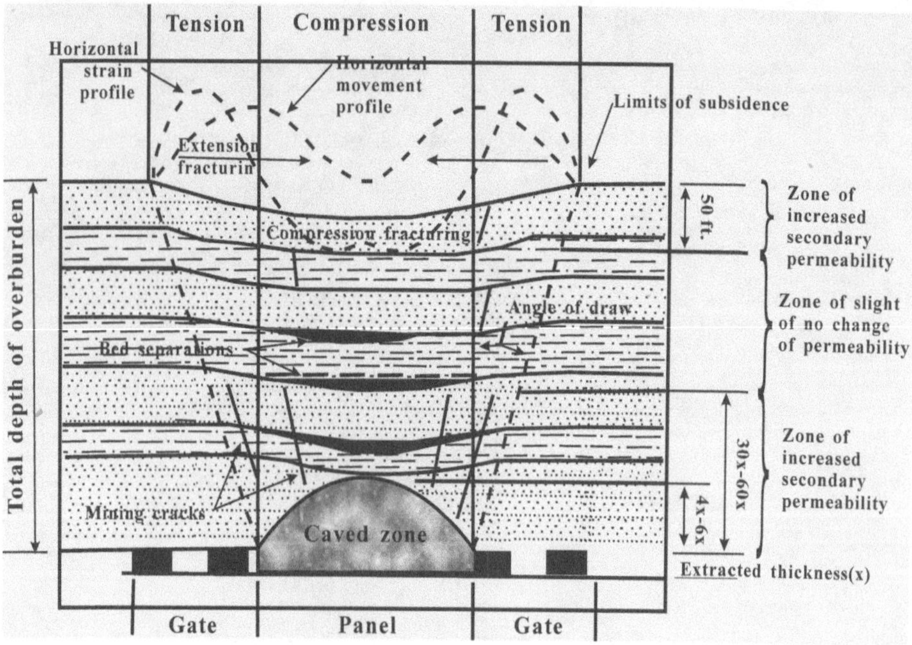

Fig. 11.10. General rock failure and subsidence related to longwall mining (Coe and Stowe 1984, Minns et al. 1995).

Above the caved zone, a transitional zone of highly fractured rock reportedly reaches as much as 30 to 60 times of the extracted thickness above the base of the mined area. This zone is called fractured zone characterized by extensive vertical fracturing and some massive block-type caving. Wells completed in either of the caved zone or fractured zone normally fail because water can rapidly drain directly to the mine works. Little recovery of water levels can be expected until the mine is allowed to flood after the completion of mining (Minns et al. 1995). The calculation of the heights of the caved zone and fractured zone can be obtained in Chap. 9, Sec. 9.2.2.

If the mine is at sufficient depth, an additional zone may exist above the extensively fractured bedrock in the subsidence trough. The majority of rock movement in this zone apparently occurs as minor horizontal slippage between strata. As a result, the strata in this zone tend to act as a "composite beam" (Minns et al. 1995). Near-surface strata (generally at depths up to about 17 m) are susceptible to fracturing and movement during subsidence.

11.4 Sedimentary structure and mining induced stress redistributions

The mechanical properties and behaviors in the sedimentary rock mass are closely related to the rock mass lithology and structure, because the rock mass is an aggregate of the rock matrix and a variety of discontinuities, such as faults, joints and bedding planes. Therefore, the effects of the lithology and structure of the rock mass need to consider for the ground control and stress redistributions due to longwall mining.

In this section, the physical simulation experiments by using a similarity model are applied to investigate the rock mass structure effects on stress redistributions induced by mining. Coal seam #13-1 in Xinji colliery in Huainan Coalfield of China is chosen for the physical model.

11.4.1 Lithology changes and mining-induced stress distribution

The thickness of coal seam #13-1 varies from 6.2 to 10 m. The average burial depth of the coal seams is 293 m. The immediate roof of the coal seam consists of grey and black mudstone or sandy mudstones. The immediate roof is rich in joints and belongs to blocky fissured rock. The coal seam roof (above the immediate roof) consists of fine- and medium-grained quartz sandstones containing many bedding planes. The thickness of the roof varies greatly, and the sedimentary faces of the roof changes frequently in the study area, as shown in Fig. 11.11. In this model, the length in the strike direction is 375 m, and mining height of coal seam #13-1 is 8.5 m. In the roof a fine-grained sandstone, which underlies lowermost in the roof of the coal seam, disappears in the middle of the model, as shown in the right hand side of the model. The second sandstone close to the seam in the roof is stable in thickness (2.09 to 3.61 m). It is the main roof of the coal seam.

In this physical model three stress sensors are examined to observe the stress redistribution due to coal mining with effects of the lithology changes.

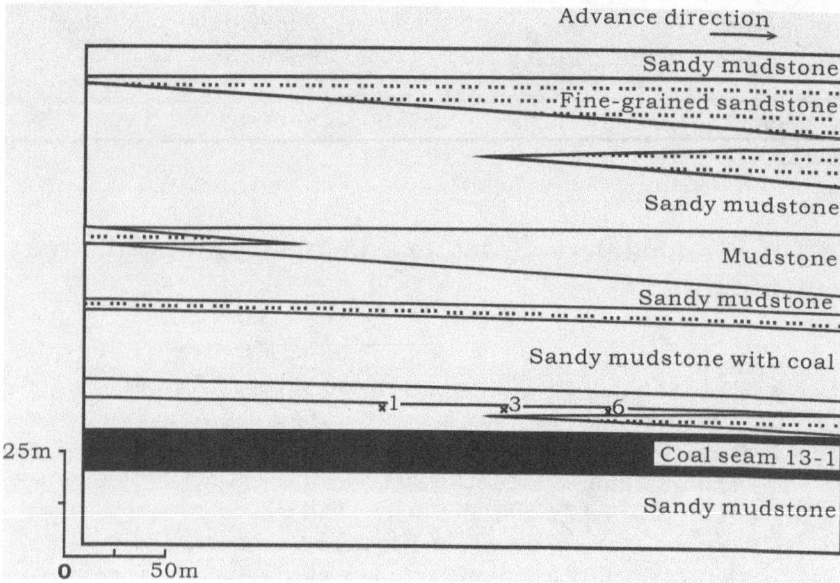

Fig. 11.11. A geologic section of the simulation model showing the changes of lithology in the overburden formations and the locations of stress sensors.

Figure 11.12 shows the mining induced vertical stress changes (the ratio of the induced and the maximum stresses) in three stress sensors located in different lithologies. It can be seen that stress distributions in different locations (with different lithologies) are different. The stress has a largest increase in the rock mass with sandstone as the immediate roof (see No. 6 sensor in Fig. 11.12). When the immediate roof is comprised of mudstones, the mining induced stress is low (refer to No. 1 and No. 3 sensors in Fig. 11.12). Therefore, the ground control and stress distributions are affected by the lithology.

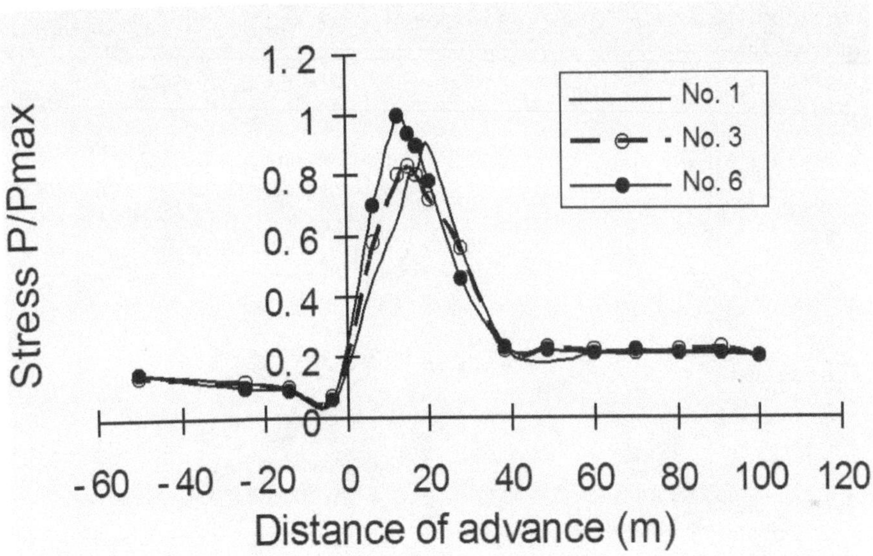

Fig. 11.12. Observed mining induced vertical stress to in-situ vertical stress ratio in three stress sensors (the positive distance represents pre-mining state, and the negative means post-mining).

11.4.2 Joints and mining-induced stress distribution

The physical models were used to simulate joint effects on longwall mining induced stress redistribution. According to the joint density in the rock mass, three rock mass structures are classified within a thickness of 30 to 40 m in the roof of the coal seam. The three rock mass structures are intact rock mass, blocky rock mass and fissured rock mass. There are almost no joints in the intact rock mass, and the joint spacing is about 5-6 cm in the blocky rock. In the fissured rock mass the joint spacing is even smaller and varies from 2 to 3 cm.

It is observed from Fig. 11.13 that in the rock masses with intact structure the abutment pressure (maximum stress) has the maximum magnitude compared to the rock masses with the blocky and fissured structures, and the location where the coal seam has the maximum stress is the closest to the coal seam wall/pillar (with smallest yield zone). This is primarily due to that the intact rock mass has the highest compressive strength compared to blocky and fissured rocks and is able to bear largest overburden stress (Peng and Meng 2002).

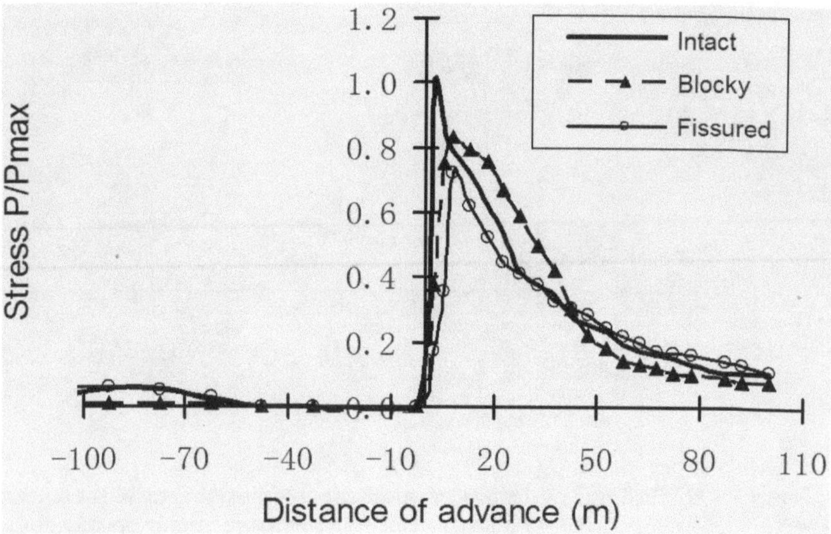

Fig. 11.13. Observed mining induced vertical stress to the maximum stress ratio in three models with different rock masses (the positive distance represents pre-mining state, and the negative means post-mining).

References

Brady BHG, Brown ET (1995) Rock mechanics for underground mining. George Allen & Unwin

Coe CJ, Stowe SM (1984) Evaluating the impact of longwall mining on the hydrologic balance, in Graves. Symp on Surface Mining, Hydrology, Sedimentology, and Reclamation, Univ of Kentucky, Lexington, pp 395-403

Dong F, Song H, Guo Z, Lu S, Liang S (1994) Support technique of strata relaxation zone. J China Coal Society, 19:21-31 (in Chinese)

Ladanyi B (1974) Use of the long-term strength concept in the determination of ground pressure on tunnel linings. Advance Rock Mechanics, Proc 3rd Congress ISRM, Vol 2, Part B, National Academy of Sciences, Washington DC, pp1150-1156

Minns SA, Kipp JA, Carey DI, Dinger JS, Sendlein LVA (1995) Effects of longwall mining on hydrogeology, Leslie County, Kentucky. Part 1: Pre-mining conditions. Univ of Kentucky, Lexington

Peng S, Meng Z (2002) Theory and practice of mining engineering geology. Geological Press (in Chinese)

Peng S, Yang G, Chugh YP (1996) Longwall chain pillars design for weak floor conditions. Rock Mechanics, Tools and Technique, A A Balkema1825:1831

Peng S, Ling B, Zheng G, Liang Y, Jin W, Geng Z, Ding H (2002) Study on roadway deformation and strata movement in mining bent and sunk zone. J China Coal Society 27(1):21-25 (in Chinese)

Peng S, Wang X, Liu X, Zhao S (2001) Experimental study of creep behaviors of the roadway in weak coal seam and rocks. J China Coal Society 26(2):149-152 (in Chinese)

Yavuz H (2004) An estimation method for cover pressure reestablishment distance and pressure distribution in the goaf of longwall coal mines. Int J Rock Mech Min Sci 41:193–205

Wilson AH (1977) The effect of yield zones on the control of ground. Proc 6th Int Strata Control Conf Paper 3

Wilson AH (1981) Stress and stability in coal ribsides and pillars. Proc 1st Annl Conf on Ground Control in Mining, West Virginia Univ (Peng SS ed) pp1-12

Zhang J, Bai M, Roegiers JC (2006) On drilling directions for optimizing horizontal well stability using dual-porosity poroelastic approach. J Petrol Sci Eng 53:61-76

Zhang J, Shen B (2004) Coal mining under aquifers in China: a case study. Int J Rock Mech Min Sci 41(4):629-639

Zhang J, Zhang Y, Liu T (1997) Rock mass permeability and coal mine water inrush. Geological Publ House Beijing, p36 (in Chinese)

Index